Landslide Science for a Safer Geoenvironment

Landslide Science for a Safer Geoenvironment

Editor

Emiliano Oriol

scitus academics

Landslide Science for a Safer Geoenvironment
Edited by **Emiliano Oriol**

ISBN: 978-1-68117-544-7
Library of Congress Control Number: 2016942944

SCITUS Academics LLC,
www.scitusacademics.com
Box No. 4766, 616 Corporate Way,
Suite 2, Valley Cottage,
NY 10989

Printed in United States of America on Acid Free Paper

Preface

When soil, rock, and other earth debris can no longer hold it together and gives way to gravity, landslides happen. A landslide is the movement of rock, debris or earth down a slope. They result from the failure of the materials which make up the hill slope and are driven by the force of gravity. Landslides are known also as landslips, slumps or slope failure. The downward force of a landslide can move slowly or quickly with disasterous effects. Landslides can be triggered by natural causes or by human activity. They range from a single boulder in a rock fall or topple to tens of millions of cubic metres of material in a debris flow. They can also vary in their extent, with some occurring very locally and impacting a very small area or hill slope while others affect much larger regional areas. The distance travelled by landslide material can also differ significantly with slides travelling from a few centimetres to many kilometres depending on the volume of material, water content and gradient of the slope. Landslides can even occur underwater, causing tidal waves and damage to coastal areas. These landslides are called submarine landslides.

Landslide Science for a Safer Geoenvironment presents comprehensive overview of progress in landslide science, landslide dynamics, landslide monitoring, and landslide risk assessment. Such alternative strategies include the use of building codes, land use planning controls, preventing water leakage, early warning systems and insurance schemes. This book addresses these developments and provides a multidisciplinary approach.

Table of Contents

Chapter 1

The effectiveness of high-resolution LiDAR data combined with PSInSAR data in landslide study

Andrea Ciampalini*, Federico Raspini, William Frodella, Federica Bardi, Silvia Bianchini, Sandro Moretti

Department of Earth Sciences, University of Firenze, via La Pira, 4, 50121, Florence, Italy

ABSTRACT

The spatial resolution of digital elevation models (DEMs) is an important factor for reliable landslide studies. Multi-interferometric techniques such as persistent scatterer interferometric synthetic aperture radar (PSInSAR) are used to evaluate the landslide state of activity and its ground deformation velocity, which is commonly measured along the satellite line of sight (LOS). In order to compare velocities measured by different satellites in different periods, their values can be projected along the steepest slope direction, which is the most probable direction of real movement. In order to achieve this result, DEM-derived products are needed. In this paper, the effectiveness of different DEM resolutions was evaluated in order to project ground deformation velocities measured by means of PSInSAR technique in two different case studies in the Messina Province (Sicily, southern Italy): San Fratello and Giampilieri. Three DEMs were used: (i) a 20-m resolution DEM of the Italian Military Geographic Institute (IGM), (ii) a 2-m resolution DEM derived from airborne laser scanning (ALS) light detection and ranging (LiDAR) data for the San Fratello 2010 landslide, and (iii) a 1-m resolution DEM derived from ALS LiDAR data for the area of Giampilieri. The evaluation of the applied method effectiveness was performed by comparing the DEMs elevation with those of each single permanent scatterer (PS) and projecting the measured velocities along the steepest slope direction. Results highlight that the higher DEM resolution is more suitable for this type of analysis; in particular, the PS located nearby the watershed divides is affected by geometrical problems when their velocities are projected along the steepest slope.

KEYWORDS

Landslide; PSInSAR; DEM; Velocity projection; LiDAR

1. INTRODUCTION

Spaceborne synthetic aperture radar (SAR) images are currently used to detect, monitor, and forecast active geological processes leading to ground deformation. For example, SAR data have been extensively used to study subsidence phenomena (Tomás et al. 2005; Raspini et al. 2012; Rosi et al. 2014). tectonic (Massironi et al.2009; Vilardo et al. 2009; Lagios et al. 2013). earthquakes (Lagios et al. 2012; Luo et al. 2013). and volcanic activity (Peltier et al. 2010). Among ground deformation phenomena, landslides represent one of the most studied natural hazards by using this technique (Herrera et al. 2009; Notti et al. 2010; Meisina et al. 2013; Tofani et al.2013; Oliveira et al. 2014). Differential synthetic aperture radar (DInSAR) represents the basic technique which exploits the phase difference between two SAR images acquired in different periods producing interferograms (Gabriel et al. 1989). More recently, several different multi-interferometric techniques have been developed, such as persistent scatterer interferometric synthetic aperture radar (PSInSAR) (Ferretti et al. 2001) and SqueeSAR (Ferretti et al. 2011). Small Baseline Subset (Berardino et al.2002). Stable Point Network (Arnaud et al. 2003) and Coherent Pixel Technique (Mora et al. 2003). These techniques use an extensive archive of satellite radar data (dating back to 1992) in order to identify networks of persistently scattering (i.e., radar reflecting) features, such as buildings and bridges, or natural features, such as rocky outcrops, against which relatively precise motion measurements are calculated retrospectively over the time spanned by the data archive. Landslide mapping and monitoring can greatly benefit from spaceborne InSAR data analysis because of its great cost-benefit ratio (especially by using archival data), non-invasiveness, wide area coverage, and high precision (Crosetto et al. 2010; Frattini et al. 2013; Wasowski and Bovenga 2014). Spaceborne multi-interferometric SAR techniques measure ground motion, with millimeter accuracy, representing a useful tool to detect and characterize slow surface phenomena, rapidly providing products such as landslide inventory and state of activity maps (Bovenga et al. 2006; Colesanti and Wasowski2006; Cascini et al. 2009; Hung et al. 2011; Ciampalini et al. 2012; Raspini et al. 2012; Bianchini et al. 2013; Del Ventisette et al. 2013; Bianchini et al. 2014).

Several studies have investigated the role of the digital elevation model (DEM) spatial resolution and its impact on landslide investigations (Glenn et al. 2006; Kasai et al.2009; Keijesers et al. 2011; Jabayedoff et al. 2012; Tarolli 2014) by comparing results obtained through the use of both low and high spatial resolution DEMs. During the last years, the increase in the use of very high-resolution DEM derived by light detection and ranging (LiDAR) data led to an improvement in terrain analysis, highlighting the advantages and limitations with respect to the low-resolution DEM (Classens et al.2005; Ardizzone et al. 2007; Chen et al. 2013; Fuchs et al. 2014; Tarolli 2014). For example, LiDAR-derived DEM has been extensively used in landslide studies (Jabayedoff et al. 2012 and reference therein). for the characterization of channel networks (Montgomery and Foufoula-Georgiu 1993; Tarolli and Dalla Fontana 2009; Cavalli et al. 2008), for river morphology analysis (Jones et al. 2007; Marcus and Fonstad 2010). active tectonic deformation recognition (Cunningham et al. 2006; Oskin et al. 2007). and in the detection of morphological changes in volcanic environments (Davila et al. 2007; Csatho et al. 2008). Considering the abovementioned works, the effectiveness of high- or low-resolution DEM in landslide study depends on the purpose of the research. On the other hand, several studies on landslide susceptibility assessment highlighted that low- to medium-resolution DEM can be more suitable (Catani et al. 2013). Normally, the optimal resolution of a DEM is directly related to the average size of the investigated hillslope process (Tarolli 2014) and also to the target of the investigation (hazard assessment, geotechnical study, risk evaluation, etc.). PSInSAR post processing analysis can benefit from its combination with DEM data for a more accurate evaluation of the landslide kinematic. For each detected permanent scatterer (PS), both displacements and ground deformation velocity are measured by the satellite sensor along its line of sight (LOS) even though the real displacement, which results slightly underestimated, actually occurs in three dimensions (Cascini et al. 2009). PSInSAR velocities in landslide studies are commonly used in the component measured along the LOS (Cigna et al. 2012; Meisina et al. 2008; Oliveira et al. 2014; Uzielli et al. 2015). A more reliable velocity measure can be obtained by projecting the velocity measured in the LOS direction, along the steepest slope considered as the most probable real movement direction (Notti et al. 2014; Bianchini et al. 2014). The velocity projection along the steepest slope uses the products derived from a DEM of the analyzed area, besides the satellite acquisition parameters (azimuth, incidence angle, and directional cosines) of the LOS. In order to achieve this result, a low- to medium-resolution DEM or a resample of a detailed DEM is recommended (Notti et al. 2014) with the aim of reducing the effect of terrain roughness and the presence of counterslopes. This work evaluates the usefulness of DEMs characterized by different spatial resolutions, during the multi-temporal analysis of PSInSAR post processing phase, in order to improve the effectiveness of this type of data in landslide kinematic analysis. The DEM resolution evaluation was tested in the Messina Province (Sicily, Italy), an area strongly affected by landslide phenomena.

2. STUDY AREAS

The Messina Province (Sicily, Italy) is an area prone to landslides, due to the combination of different triggering (seismicity, exceptional rainfall events, active tectonics, and volcanic activity) and predisposing factors (lithology, land use, morphology) (Billi et al. 2008; Ciampalini et al. 2015). Between 2009 fall and 2010 winter on both the Nebrodi and Peloritani mountain ranges, exceptional rainfall events led to more than a thousand different types of mass movements: debris flows, complex, rotational, and translational landslides, rock falls (following the classification of Varnes 1978). which caused 37 victims, and intense damages to urbanized areas, infrastructures, and cultivated and pasture lands.

2.1 The San Fratello site

The first chosen area is San Fratello (Figure 1), located along the Messina Province Tyrrhenian coastline, a town historically affected by landslide phenomena (Bardi et al.2014; Ciampalini et al. 2014). It is located at an altitude of 675 m asl. The town is crossed by an important local roadway (SS 289) which connects several villages scattered within the Nebrodi mountains to the coastal highway. The inhabited area is located along the north–south-oriented divide between two watersheds, the Furiano Creek valley to the west and the Inganno Creek valley to the east. The eastern slope altitude varies between 250 m and 675 m asl. From the Inganno Creek valley up to the divide, the slope has gently rises from 250 to 620 m asl, with an average slope of 14°. From 620 m asl to the top of the divide, the average slope increase to 24° (Figs. 2 and3). This morphology is probably linked to the presence of older landslides which modeled the slope. The most recent landslide event occurred on 14 February 2010, affecting a large sector of the eastern hillside, damaging several buildings and infrastructures (Bianchini et al. 2014). This landslide, about 1.8 km in length, involved an area of about 1 km^2, spanning from the town area eastern sector toward the Inganno Creek valley. This complex landslide produced also multiple failures, traction cracks, and counterslopes. In particular, the landslide is characterized by a rotational movement in its upper sector, evolving in an earth flow in its lower portion. From a geological point of view, the area of San Fratello is part of the northeastern sector of the Apennine–Maghrebian orogenic belt. In particular, the area is characterized by the presence of a sequence of terrigenous to calcareous sedimentary sequences. The southern part of the village is made of terrigenous terrains, lower Cretaceous in age clayey sequences (Argille Scagliose unit). In the northern portion of the village, carbonate complexes, represented by Liassic limestone platform sequences crop out, overlapped by a terrigenous Late Eocene–Oligocene Flysch (Frazzano` Flysch). The top of the sequence is represented by the Cretaceous pelagic dolostones and limestones (San Marco D'Alunzio unit), which crops out in the N–NW San Fratello village area (Nigro and Sulli 1995; Lavecchia et al. 2007). The landscape of the study area and its slope instability are strongly influenced by this geological framework because of the overlapping of hard brittle lithologies on the soft clayey formations. Furthermore, the San Fratello area is characterized by a 10-m-thick cover of silty clayey sediments which increases the formation of slope instability phenomena.

2.2 The Giampilieri site

The second chosen study area is Giampilieri, located between the Messina Province Ionian coast and the Peloritani mountain ridge (Ardizzone et al. 2012; Del Ventisette et al. 2012) (Figure 1), which on 1 October 2009, was hit by an exceptional storm (225 mm of rainfall recorded in 8 h). Giampilieri is located within the Giampilieri Creek valley on a southwest-oriented slope, which reaches its maximum altitude (516 m) about 2 km at northeast with respect to the village (Figure 4). The urban area is developed between 128 and 190 m asl, and it is connected with the coastal highway by a single roadway axis, climbing upward along very steep slopes. In correspondence to the village, the slope rises up to 175 m asl, with an average angle of 10.5°. The slope increases rapidly until 180 m asl, with an average angle of 28° rising up to 44° between 180 m and 270 m asl. The highest part of the slope is characterized by a decrease of the angle (32°) (Figure 4). The persisting rainfall triggered more than 600 slope failures, mainly shallow soil slides, debris flows, and debris avalanches, on an area of about 50 km^2. Giampilieri (Figure 1) was the most damaged village, where landslides and floods caused 37 fatalities (including 31 deaths and 6 missing persons), 122 injured people and 2019 evacuated people (Ciampalini et al. 2015). The geological framework of the Giampilieri area is characterized by the presence of strongly weathered metamorphic rocks, topped by 1.5–2-m-thick weathered soil. The area northwestern sector is made of low-grade phyllites and metasandstones, whereas the southeastern sector of the village is made of medium-grade amphibolites of the Aspromonte unit (Del Ventisette et al. 2012). The contact between the amphibolites and the phyllites is represented by a fault. Both the lithologies appear deeply fractured and characterized by a cleavage system which reduces the geotechnical characteristic of the rocks.

3. METHODOLOGY

In the last decades, different SAR satellite missions have been used to study ground deformation phenomena, providing the possibility to reconstruct the past deformation of the areas of interest. The European C-band missions ERS 1/2, environmental satellite (ENVISAT), routinely acquired images between 1992 and 2010. These data can be coupled with the currently operating C-band sensors, such as RADARSAT 1/2 and X-band sensors (TerraSAR-X and COSMO-SkyMed (constellation of small satellites for the Mediterranean basin Observation)), allowing the analysis of both past and recent ground displacements of the observed scenes, in a time interval of more than 20 years (Bianchini et al. 2013). PSInSAR (and its evolution SqueeSAR) provides deformation time series on grids of stable reflective point-wise targets called PS or diffuse scatterers (DS), which correspond to hand-made artifacts, such as buildings, railways or highways, and/or to natural targets (rocky outcrops), characterized by a coherent electromagnetic behavior in time (Ferretti et al. 2001;2011). The displacements recorded for each PS or DS are measured considering a stable ground point (reference point) of known coordinates. A ground deformation multi-temporal analysis can use different sensors acquiring in different periods. This means that ground deformation velocity of each PS dataset is measured along a different LOS. In order to compare different sensors data, characterized by different look angles and wavelengths, and thus different LOS, Notti et al. (2014) suggested to project the velocity, measured along the LOS, along the steepest slope (Eq. 1).

$$V_{\text{Slope}} = V_{\text{Los}}/C \tag{1}$$

where V_{Slope} is the velocity projected along the steepest slope, V_{Los} represents the velocity measured along the satellite LOS, and C is the coefficient calculated as follows (Eq. 2):

$$C = (\text{nlos COS}(S)\ \text{SIN}(A-1.571)) + (\text{elos}\ (-1\ \text{COS}(S)\ \text{COS}(A-1.571)) + (\text{hlos SIN}(S))) \tag{2}$$

where S is the Slope, A is the Aspect both derived from the DEM and nlos, and hlos and elos are the direction cosines of the satellite LOS.

The proposed formula uses DEM-derived products such as Slope and Aspect, therefore this procedure obviously needs a DEM. Three DEMs having different resolutions (Table 1) were used to project the PS velocities along the steepest slope. The first one is the 20-m resolution DEM, created in 2002 by Italian Military Geographic Institute (IGM) and derived from the interpolation of the contour lines and the geodetic points of the, which covers the whole Sicily area. For the San Fratello landslide area, a high-resolution airborne laser scanning (ALS) DEM, acquired soon after the 2010 event, was used. This DEM is characterized by spatial resolution of 2 m and 4 p/m^2. Even for the Giampilieri area, an ALS DEM, acquired during 2009, was used to produce a 1-m resolution DEM with 8 p/m^2.

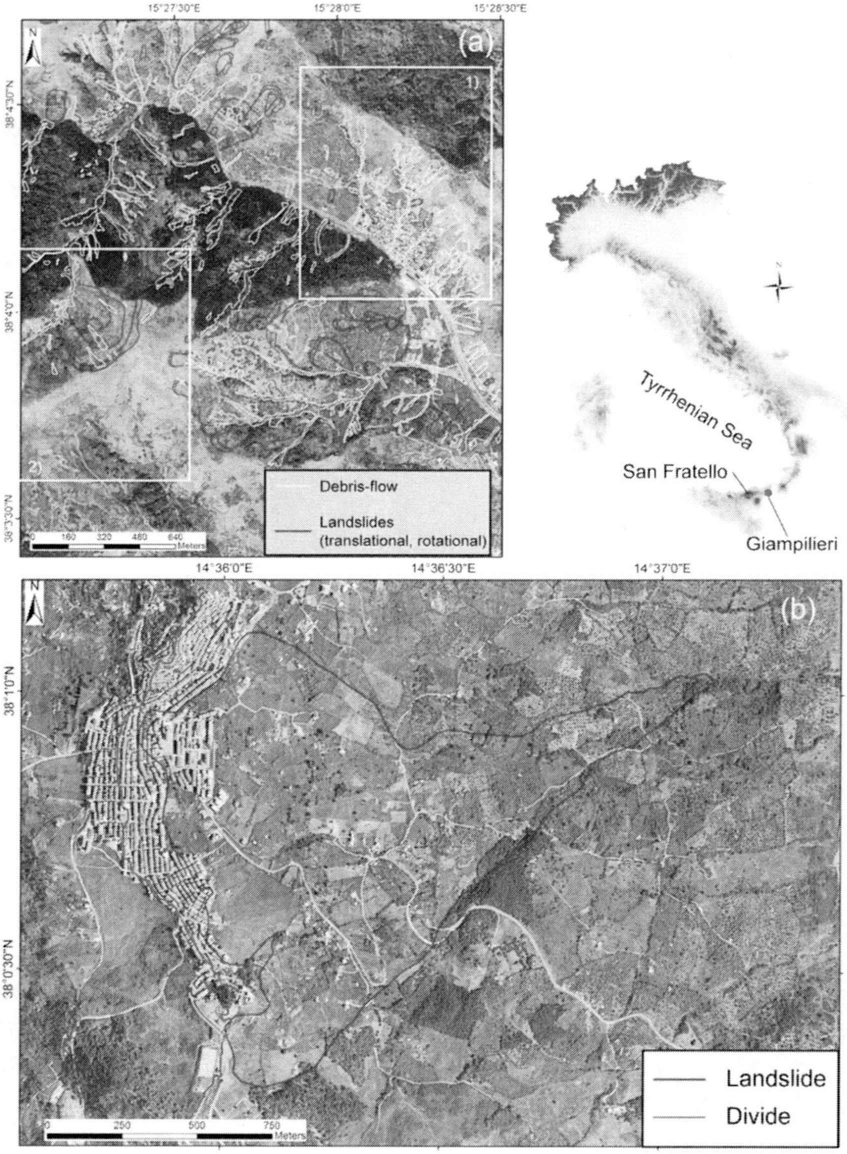

Figure 1. Location of the study areas: **a** Giampilieri and **b** San Fratello. The landslide inventory map in **a** is from Ardizzone et al. 2012. The landslide boundary in **b** is from Bardi et al. 2014. *White rectangles* (*1* and *2*) in **a**correspond to the area showed in Figure 9

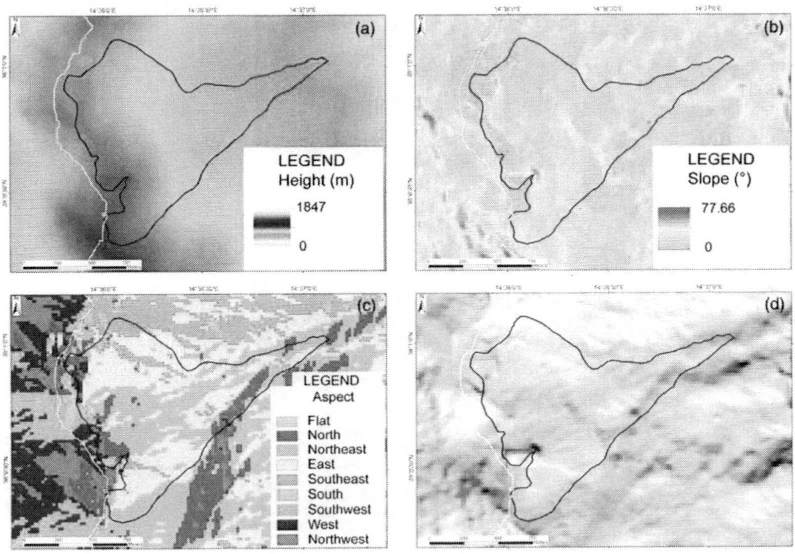

Figure 2. Example of the 20-m resolution DEM-derived products for the area of San Fratello. **a** DEM, **b** slope, **c** aspect, and **d** hillshade. The *white line* is the watershed divide; the *black line* corresponds to the 2010 landslide boundary

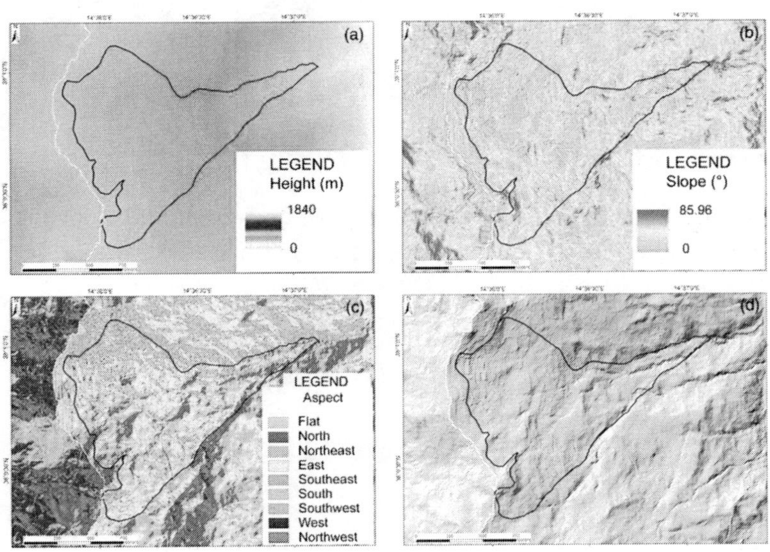

Figure 3. Example of the 2-m resolution DEM derived products for the area of San Fratello. **a** DEM, **b** slope, **c** aspect, and **d** hillshade. The *white line* is the watershed divide; the *black line* corresponds to the 2010 landslide boundary

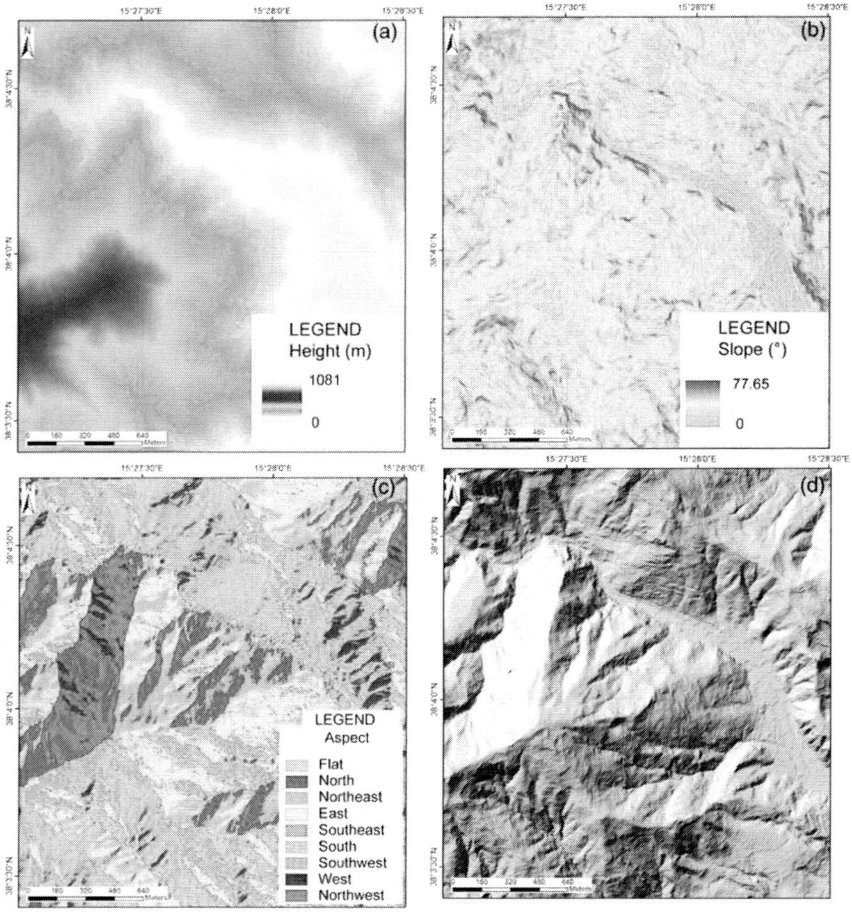

Figure 4. Example of the 2-m resolution DEM-derived products for the area of Giampilieri. **a** DEM, **b** slope, **c** aspect, and**d** hillshade

Table 1. Characteristic of the used PSInSAR datasets

Area	Sensor	Geometry	Time interval	No. images	Density (PS/km^2)	Resolution (m)
San Fratello	CSK	Descending	16/01/2011–02/05/2012	32	400.62	3
Giampilieri	ENVISAT	Ascending	22/01/2003–20/05/2009	55	414.88	30

PS heights included in the PS datasets have been retrieved using the shuttle radar topography mission (SRTM) DEM (WGS84) calibrated on the basis of the reference point which usually has an accurate GPS position. The orthometric height of each PS has been deduced subtracting a correction factor (around 43 m) which considers the ellipsoid height from the SRTM height. In order to evaluate the effectiveness of the DEM resolution, first of all, the height above the sea level of each PS, retrieved by the PS dataset, was compared to the height extracted from each available DEM by using a linear regression (Figure 5). The extraction of the PS heights from each available DEM was performed considering the height of each PS data center in GIS environment.

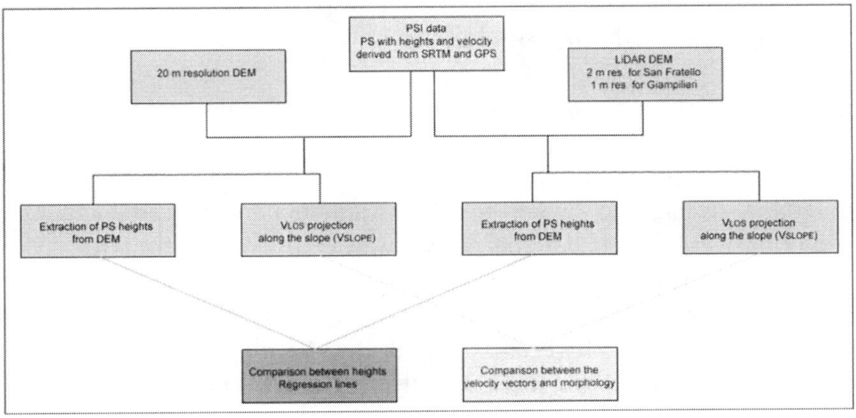

Figure 5. Flow chart illustrating the adopted methodology

The second step regards the projected velocities and their directions with respect to the North and to the slope morphology (Figure 5). In particular, the projected velocities direction was evaluated with respect to the slope orientation (Aspect). When a low-resolution DEM is used, geometrical problems can be detected especially in correspondence with watershed divides, where several velocity vectors should be oriented upslope. A visual inspection was used to detect the area affected by geometrical problems. In order to evaluate the effectiveness of different DEM resolution in PSInSAR post processing analysis, phase 2 PS dataset was used, one for each investigated area. In the San Fratello landslide area, PS data acquired with COSMO-SkyMed sensor were used to evaluate the post event ground deformation (Table 2). ENVISAT data were used to assess the pre-event deformation phenomena in correspondence of the area of Giampilieri (Table 1).

Table 2. Characteristic of the used DEMs

Area	Type	Date	Resolution
Sicily	Interpolation	2002	20 m
San Fratello	LIDAR	2010	2 m–4 p/m^2
Giampilieri	LIDAR	2009	1 m–8 p/m^2

4. RESULTS AND DISCUSSION

4.1 The San Fratello landslide

In the San Fratello area, the results obtained by using the 20- and 2-m resolution DEMs were compared. The test site location is very useful because of its position, located across a divide between two watersheds. Along the divide, the higher the approximation introduced by a low-resolution DEM, the lower the accuracy in obtaining an accurate divide position. An example of the DEM-derived products is reported in Figs. 2 and 3. After a simple visual comparison, the 2-m resolution DEM (Figure 3) can be considered more useful to characterize the slope morphology. In particular, both the derived Slope and hillshade clearly show several elements on the eastern slope of San Fratello such as the presence of a channel network, counterslopes, and secondary scarps that cannot be detected by using the 20-m resolution DEM (Figure 2). Furthermore, the comparison between the two obtained Aspects highlights the presence of an area, about 200 m^2 in extension, located across the divide showing different orientation (Figure 2c). In the 20-m resolution-derived Aspect, the N-S-oriented divide does not perfectly separate the eastward-oriented slope from the westward-oriented ones. A strip located eastward of the divide shows a westward orientation which does not correspond to the real slope morphology (Figure 2c). This area can affect a proper projection of the ground deformation velocities along the steepest slope. Among the data included in each PS dataset (e.g., deformation, velocity), the elevation of each PS is reported. These elevations were compared to those derived by both the available DEMs in order to compare their correlations (Figure 6). The comparison between PS elevations and those retrieved from the 20-m resolution DEM shows that the correlation between elevations is characterized by an $R^2 = 0.59$. This correlation is based on 4098 PS. A strong improvement in the correlations between the elevations was observed comparing the PS heights to the elevations extracted from the 2-m resolution DEM (Figure 7). In this case, the R^2 is

0.99 suggesting that this DEM is strongly reliable. This correlation is also based on the same 4098 PS. Following the equation proposed by Notti et al. (2014). the PS velocities were projected along the steepest slope using both the 2- and 20-m resolution DEMs. Usually, the directions of the projected PS velocity, by using both DEMs, are in agreement with the average slope direction, except for the area characterized by a higher terrain roughness, where the 2-m resolution DEM resulted more accurate. This agreement strongly decreases in correspondence with the divide, especially in the previously mentioned area where the aspect derived from the 20-m resolution DEM shows a different orientation with respect to the real morphology. In this area, the PS velocities projected using the 20-m resolution DEM are directed upward (Figure 8a), which is a very unlikely condition. On the contrary, projected PS velocities obtained by using the 2-m resolution DEM are in agreement with the main slope direction (Figure 8b). Considering the unreliable projected PS velocities, it is possible to detect the area where the 20-m resolution DEM is affected by geometric problems (Figure 8c), probably related to the approximation due to its lower resolution which increased nearby to the divide. This area includes also part of the 2010 San Fratello landslide crown. The total amount of PS located in this area is 1609 (39.26 % of the available PS) which must not be considered for a correct ground deformation analysis using the projected PS velocities. A subset of the whole PS population was extracted considering those PS located inside the detected area in order to investigate the relationship between PS elevations and DEMs elevations. A decrease of the correlation can be observed both for the 20- and 2-m resolution DEM but, while this decrease is still acceptable ($R^2 = 0.86$) for the 2-m DEM (Figure 7a), for the 20-m DEM, it is very high ($R^2 = 0.03$) (Figure 9b). These correlations have been assessed using a total of 1609 PS.

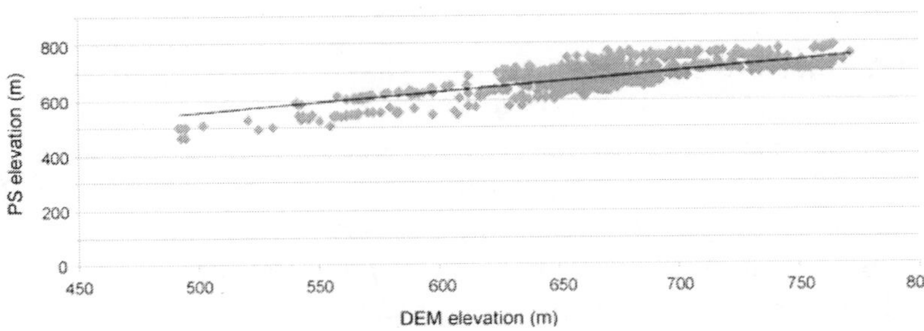

Figure 6. Correlation between PS heights and 20-m resolution DEM elevations in the San Fratello area

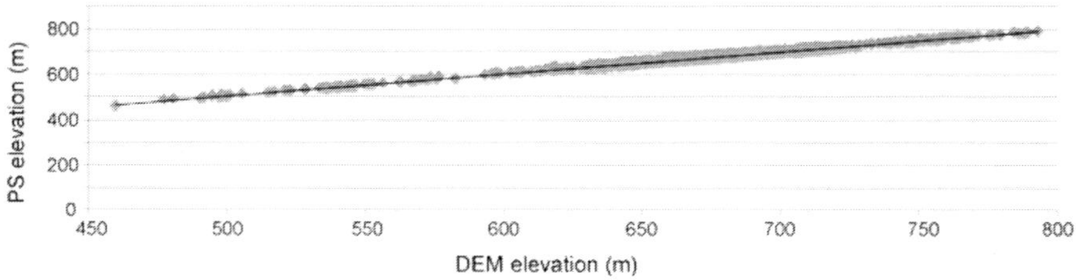

Figure 7. Correlation between PS heights and 2-m resolution DEM elevations in the San Fratello area

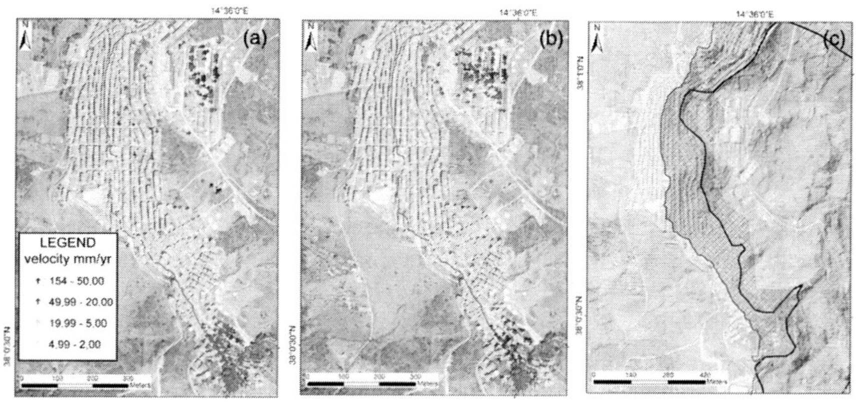

Figure 8. Projection of the PS velocities using the 20-m resolution DEM for the crown of the San Fratello landslide (**a**) and the 2-m resolution DEM (**b**). The *blue line* represents the divide; the *black line* represents the landslide boundary. **c** delimitation of the area affected by geometrical problems

4.2 The Giampilieri area

The same approach described for the San Fratello site was also applied in the area of Giampilieri, in order to evaluate the differences between the 20-m resolution DEM and the available 1-m resolution DEM. In this case, PS data show a local lower density because they were elaborated by using the C-band sensor ENVISAT, which is characterized by lower spatial and temporal resolution with respect to the X-band sensor COSMO-SkyMed. The comparison among the PS heights and the elevations measured with the available DEM show a good correlation for both the 20-m and the 1-m resolution DEMs, but, also in this case, the DEM having the higher resolution shows a better $R^2 = 0.9992$, whereas R^2 for the 20-m resolution DEM is equal to 0.9471 (Figure 10). Even though also the 20-m resolution DEM shows a very good agreement between the PS heights and the related DEM elevations, when PS velocities were projected along the steepest slope direction, several geometric problems can be detected within the study area (Figure 11). For example, north of the village of Giampilieri, several projected PS velocities, using the 20-m resolution DEM, are directed upslope (Figure 11a); on the contrary, all the projected PS velocities using the 1-m resolution DEM are correctly oriented along the slope direction (Figure 11b). The usefulness of the high-resolution DEM (1 m in this case) for the PS velocities projection can be appreciated especially for the areas characterized by complex relief. For example, in case of mountainous areas with narrow valleys and high difference in altitudes between the valley bottom and the divides, the higher is the DEM resolution, the more accurate will be the V_{slope} direction (see Figure 11c, d, where most of the PS velocities projected using the 20-m resolution DEM show a very different direction with respect to those of the real morphology).

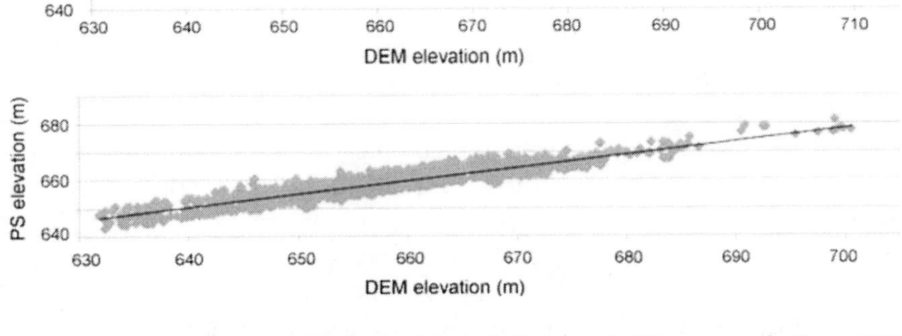

Figure 9. Correlation between PS heights and 20-m resolution DEM elevations (**a**); correlation between PS heights and 2-m and resolution DEM elevations (**b**) for the PS located inside the area detected in Figure 6c

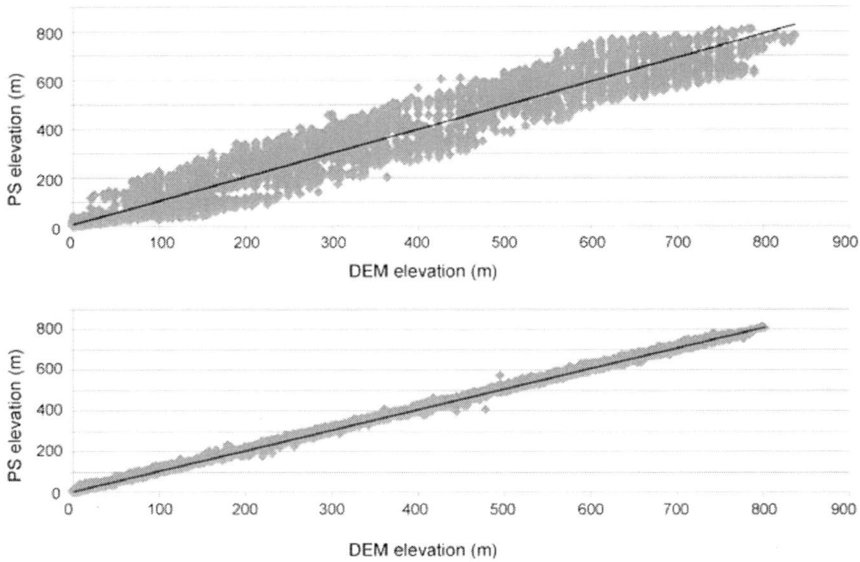

Figure 10. Correlation between PS heights and 20-m resolution DEM elevations (**a**); correlation between PS heights and 1-m and resolution DEM elevations (**a**)

5. CONCLUSION

In this work, the reliability of high spatial resolution DEMs was tested in order to provide new opportunities and challenges for mapping landslide morphology with high accuracy, both through visual interpretation and statistical analysis.

In particular, the effectiveness of two DEMs, characterized by different resolutions, was discussed in landslide study as tools for an accurate PSInSAR post processing analysis. The comparison was made considering a high-resolution LiDAR DEM and a low-resolution DEM derived from the interpolation of contour lines. The test was performed in two different sites. For each test site, a LiDAR DEM was available. PS data measure ground deformation velocities along the LOS of the used satellite sensor. These measures can underestimate the real velocity which usually occurs along the steepest slope. Furthermore, when PS datasets are acquired by using different sensors and in different periods on the same area, they are also acquired by different LOS. In order to compare the ground deformation velocities of different PS datasets, they must be re-projected along a common direction. The steepest slope direction is commonly considered as the most reliable and thus it can be used to project the velocities. This process needs for a DEM to calculate the characteristics of the slope where the PS are located, in particular, the formula used for the projection needs for the local value of both the slope gradient (Slope derived from a DEM) and the slope orientation (Aspect derived from a DEM).

The evaluation of the effectiveness of DEMs having different resolution in the projection of the ground deformation velocities along the steepest slope was carried out. First, by comparing the height of all the available PS with the relative DEM elevation for both the 20- and 2-m resolution DEM. Second, the velocities were projected along the steepest slope by using again both of the DEMs and identifying the presence of areas where the projected velocities show an unreliable direction (upslope). This problem particularly affects those PS located in the upper part of the slope, close to the divide, where the approximation that characterizes the low-resolution DEM does not allow a proper projection of the PS velocities along the steepest slope.

Despite the used DEMs were created in different time periods, the variation between the terrain surface heights due to possible tectonic activity is negligible. The investigated areas are tectonically active, but the vertical ground deformation measured in the long-term period highlights an overall stability during the last 15 years (Del Ventisette et al. 2013). Although the original topographic maps used to create the IGM 20-m resolution DEM can be very old (1970–1990), the height discrepancy between the latter and the ALS DEM is very high (up to 50 m) and cannot be related to vertical deformation due to the tectonic activity. Considering the comparison between PS heights, derived from the SRTM data, calibrated with the reference point GPS position, and the ALS DEMs heights, a very reduced discrepancy has been recognized. This comparison can be considered as a doubleway in order to estimate the accuracy of both the data.

High-resolution topographic data have not only the potential to detect morphological characteristic (counterslopes, secondary scarps, etc.) within a landslide, but this work also highlights that a correct PSInSAR post processing analysis can benefit from the use of a high-resolution DEM, with special regards to the landslide kinematics understanding. On the contrary, the use of a medium- to low-resolution DEM can lead to geometric problems in the projection of the PS velocities, especially nearby the watershed divides and in the areas characterized by a complex morphology.

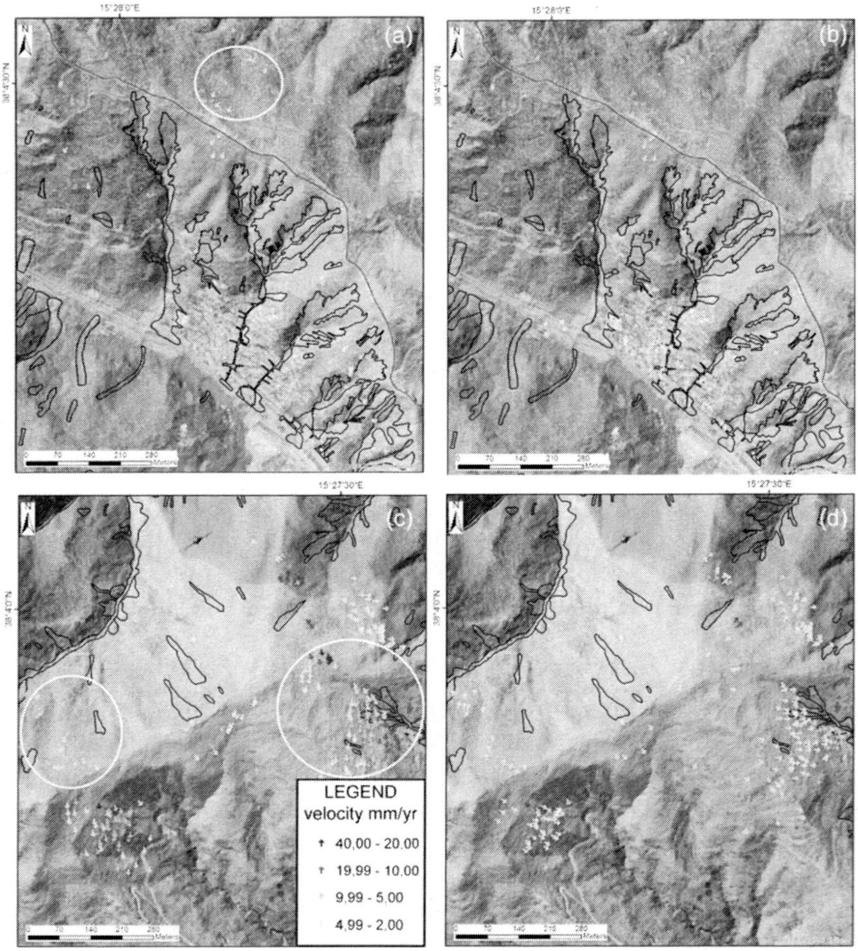

Figure 11. a PS velocities projected using the 20-m resolution DEM north of the Giampilieri village (the with circle include the PS having upslope direction); **b** PS velocities projected using the 2-m resolution DEM north of the Giampilieri; **c** PS velocities projected using the 20-m resolution DEM south of the Giampilieri; **d** PS velocities projected using the 2-m resolution DEM south of the Giampilieri. The *blue line* in **a** and **b** represents the divide; the *black lines* represent the landslide boundaries of the 2009 event occurred in the area

ACKNOWLEDGMENTS

This study was funded by the Seventh Framework Programme of European Commission: Landslide Modelling and Tools for vulnerability assessment preparedness and recovery management (LAMPRE). We thank two anonymous reviewers for their suggestions which greatly improved this work. The COSMO-SkyMed data have been processed by Tele-Rilevamento Europa (TRE).

REFERENCES

1. Ardizzone F, Cardinali M, Galli M, Guzzetti F, Reichembach P (2007) Identification and mapping of recent rainfall-induced landslides using elevation data collected by airborne lidar. Nat Hazards Earth Sist Sci 7:637–650

2. Ardizzone F, Basile G, Cardinali M, Casagli N, Del Conte S, Del Ventisette C, Fiorucci F, Garfagnoli F, Gigli G, Guzzetti F, Iovine G, Mondini AC, Moretti S, Panebianco M, Raspini F, Reichembach P, Rossi M, Tanteri L, Terranova O (2012) Landslide inventory map for the Briga and the Giampilieri catchments, NE Sicily, Italy. J Maps 8:176–180

3. Arnaud A, Adam N, Hanssen N, Inglada J, Duro J, Closa J, Eineder M (2003) ASAR ERS interferometric phase continuity. Geoscience and Remote Sensing Symposium, 2003. IGARSS Proceedings. 2003 I.E. International 1133–1135

4. Bardi F, Frodella W, Ciampalini A, Bianchini S, Del Ventisette C, Gigli G, Fanti R, Moretti S, Basile G, Casagli N (2014) Integration between ground based and satellite SAR data in landslide mapping: the San Fratello case study. Geophys J Roy Astron Soc 223:45–60

5. Berardino P, Fornaro G, Lanari R, Sansosti E (2002) A new algorithm for surface deformation monitoring based on small baseline differential SAR interferograms. IEEE Transaction on Geoscience and Remote Sensing 40:2375–2383

6. Bianchini S, Herrera G, Mateos RM, Notti D, Garcia I, Mora O, Moretti S (2013) Landslide activity maps generation by means of persistent scatterer interferometry. Remote Sensing 5:6198–6222

7. Bianchini S, Ciampalini A, Raspini F, Bardi F, Di Traglia F, Moretti S, Casagli N (2014) Multi-temporal evaluation of landslide movements and impacts on buildings in San Fratello (Italy) by means of C-band and X-band PSI data. Pure Appl Geophys.

8. Billi A, Funiciello R, Minelli L, Faccenna C, Neri G, Orecchio B, Presti D (2008) On the cause of the 1908 Messina tsunami, southern Italy. Geophysical Research Letters 35

9. Bovenga F, Nutricato R, Refice A, Wasowski J (2006) Application of multi-temporal differential interferometry to slope instability detection in urban/peri-urban areas. Eng Geol 3–4:218–239

10. Cascini L, Fornaro G, Peduto D (2009) Analysis at medium scale of low-resolution DInSAR data in slow-moving landslide-affected areas. ISPRS J Photogram Rem Sens 64:598–611

11. Catani F, Lagomarsino D, Segoni S, Tofani V (2013) Landslide susceptibility estimation by random forests technique: sensitivity and scaling issues. Natural Hazards and Earth System Sciences 13:2815–2831

12. Cavalli M, Tarolli P, Marchi L, Dalla Fontana G (2008) The effectiveness of airborne LiDAR data in the recognition of channel-bed morphology. Catena 73:249–260

13. Chen Q, Liu X, Liu C, Ji R (2013) Impact analysis of differential spatial resolution DEM on object-oriented landslide extraction from high resolution remote sensing images. Ninth International on Natural Computation (ICNC), 23–25 July 2013, Shenyang, China.

14. Ciampalini A, Cigna F, Del Ventisette C, Moretti S, Liguori V, Casagli N (2012) Integrated geomorphological mapping in the north-western sector of Agrigento (Italy). J Maps 8:136–145

15. Ciampalini A, Bardi F, Bianchini S, Frodella W, Del Ventisette C, Moretti S, Casagli N (2014) Analysis of building deformation in landslide area using multisensor PSInSARTM technique. Int J of App Earth Observ Geoinf 33:166–180

16. Ciampalini A, Raspini F, Bianchini S, Frodella W, Bardi F, Lagomarsino D, Di Traglia F, Moretti S, Proietti C, Pagliara P, Onori R, Corazza A, Duro A, Basile G, Casagli N (2015) Remote sensing as tool for development of

landslide databases: the case of the Messina Province (Italy) geodatabase. Geomorphology.

17. Cigna F, Bianchini S, Casagli N (2012) How to assess landslide activity and intensity with persistent scatterer interferometry (PSI): the PSI-based matrix approach. Landslide 10:267–283

18. Classens L, Heuvelink GBM, Schoorl JM, Veldkamp A (2005) DEM resolution effects on shallow landslide hazard and soil redistribution modeling. Earth Surf Process Landf 30:461–477

19. Colesanti C, Wasowski J (2006) Investigating landslides with space-borne synthetic aperture radar (SAR) interferometry. Eng Geol 3–4:173–199

20. Crosetto M, Monserrat O, Iglesias R, Crippa B (2010) Persistent scatterer interferometry: potential, limits and initial C- and X-band comparison. Photogramm Eng Remote Sens 76:1061–1069

21. Csatho B, Schenk T, Kyle P, Wilsond T, Krabille WB (2008) Airborne laser swath mapping of the summit of Erebus volcano, Antarctica: applications to geological mapping of a volcano. J Volcanol Geoth Res 177:531–548

22. Cunningham D, Grebby S, Tansey K, Gosar A, Kastelic V (2006) Application of airborne lidar to mapping seismogenetic faults in forested mountainous terrain, southeastern Alps, Slovenia. Geophys Res Lett 33:L20308

23. Davila N, Capra L, Gavilanes Ruiz JC, Varley N, Norini G, Vazquez AG (2007) Recent lahars at Volcán de Colima (Mexico): drainage variation and spectral classification. J Volcanol Geotherm Res 165:127–141

24. Del Ventisette C, Garfagnoli F, Ciampalini A, Battistini A, Gigli G, Moretti S, Casagli N (2012) An integrated approach to the study of catastrophic debris-flow: geological hazard and human influence. Nat Hazards Earth Syst Sci 12:2907–2922

25. Del Ventisette C, Ciampalini A, Manunta M, Calò F, Paglia L, Ardizzone F, Mondini AC, Reichembach P, Mateos RM, Bianchini S, Garcia I, Füsi B, Deak ZV, Radi K, Graniczny M, Kowalski Z, Piatkowska A, Przylucka M, Retzo H, Strozzi T, Colombo D, Mora O, Sanches F, Herrera G, Moretti S, Casagli N, Guzzetti F (2013) Exploitation of large archives of ERS and

ENVISAT C-band SAR data to characterize ground deformations. Remote Sensing 5:3896–3917

26. Ferretti A, Prati C, Rocca F (2001) Permanent scatterers in SAR interferometry. IEEE Trans Geosci Remote Sens 39:8–20

27. Ferretti A, Fumagalli A, Novali F, Prati C, Rocca F, Rucci A (2011) A new algorithm for processing interferometric data-stacks: SqueeSAR. IEEE Trans Geosci Remote Sens 49:3460–3470

28. Frattini P, Crosta GB, Allievi J (2013) Damage to buildings in large slope rock instabilities monitored with the PSInSAR™ technique. Remote Sensing 5:4753–4773

29. Fuchs M, Torizin J, Kühn F (2014) The effect of DEM resolution on the computational of the factor of safety using an infinite slope model. Geophys J Roy Astron Soc 224:16–26

30. Gabriel AK, Goldstein RM, Zebker HA (1989) Mapping small elevation changes over large areas: differential radar interferometry. Journal of Geophysical Research: Solid Earth 94:9183–9191

31. Glenn NF, Streutker DR, Chadwick DJ, Thackaray GD, Dorsch SJ (2006) Analysis of LiDAR-derived topographic information for characterizing and differentiating landslide morphology and activity. Geophys J Roy Astron Soc 73:131–148

32. Herrera G, Davalillo JC, Mulas J, Cooksley G, Montserrat O, Pancioli V (2009) Mapping and monitoring geomorphological processes in mountainous areas using PSI data: Central Pyrenees case study. Nat Hazards Earth Syst Sci 9:1587–1598

33. Hung W-C, Hwang C, Chen Y-A, Chang C-P, Yen J-Y, Hooper A, Yang C-Y (2011) Surface deformation from persistent scatterers interferometry and fusion with leveling data: a case study over the Choushui River Alluvial Fan, Taiwan. Remote Sens Environ 115:957–967

34. Jabayedoff M, Oppikofer T, Abellán A, Derron MH, Loye A, Metzger R, Pedrazzini A (2012) Use of LIDAR in landslide investigations: a review. Nat Hazards 61:5–28

35. Jones AF, Brewer PA, Johnstone E, Macklin MG (2007) High resolution interpretative geomorphological mapping of river valley environments using airborne lidar data. Earth Surf Process Landf 21:1574–1592

36. Kasai M, Ikeda M, Asahina T, Fujisawa K (2009) LiDAR-derived DEM evaluation of deep-seated landslides in a steep and rocky region of Japan. Geophys J Roy Astron Soc 113:57–69

37. Keijesers JGS, Schoorl JM, Chang K-T, Chiang S-H, Claessens L, Veldkamp A (2011) Calibration and resolution effects on model performance for predicting shallow landslide locations in Taiwan. Geophys J Roy Astron Soc 133:168–177

38. Lagios E, Papadimitriou P, Novali F, Sakkas V, Fumagalli A, Vlachou K, Del Conte S (2012) Combined seismicity pattern analysis, DGPS and PSInSAR studies in the broader area of Cephalonia (Greece). Tectonophysics 534–525:43–58

39. Lagios E, Sakkas V, Novali F, Bellotti F, Ferretti A, Vlachou K, Dietrich V (2013) SqueeSAR™ and GPS ground deformation of Santorini Volcano (1992–2012): tectonic implications. Tectonophysics 594:38–59

40. Lavecchia G, Ferrarini F, De Nardis R, Visini F, Barbano MS (2007) Active thrusting as possible seismogenic source in Sicily (Southern Italy): some insights from integrated structural-kinematic and seismological data. Tectonophysics 445:145–167

41. Luo S, Fu L, Zhu S, He Q, Wan W, Yang B (2013) Processes of the displacement field change of the 2009 April 6 Mw 6.3 L'Aquila earthquake using persistent scatterer and small baseline methods. Earthquake Science 26:293–299

42. Marcus WA, Fonstad MA (2010) Remote sensing of rivers: the emergence of a subdiscipline in the river sciences. Earth Sur Proc Land 35:1867–1872

43. Massironi M, Zampieri D, Bianchi M, Schiavo A, Franceschini A (2009) Use of PSInSAR™ data to infer active tectonics: clues on the differential uplift across the Giudicarie belt (Central-Eastern Alps, Italy). Tectonophysics 476:297–303

44. Meisina C, Zucca F, Notti D, Colombo A, Cucchi A, Savio G, Giannico C, Bianchi M (2008) Geological interpretation of PSInsAR data at regional scale. Sensors 8:7469–7492

45. Meisina C, Notti D, Zucca F, Ceriani M, Colombo A, Poggi F, Roccati A, Zaccone A (2013) The use of PSInSAR™ and SqueeSAR™ techniques for updating landslide inventories. C. Margottini, P. Canuti, K. Sassa (Eds.), Landslide Science and Practice, Springer, Berlin Heidelberg, pp. 81–87

46. Montgomery DR, Foufoula-Georgiu E (1993) Channel network source representation using digital elevation models. Water Resour Res 29:3925–3934

47. Mora O, Mallorqui JJ, Broquetas A (2003) Linear and nonlinear terrain deformation maps from a reduced set of interferometric SAR images. IEEE Trans Geosci Remote Sens 41:2243–2253

48. Nigro F, Sulli A (1995) Plio-Pleistocene extensional tectonics in the Western Peloritani area and its offshore (northeastern Sicily). Tectonophysics 252:295–305

49. Notti D, Davalillo JC, Herrera G, Mora O (2010) Assessment of the performance of X-band satellite radar data for landslide mapping and monitoring: upper Tena Valley case study. Nat Hazards Earth Sist Sci 10:1865–1875

50. Notti D, Herrera G, Bianchini S, Meisina C, Davalillo JC, Zucca F (2014) A methodology for improving landslide PSI analysis. Int J Remote Sens 35:2186–2214

51. Oliveira SC, Zêzere JL, Catalão J, Nico G (2014) The contribution of PSInSAR interferometry to landslide hazard weak rock-dominated areas. Landslides 12:703–719

52. Oskin ME, Le K, Strane MD (2007) Quantifying fault-zone activity in arid environments with high-resolution topography. Geophys Res Lett 34:L23S05

53. Peltier A, Bianchi M, Kaminski E, Komorowsky J-C, Rucci A, Staudacher T (2010) PSInSAR as a new tool to monitor pre-eruptive volcano ground deformation: validation using GPS measurement on Piton de la Fournaise. Geophys Res Lett 37, L12301

54. Raspini F, Cigna F, Moretti S (2012) Multi-temporal mapping of land subsidence at basin scale exploiting persistent scatterer interferometry: case study of Gioia Tauro plain (Italy). J Maps 8:514–524

55. Rosi A, Agostini A, Tofani V, Casagli N (2014) A procedure to map subsidence at the regional scale using the persistent scatterer interferometry (PSI) technique. Remote Sensing 6:10510–10522

56. Tarolli P (2014) High-resolution topography for understanding Earth surface processes: opportunities and challenges. Geophys J Roy Astron Soc 216:295–312

57. Tarolli P, Dalla Fontana G (2009) Hillslope to valley transition morphology: new opportunities from high resolution DTMs. Geophys J Roy Astron Soc 113:47–56

58. Tofani V, Raspini F, Catani F, Casagli S (2013) Persistent scatterer interferometry (PSI) technique for landslide characterization and monitoring. Remote Sensing 5:1045–1065

59. Tomás R, Márquez Y, Lopez-Sanchez JM, Delgado J, Blanco P, Mallorquí JJ, Martínez M, Herrera G, Mulas J (2005) Mapping ground subsidence induced by aquifer overexploitation using advanced differential SAR inteferometry: Vega Media of the Segura River (SE Spain) case study. Remote Sens Environ 98:269–283

60. Uzielli M, Catani F, Tofani V, Casagli N (2015) Risk analysis for the Ancona landslide-I: characterization of landslide kinematics. Landslides 12:69–82

61. Varnes DJ (1978) Slope movements and types and processes. Landslides Analysis and Control, Transportation Research Board Special Report 176:11–33

62. Vilardo G, Ventura G, Terranova C, Matano F, Nardò S (2009) Ground deformation due to tectonic, hydrothermal, gravity, hydrogeological, and anthropic processes in the Campania Region (Southern Italy) from permanent scatterers synthetic aperture radar interferometry. Remote Sens Environ 113:197–212

63. Wasowski J, Bovenga F (2014) Investigating landslides and unstable slopes with satellite multi temporal interferometry: current issues and future perspectives. Eng Geol 174:103–138

Chapter 2

Forecasting rock slope failure: how reliable and effective are warning systems?

Martina Sättele[1], Michael Krautblatter[2], Michael Bründl[1], Daniel Straub[3]

[1]WSL Institute for Snow and Avalanche Research SLF, ? Avalanche Dynamics and Risk Management, Flüelastrasse 11, 7260, Davos Dorf, Switzerland

[2]Technische Universität München, Landslide Research, Arcisstrasse 21, 80 333, Munich, Germany

[3]Technische Universität München, Engineering Risk Analysis Group, Arcisstrasse 21, 80 333, Munich, Germany

ABSTRACT

Warning systems are increasingly applied to reduce damage caused by different magnitudes of rockslides and rockfalls. In an integrated risk-management approach, the optimal risk mitigation strategy is identified by comparing the achieved effectiveness and cost; estimating the reliability of the warning system is the basis for such considerations. Here, we calculate the reliability and effectiveness of the warning system installed in Preonzo prior to a major rockfall in May 2012. "Reliability" is defined as the ability of the warning system to forecast the hazard event and to prevent damage. To be cost-effective, the warning system should forecast an event with a limited number of false alarms to avoid unnecessary costs for intervention measures. The analysis shows that to be reliable, warning systems should be designed as fail-safe constructions. They should incorporate components with low failure probabilities, high redundancy, have low warning thresholds, and additional control systems. In addition, the experts operating the warning system should have limited risk tolerance. In an additional hypothetical probabilistic analysis, we investigate the effect of the risk attitude of the decision makers and of the number of sensors on the probability of detecting the event and initiating a timely evacuation, as well as on the related intervention cost. The analysis demonstrates that quantitative assessments can support the identification of optimal warning system designs and decision criteria.

KEYWORDS

Rockfall; Inverse velocity; Warning system; Reliability; Effectiveness

1. INTRODUCTION

Mountain regions are prone to gravitationally driven natural hazard processes. Mid- (>10^4 m^3) and high-magnitude (>10^6 m^3) rockfalls and rockslides can cause much damage due to on-site and off-site effects such as inundation, dam failure, and subsequent debris flow activity in released debris (Hewitt et al. 2008). In some areas, especially those affected by melting permafrost and rapid deglaciation, there is increasing evidence for rising frequencies of rock slope failure (Ravanel and Deline2011; Huggel et al. 2012; Krautblatter et al. 2013). Out of 38 recorded landslide disasters (1000–1999 AD) that have each caused more than 1000 casualties worldwide, 75 % were due to massive high-magnitude rock slope failure (Evans 2006). But also, mid-magnitude rockfalls pose a great risk to densely populated alpine areas. This is because of their sudden release, fast propagation, long and (often unpredictable) runout, and impact energy far beyond the retention capacity of most rockfall nets or other protection measures (Zimmer et al. 2012). They often result in total destruction of buildings and infrastructure as has recently been demonstrated by a 3000-m^3 rockfall on the major Gotthard Railway (Switzerland) on June 5, 2012, which caused one casualty and closed the railway line for more than a month.

To protect individuals and mobile objects close to potential release areas, warning systems are increasingly applied as a risk mitigation measure. They can generate timely information and allow at-risk people to act and to reduce their impacts (UNISDR2007). Warning systems are especially useful when the capacity of structural rockfall protection measures is likely to be exceeded; they are typically associated with low life cycle cost, a high flexibility, and with minimal impact on the environment (Hattenberger and Wöllik 2008). Modern warning systems for forecasting mid- and high-magnitude rockfall events consist of technical components, models, and human decision making (Ganerød et al. 2008; Sturny and Bründl 2014). They make use of typical velocity patterns of rockfall processes for detecting events.

To this end, displacements are monitored, often with total stations and extensometers. In the operation of rockfall warning systems (RWS), two phases can be distinguished, which are related to the progress of a rockfall failure. In the first phase, to which we refer to as "detachment phase," warning systems constantly monitor movement patterns of the unstable rock mass to send warning information instantaneously when a predefined threshold is exceeded. Higher-magnitude rock slope failures characteristically evolve slowly, typically weeks to several years, until a critical level of detachment is developed. Failing slopes show early signs of deformations such as tension cracks, movement, and increasing rockfall activity (Hungr and Evans 2004). This progressive failure includes daily fluctuations and depends on temperature, rain, snow melt, and long-term stress–strain behavior of slopes which control fracture propagation. Rainfall, earthquakes and snow melting, weathering, and aging can be important triggers or driving factors that weaken the rock mass during the first phase (Lacasse et al. 2008). Rainfall and snow melt directly influence the pore water pressure, which can be a major driver for rock displacement and the final failure (Loew et al. 2012). To include these contributing factors in decision procedures associated with warning systems, they are often equipped with rain gauges and sensor measuring pore water pressures.

In the second phase, which we refer to as the "acceleration phase", the final failure occurs after an acceleration in which rock bridges are destroyed, often preceded by subcritical fracture propagation and stress corrosion, especially in a brittle rock mass (Petley et al. 2005a, b; Petley and Petley 2006). In this acceleration phase, experts should analyze sensor data and apply models to forecast the event time and execute appropriate intervention measures. The time of failure can be forecasted using measured velocity (Fukuzono 1990). Figure 1 illustrates how the inverse velocity (1/v) is plotted against time in a linear 1/v model to obtain an event forecast (Saito 1969; Hashimoto et al. 1982).

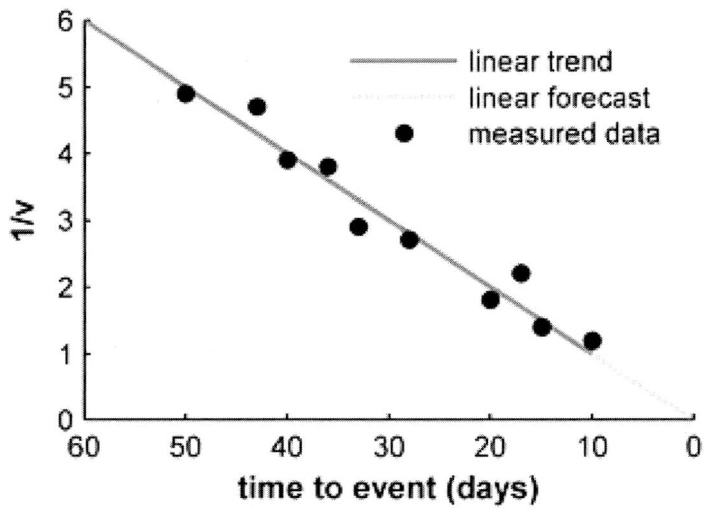

Figure 1. Sketch of a 1/v model to predict the event date of brittle rock slope failure based on a measured kinematic time series

Many authors emphasize the applicability of the linear 1/v model, especially for brittle rock slope failures, i.e., forming of a rupture surface and first time failure in cohesive materials whereas asymptotic trends can be often observed in pre-existing shear surfaces or ductile deformations in the basal region (Petley and Petley 2006). Successful forecasts of a mid-magnitude icefall ($1*10^5$ m^3) and a hazardous rockfall (3000–5000 m^3) in the early 1990s are described by Wegmann et al. (2003). Krähenbühl (2006) presents an application of the 1/v model to forecast two successive rockfall events (40,000 and 200,000 m^3) in the alpine region of Switzerland. Accurate forecasts could also be made for three major rockfall events (1, 2, and 18 million m^3), which occurred in two open pit gold mines in northeastern Nevada (Rose and Hungr 2007); the largest event was even forecasted 3 months prior to the event.

In contrast to these successful forecasts, a number of catastrophic events have occurred where prediction has failed. An infamous example is the Vajont landslide (270 million m^3) in 1963, which triggered a 245-m-high flood wave in a water reservoir that consequently overtopped the dam and killed almost 2000 people in villages downstream (Kilburn and Petley 2003). Event analyses demonstrated the difficulty in forecasting the Vajont landslide with the 1/v model and confirmed that the final failure depends on the characteristics of the rock slope failure (brittle, ductile) and on the type of movement. Although clear accelerations were monitored over years at Vajont, they were misinterpreted, and the failure occurred more rapid than expected in a period of 45 s. Experts assumed that velocities would slow down following a nonlinear trend as two previous, smaller events did (Kilburn and Petley 2003; Petley and Petley 2006). Even 50 years after the events, there is still uncertainty as to why the velocity patterns changed during the multiple deformation phases from 1960 to 1963 (Barla and Paronuzzi 2013).

Accurate event forecasts are still a major challenge. Finding the balance between an early and safe evacuation and cost caused by unnecessary intervention measures remains a critical task for decision makers, who can be landslide experts, politicians, or a combination of both. To support decision makers in selecting an optimal risk mitigation strategy, the effectiveness and cost of alternative measures should be quantified and compared. However, existing frameworks for evaluating the effectiveness of mitigation measures focus on structural measures such as dams, rockfall nets, and galleries (Fell et al. 2005; Penning-Rowsell et al. 2005; Romang et al.2011; SafeLand 2012). The effectiveness of rockfall warning systems and warning systems for other types of natural hazard has so far not been considered systematically or quantitatively. This study aims to address this by quantitatively analyzing the reliability and effectiveness of a rockfall warning system installed for slowly evolving mid- and high-magnitude rockfalls. First, we assess the reliability of the warning system installed in Preonzo, Switzerland, to detect a past rockfall event on May 15, 2012. The reliability is expressed as the probability that an evacuation is set up timely before the event, and the analysis is conducted for both the detachment and the acceleration phase. Second, we investigate the influence of human decision-making and the number of sensors on the system effectiveness and the intervention cost using hypothetical predictive analysis.

2. EVALUATION CRITERIA FOR WARNING SYSTEMS

Warning systems are part of an integrated risk management aimed at increasing the preparedness for potential risks created by natural hazards and improving the capability to respond to and recover from an event. To compare preventive warning systems with alternative risk mitigation measures and to identify an optimal risk mitigation strategy, the effect on risk reduction and related cost should ultimately be considered in cost-benefit analyses (Penning-Rowsell et al. 2005; Safe Land 2012; Špačková and Straub 2015). For warning systems, these evaluation criteria should address specific needs. Where the cost associated with the hazard consequence is explicitly quantified, a cost-effectiveness analysis can be carried out in which the cost of mitigation measures is compared against their effectiveness.

The effectiveness of a warning system is a function of the reduction in risk it achieves. The effectiveness E_S can be defined through the relative risk reduction achieved with the warning system (Sättele et al. 2015). With R being the overall risk without the warning system and $R^{(S)}$ the risk with the warning system installed, the effectiveness is as follows:

$$E_S = 1 - \frac{R^{(S)}}{R} \tag{1}$$

The risk R_{ij} associated with an exposed object i and a hazard scenario j can be calculated as follows (e.g., (Fuchs 2006; Bründl et al. 2009)):

$$R_{ij} = p_j \times pe_{ij} \times v_{ij} \times A_i \tag{2}$$

Here, p_j is the probability of occurrence of a scenario j, pe_{ij} is the exposure probability of object i in scenario j, v_{ij} is the vulnerability of object i in scenario j and A_i is the value of object i. The overall risk R is calculated by summing over all exposed objects i and hazard scenarios j:

$$R = \sum_{j=1}^{n_{scen}} \sum_{i=1}^{n_{obj}} R_{ij} \tag{3}$$

Warning systems reduce the risk primarily by decreasing the exposure probability $pe_{ij}$of persons and mobile objects i in a hazard scenario j. Therefore, this paper addresses two questions: (i) How can the reduced exposure probability achieved with a rockfall warning system be quantified? (ii) And how can a rockfall warning system be optimized to minimize pe_{ij} with minimal cost?

Reliability analysis has been identified as an accurate basis for a comprehensive system evaluation of structural risk mitigation measures (Margreth and Romang 2010; Balbi et al. 2014) and can be applied to determine the exposure probability pe_{ij} of warning systems (Sättele et al. 2015). The reliability of a system is defined as its ability to fulfill the designated performance over a certain time (IEEE 2010). RWS for high- and mid-magnitude rockfalls are reliable if they detect dangerous events in time and lead to measures that reduce the exposure probability pe_{ij}. The reliability of a rockfall warning system can be quantified through the probability of detection, i.e., the probability that a warning system detects a hazardous event and adequate measures are taken and the probability of false alarms, i.e., the probability that measures are taken when they were unnecessary. A large number of false alarms can reduce the effectiveness, because frequent false alarms reduce the probability that people comply to an issued warning (Breznitz 1989; Sättele et al. 2015). In addition, the available lead time and the consequences of missed events must be taken into account as it contributes to the effectiveness (Paté-Cornell 1986; Schröter et al. 2008; Busslinger2009; UNEP 2012). A missed event means that an alarm is not given in time, as it happened in Vajont in 1963, when the destructive rockfall occurred before persons at risk were evacuated.

A reliability analysis for warning systems must address both the technical reliability and the inherent reliability of the system (Sättele et al. 2015). The technical reliability accounts for failures of technical system components due to aging and external causes such as lightning and destruction. It considers the configuration of components within the system and addresses redundancies as well as dependencies of the system components. It results in the probability of technical system failure. The inherent reliability of a warning system describes the general ability of the system to detect an event. It is primarily a function of the warning thresholds, the model forecast accuracy of models, and human decision-making.

The technical and inherent reliability of a rockfall warning system should be analyzed separately for the detachment and acceleration phase of rockfall processes. During the detachment phase, the technical reliability remains constant and can be modeled using classical methods such as fault, event or decision tree analysis, and failure mode and effect analysis (Stewart and Melchers 1997). Here, we model this technical reliability of warning systems with Bayesian networks (BN), which are increasingly applied to probabilistically model probabilities of systems failures (Bensi et al. 2012) and natural risks (Aguilera et al. 2011) such as rockfall hazard rating

(Straub 2005). The Bayesian network is a probabilistic graphical modeling tool, in which nodes represent random variables and arcs describe the stochastic dependencies among the nodes (Jensen and Nielsen 2007). The inherent reliability during the detachment phase depends initially on the warning thresholds, the measured sensor data, and their positioning in the field. All these aspects can also be modeled in the Bayesian network (see "Inherent reliability" section).

In the acceleration phase, the technical reliability of the system is no longer constant, but decreases over time, because sensors reach their mechanical limits or are likely to be destroyed by the failure. Additionally, power support and communication lines may be interrupted on purpose to prevent additional consequences in case of a direct hit. Thus, the power supply of the rockfall warning system becomes less reliable close to the event. We propose to address these increasing failure probabilities of technical failures through inhomogeneous Poisson processes (see "Technical reliability" section). The inherent reliability during the acceleration phase addresses the ability of the warning system and the decision makers to implement an evacuation before the event. The forecast ability depends on the ability of the 1/v model to forecast the event time and on the risk tolerance of the decision maker. To assess the inherent reliability achieved with the rockfall warning system, event forecasts are calculated and summarized graphically (see "Inherent reliability" section).

3. THE PREONZO ROCK SLIDE WARNING SYSTEM

3.1 Description of the event in May 2012

An unstable rock mass which threatens the community of Preonzo (Switzerland) has been recognized for decades. Slope failures occurred in the 18th century followed by temporary stabilization. A new tension crack was discovered by residents in 1989. Since 1991, the movement has been increasing significantly, and in May 2002, about 150,000 m^3 of the rock mass failed in the southern area of the Alpe di Rioscioro. This event, together with a minor event in May 2010, did not cause loss or damage to the downslope factories and roads (Willenberg et al. 2009; Loew et al. 2012).

Here, we investigate the latest event, which took place on May 15, 2012. The entire spring period of 2012 was characterized by unusually high displacement rates. At the beginning of May, local authorities were informed about critical displacements from the geologist operating the warning system. Immediately, a crisis team was established comprising local authorities, safety officers, and geologists operating the rockfall warning system. After several days of heavy rainfall, the velocity of the rock mass increased significantly, and on May 6, the crisis team evacuated people from the underlying factories and closed the nearest road. On May 8, the rain stopped, the velocity decreased, and intervention measures were discontinued to avoid losses from business interruption. On May 12, the velocity increased again, and a second evacuation was initiated. In the early morning of May 15, approximately 300,000 m^3detached from the rock face. Fortunately, the rock mass stopped on the slope and did not damage any infrastructure or injure any person.

3.2 Components of the Preonzo rockfall warning system

Between 1999 and May 2012, a rockfall warning system was installed at the Preonzo site. The system was operated and maintained by the local geologists responsible for natural hazard management in the Canton Ticino. From 1999 onward, five extensometers continuously measured the rock movements to detect accelerations and to automatically generate warning information. To increase the system reliability, an automated total station with 14 reflectors was set up in summer 2010 (Loew et al.2012). The configuration of the Preonzo warning system has three units for monitoring, data interpretation, and information dissemination (Sättele et al. 2012).

The monitoring unit incorporated two main sensor technologies. In the tension crack of the northern section, five extensometers monitored the displacement (Figure 2a). The extensometer measurements (every 15–60 min) were controlled by a remote data logger. Sensors and the logger communicated via protected cable connections. The power supply was provided by a battery and was recharged by a solar panel. In addition to the extensometers, an automated total station measured the distances to 14 reflectors on the front face of the slope (Figure 2b). The total station in the valley was connected to a computer that initiated measurements every 20 min. Power was provided by the power network, and the system was located in a heated cabin, built on a concrete foundation to minimize movement. In addition, a rain gauge was installed close to the tension cracks.

Figure 2.a Positioning of the five extensometers in the tension cracks and **b** 14 reflectors in the front face which were regularly monitored by an automated total station

The data interpretation unit included two main decision levels: one in the detachment and one in the acceleration phase. During the detachment phase, warnings were automatically generated when predefined warning thresholds were exceeded; in the acceleration phase, the crisis team analyzed the data to decide on intervention measures. To generate automated warning information, three threshold levels were defined for each extensometer in the tension crack (3 mm/day, 5 mm/day, and 5 mm/h) and one threshold for the total station (50 mm/day). The threshold for the total station was higher, because measured displacements were generally higher at the front face. Whenever a threshold for an extensometer was exceeded, the remote data logger issued warning information via the mobile network to the geologists. Independently, the computer connected to the total station in the valley issued a warning if the threshold was exceeded. While data from the remote data logger was transmitted via mobile network, data from the total station was sent via cable connection. Whenever the system operators received automated warning information, they analyzed sensor data to decide on further responses. The rain gauge serves as additional information source and decision add but is not associated with a threshold. All data were collected, processed, and visualized on a central server, and the event date was forecasted by the application of the 1/v model. The final decision about an evacuation was made by the crisis team and was based on the calculated forecasts. The warning system included a control system, which observed the availability of individual sensors, the status of the remote batteries, and the availability of the mobile network at regular intervals. Consequently, system failures could be detected intermediately to implement alternative temporary risk mitigation measures.

The information dissemination unit of the Preonzo warning system consisted of intervention plans, which summarized mitigation measures and responsibilities. The intervention was planned and coordinated by the crisis team to protect the underlying factories and roads. The evacuation of the factories could be initiated through activation of acoustic signals and was organizationally led by the police, who were also responsible for closing the roads below.

3.3 Quantifying the reliability of the Preonzo warning system

This section describes the reliability analysis of the warning system installed before and during the event in 2012, considering both the technical and the inherent reliability. The reliability analysis is presented separately for the detachment and acceleration phase. In each phase, factors that determine the system reliability are described, selected methods used to assess the reliability are presented, and main results are summarized.

3.4 System reliability during the detachment phase

During the detachment phase, the warning system monitored the displacement to alert local geologists about relevant accelerations. Whether or not the geologists received timely information depended on the reliability of the technical system configuration and on the predefined warning thresholds.

3.4.1 Technical reliability

Accurate automated warning information can be generated only when technical system components work properly. In Preonzo, the warning system is equipped with a control system that sends information when system components fail. In this situation, experts are alerted to assess the situation on-site. Due to this fail-safe system configuration, technical failure of the system will not lead to events being missed. Nevertheless, to avoid high cost due to unnecessary interventions because of frequent alarms, the technical reliability is relevant and should be maximized.

The Bayesian network (BN) to model the technical reliability of the Preonzo warning system consists of two different types of nodes (Figure 3). Grey nodes describe the causal chain from measured sensor data to the warning provided to system operators. The conditional probability tables of the gray nodes define the logical relations between the different nodes and are here modeled deterministically. As an example, the node "warning issued" incorporates both serial connections ("AND") and parallel connections ("OR"). Figure 4a exemplarily depicts an OR connection and the associated conditional probability table. Here, only one single sensor has to indicate an event (state 1) to issue a warning. Figure 4b represents an exemplarily AND connection with the associated conditional probability table. Here, both the logger and the battery have to work to issue a warning.

White nodes in the Bayesian network specify the failure probabilities of system components. These depend on internal failure causes and external factors. Most components are installed in rough alpine terrain, and they are exposed to external failure sources such as lightning, animals, and extreme weather. Assuming that failures occur randomly through time, failure probabilities of individual technical system components at time t (the unavailability) can be calculated as follows (Straub 2012):

$$\Pr(F(t)) \approx \lambda \times \mathrm{E}[T_r] \tag{4}$$

where λ is the failure rate of the system component and $\mathrm{E}[T_r]$ is the expected time it takes to detect and repair a failure. This approximation is valid for small values of λ, i.e., for $\lambda \ll \frac{1}{\mathrm{E}[T_r]}$. The failure rate λ of a system component includes both the internal failure rate λ_{IF} and the rate of failures caused by external influences λ_{EF}:

$$\lambda = \lambda_{IF} + \lambda_{EF} \quad (5)$$

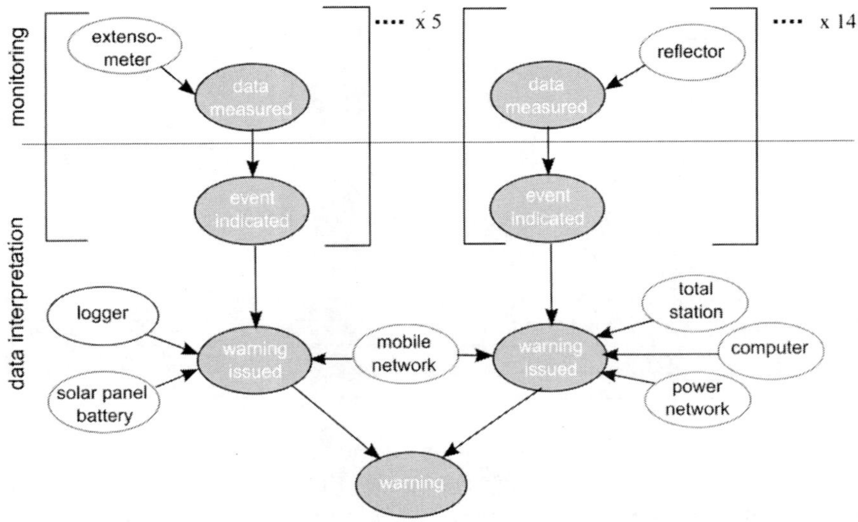

Figure 3. Bayesian network to model the technical reliability in the detachment phase. The monitoring unit incorporates 5 extensometers and 14 reflectors and the data interpretation components to initiate and transfer warnings

The failure probabilities for components of the Preonzo system were adapted from results of a similar case study (Sättele et al. 2015) and are estimated for individual components as $\Pr(F(t)) \sim 5 \times 10^{-4}$. Only the failure probability of the mobile network is significantly higher at 1.2×10^{-2} according to a study conducted by the Swiss Federal Office for Civil Protection (FOCP 2013).

The analysis shows that the technical reliability achieved with the Preonzo system is high, due to multiple and redundant sensor units. The system is equipped with two sensor technologies that both incorporate redundant sensors. The limiting factor for the technical reliability is the availability of the mobile network, which has no redundancy. This results in an unavailability of the technical system of $\Pr(F_s(t) = 1.2 \times 10^{-2}$. Despite this relatively high probability of system failure, this is not critical due to the fail-safe configuration, in which decision makers are alerted if the alarm system has a technical malfunction.

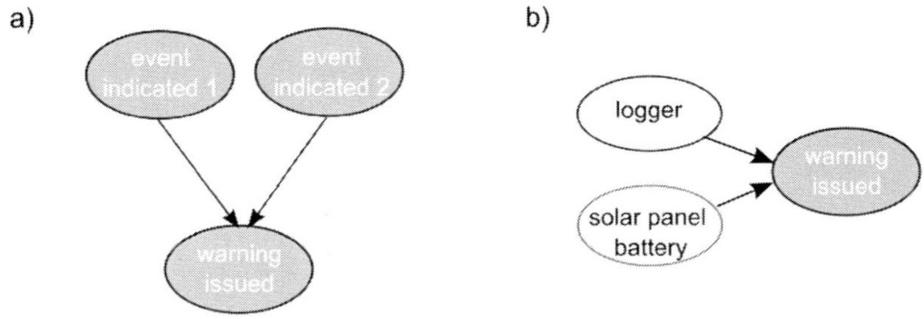

event indicated 1		yes		no	
event indicated 2		yes	no	yes	no
warning issued	yes	1	1	1	0
	no	0	0	0	1

logger		yes		no	
solar panel/ battery		yes	no	yes	no
warning issued	yes	1	0	0	0
	no	0	1	1	1

Figure 4. Illustrative example to represent the relations between the nodes of a Bayesian network: **a** "OR" connection and associated conditional probability table; **b** "AND" connection and associated conditional probability table

3.4.2 Inherent reliability

The inherent reliability of a warning system is its ability to interpret the measured sensor data, to distinguish between noise and real displacement, and to inform system operators before the event occurs. During the detachment phase, the relationship between the probability of detection and the probability of false alarms is determined by the predefined warning thresholds. To achieve a large probability of detection, thresholds for the Preonzo warning system are intentionally set low (see "Components of the Preonzo rockfall warning system" section). However, the warning thresholds are high enough to avoid perpetual false alarms due to displacement noise that arises from daily fluctuations, e.g., due to temperature changes.

Existing movement records summarize the displacement rates measured by the extensometers and the total station between August 2010 and May 2012. Dilatations measured by extensometers positioned in the northern section are higher (extensometers 3–5) than those measured by similar sensors in the southern section (Figure 5). A few hours before the event, extensometers 4 and 5 reached their technical limits and failed. The reflectors show similar displacement characteristics to the extensometers (Figure 6). Reflectors 1–6 and 8, 9 were mounted in the northern sector and showed large displacement rates. The remaining reflectors (7, 11–14) were installed in the southern section and did not indicate any discernable acceleration. Reflector 10 had already failed in the summer of 2011.

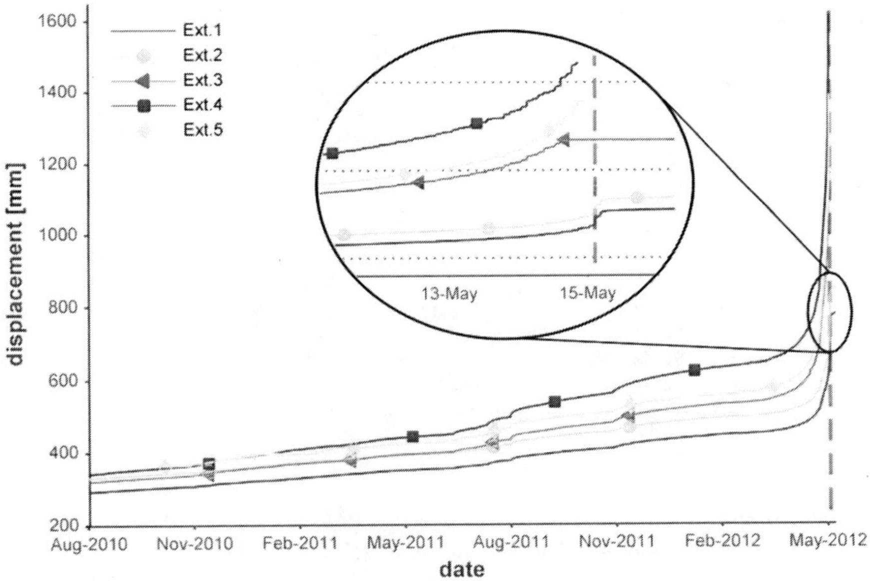

Figure 5. Displacements measured by the extensometers 1–5 in the tension cracks between August 2010 and May 2012. Extensometers positioned in the northern section (3–5) indicate higher acceleration before the event

In Figure 7, the sensor data measured by extensometer 5 from January to May 2012 is compared to the lowest warning level. The lowest warning threshold (1 mm/day) was regularly exceeded by those sensors installed in the northern section from the beginning of March onward. Starting in early April, the warning threshold was constantly exceeded, and system operators started to analyze the sensor data at more frequent intervals independent of the automated warning information. The two upper thresholds (3 and 5 mm/h) were reached several days before the event.

Other extensometers in the northern section showed similar results. The only warning level defined for the reflectors (50 mm/day) was also exceeded before the event by those reflectors installed in the northern section.

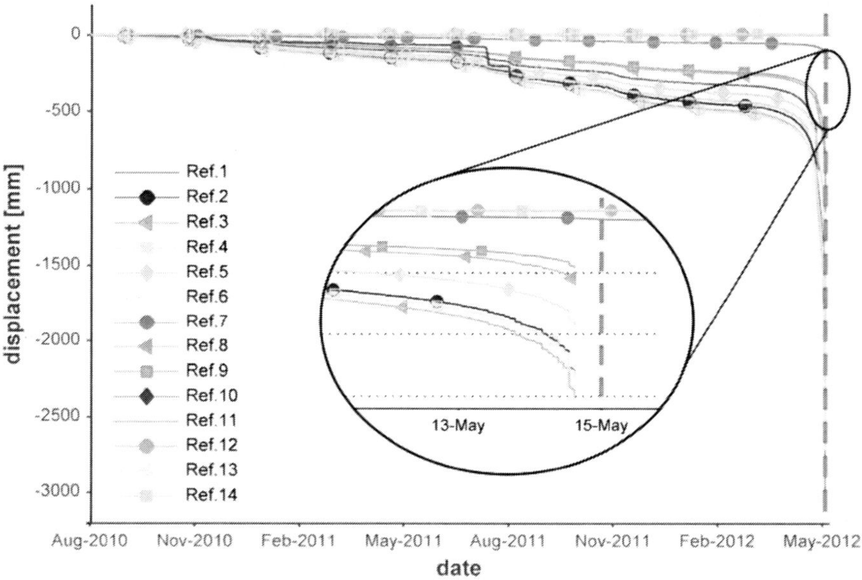

Figure 6. Displacements measured by the reflectors 1–14 installed at the front face of the slope between August 2010 and May 2012. Reflectors positioned in the northern section (1–7, 9) indicate higher acceleration before the event

The influence of warning thresholds on the system reliability can be modeled within the Bayesian network, by making the conditional probability of the nodes "event indicated" dependent on the threshold. Such an approach was followed in Sättele et al. (2015). However, during the detachment phase, a low threshold can generally be chosen, which leads to a probability of detection of one. This is because the effect of false alarms is less relevant in this phase, as the warning information is sent to system operators and not directly to endangered people. System operators should be interested in receiving information from every unexceptional displacement. In combination with the fail-safe configuration of the technical system components, the overall probability of identifying an event in the detachment phase is very close to one. The high reliability of the Preonzo warning system also leads to a reasonably small number of false alarms to system operators.

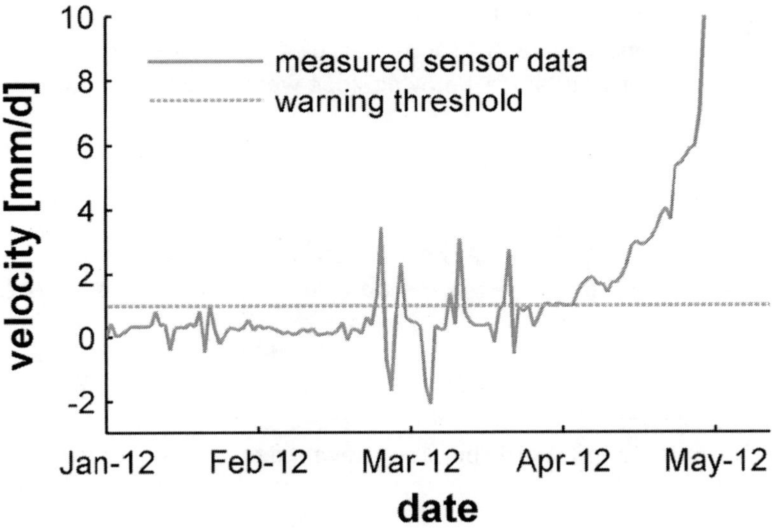

Figure 7. Velocities measured by extensometer 5. The velocities temporarily exceeded the lowest warning threshold in March and then constantly from beginning of April

3.5 System reliability during the acceleration phase

During the acceleration phase, specialists analyze the sensor data to forecast the failure time and support the crisis team in planning appropriate response. The system reliability in this phase is a function of increasing failure probabilities of technical system components, the forecast accuracy of the 1/v model, and human decision-making.

3.5.1 Technical reliability

The technical reliability in the acceleration phase influences the warning system's ability to support the crisis team in reaching an accurate event forecast. The forecast of the event time depends directly on the availability and quality of the measured sensor data. Power interruptions due to safety reasons or sensor failures due to large movements in the surrounding area occur more frequently close to the event. The effect of increasing system failure probabilities on the forecast accuracy should therefore be considered in warning system technical reliability analyses. In the northern section of the Preonzo rock face, the majority of sensors failed in the hours before the event (Figure 8). Three out of five extensometers and all reflectors positioned in the fast moving slope were destroyed.

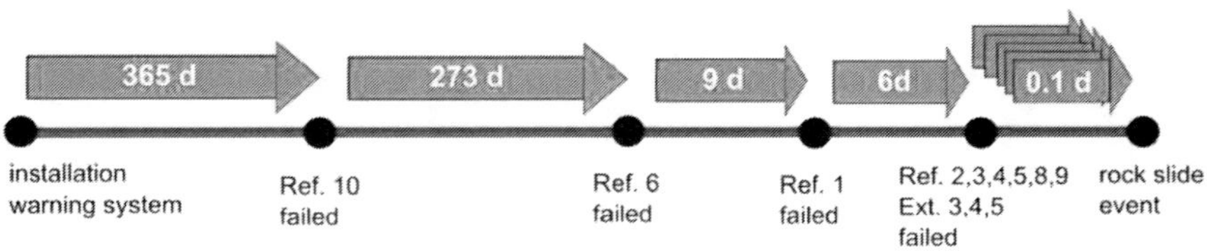

Figure 8. Failures of sensors before the event on May 15, 2012. Many sensors are destroyed shortly before the event due to the fast moving slope

To quantify the accelerating destruction of sensors in Preonzo before the event 2012, a function describing the probability of failure was fitted to the observed number of sensors that failed in the northern section (extensometer 1–5; reflectors 1–6 and 8, 9, 10). The failure probability P_f for the Δt days to the event in Preonzo is modeled as follows:

$$P_f(\Delta t) = p_{basic} + p_{end} \times \exp(-b \times \Delta t) \qquad (6)$$

where the basic failure probability of technical system components p_{basic} is 5.0×10^{-4} (section "Technical reliability"). To obtain values for p_{end} and b, a likelihood function describing the observed number of failures is established. The probability of n_F failures out of n sensors on a given day is described through the binomial distribution as follows:

$$\Pr(N_F = n_F) = \binom{n}{n_F} \left\{ P_f(\Delta t)^{n_F} \times \left[1 - P_f(\Delta t)\right]^{n - n_F} \right\} \qquad (7)$$

The parameters b and p_{end} are found by maximizing the log-likelihood function, which is defined as follows:

$$\ln L\left(p_{end}, b\right) = \sum_{t=1}^{44} \ln \Pr\left(N_F = n_F(t) \middle| p_{end}, b\right) \quad (8)$$

where $n_F(t)$ are the observed failures on days 1–44 before the event. By maximizing Eq. 7, maximum likelihood estimates for p_{end} and b are calculated as $p_{end} = 0.689$ and $b = 0.510$. In Figure 9, the percentage of sensors that failed in the northern section and the calculated failure probabilities using Eq. 6 for days 1–44 before the event are summarized.

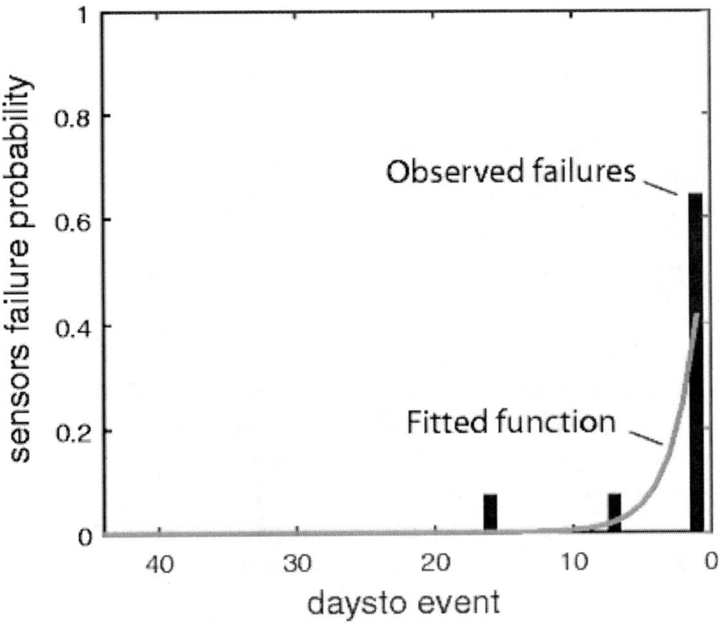

Figure 9. During the acceleration phase, an increasing number of sensors fail just before the event. The *bars* (*left axis*) show the observed percentage of sensors that fail; the *line* (*right axis*) depicts the failure probabilities according to Eq. 6

3.5.2 Inherent reliability

The inherent reliability in the acceleration phase is a function of the available sensor data, the accuracy of the 1/v model, and the expert's ability to forecast the event based on these data. The decision to evacuate depends on the risk tolerance of the crises team. Due to the dependence on the available sensor data, the inherent reliability is related to the technical reliability of the rockfall warning system. An increasing number of sensor failures reduces the forecast ability.

To assess the inherent reliability of the Preonzo system, we apply a linear 1/v model using measured sensor data to obtain event forecasts. The inverse velocity at time *t* is calculated as follows:

$$\frac{1}{v(t)} = a + b \times t \tag{9}$$

a and *b* are the modeled parameters and *t* is the time (in days). To avoid a large scatter generated by small velocity values deteriorating the accuracy of the forecast, the parameters are fitted to the measured velocity rather than to its inverse. Specifically, a least-squares fit of *a* and *b* to measured values of *v*(*t*) is carried out through the following relationship:

$$v(t) = \frac{1}{a + b \times t} \tag{10}$$

Figure 10a exemplarily illustrates the velocities measured by extensometer 1 in the 10 days before May 14 and the model fit. To obtain a forecast of the event time, the inverse velocities 1/v are calculated according to Eq. 8 for dates *t* in the future. The event date forecasted with the 1/v model is the day where the inverse velocity (1/v) cuts the *x*-axis. In Figure 10b, the inverse velocity corresponding to Figure 10a is shown. On May 14, the forecast of the event with extensometer 1 is made for May 16, 1 day later than the event actually happened.

Following the approach illustrated above for extensometer 1 and data available before May 15, event forecasts were made for every sensor and each day between April 1 and May 14. Figure 11a displays the forecasted event dates using data from different sensors installed in the northern section as a function of the date on which the prediction is made. In April, the forecasts made by different sensors vary significantly, but they become more aligned by the end of April. In this final phase, the predictions based on extensometer data show larger scatter than those based on reflector data. On May 14, the majority of sensors provided velocity data that indicated a hazardous event for the next day. Extensometers 1 and 2 are located further south, where the measured displacements were less significant. They forecast the event with a delay of 1 day. Nevertheless, 10 out of 12 sensors lead to an accurate event forecast.

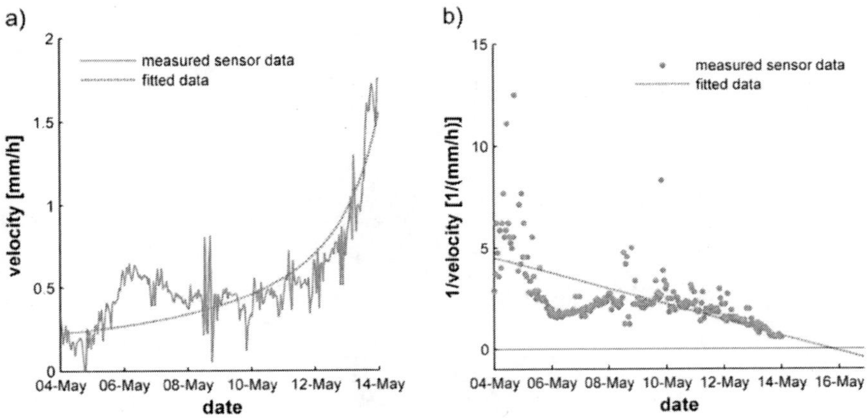

Figure 10. a Velocity recorded by extensometer 1 from May 4 to May 14 and fitted function; **b** 1/v versus time using data from extensometer 1. On May 14, the 1/v model forecasts the event to occur on May 16, which is 1 day later than the observed event

The inherent reliability in the phase of acceleration depends on the decision of the crises team, who analyze the displacement rates with respect to quantities of rain that has fallen and is expected to fall. To understand the human impact on the inherent reliability, the forecasted number of days to the event is plotted for days between April 1 and May 14 (Figure 11b) for all five extensometers and reflectors showing large displacement rates in the northern section (reflectors 1–5 and 8, 9). There are several days in April on which extensometers 3–5 and all those northern reflectors forecast the event for the following day. Here, different experts may reach different decisions. In May, the forecasts vary less, and the influence of human decision-making becomes less important. It can be assumed that even decision makers with different risk tolerances would come to the same decision in this period (which is the one that was actually taken by the crisis team in Preonzo).

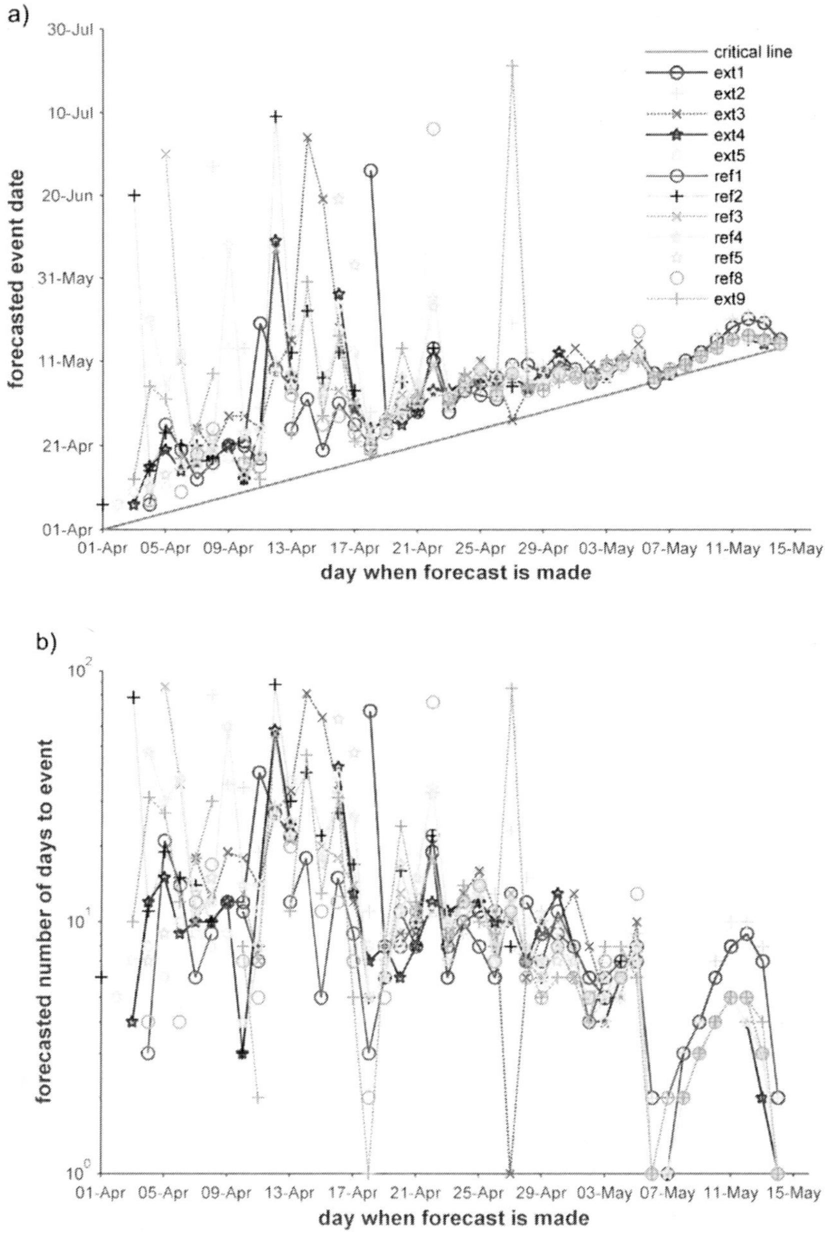

Figure 11. a Daily forecasted event date calculated with the 1/v model for the sensors in the northern section between April 1 and May 14; **b** forecasted number of days to the event calculated with the 1/v model for sensors in the northern section on every day between April 1 and May 14

3.6 Hypothetical analysis: optimizing warning systems

To compare the effectiveness and the cost for varying designs of the Preonzo system, the probability of detection achieved with the system and the expected cost arising from the intervention measures are estimated in a hypothetical analysis. Since the actual decision on intervention measures is based on expert assessments, which can include more information than only sensor data, e.g., rain measurements, the analysis is simplified and likely to underestimate the true capacities of the warning system.

To assess the effect of sensor numbers on the reliability, we vary their amount between 5 and 50 in the hypothetical analysis. To investigate the effect of human decision-making, we specify two decision makers with different risk types. A risk averse and a risk tolerant decision maker are associated with different evacuation criteria (Table 1). A technical evacuation criterion determines the minimum numbers of sensors that must be available for a forecast. Whenever fewer sensors are functioning, the crisis team initiates an evacuation and the closure of the road. The inherent evacuation criterion defines the minimum proportion of sensors that must forecast a failure for the next day in order to initiate an evacuation and road closure.

Table 1. Definition of risk types based on different evacuation criteria

Risk type		Technical evacuation criterion, evacuate when:	Inherent evacuation criterion, evacuate when:
Risk averse	Less risk tolerant decision makers	Less than 6 sensors are functioning	20 % of the sensors forecast the event for the next day
Risk tolerant	More risk tolerant decision makers	Less than 3 sensors are functioning	50 % of the sensors forecast the event for the next day

Using a Monte Carlo analysis, we estimate the probability of detection and intervention cost for the specified risk averse and tolerant decision maker and modified numbers of sensors. In the Monte Carlo analysis, the investigated Preonzo rockfall event is randomized, as is the response of the individual sensors. We use $n_S = 10,000$ random realizations (run) of the process. For each run, it is checked if and when evacuations would be initiated, based on technical and inherent evacuation criteria. In each run, the number of functioning sensors for all days between April 1 and May 14 is simulated and compared to the minimal required number to investigate whether the technical evaluation criterion is fulfilled. The number of functioning sensors on each day is simulated based on the binomial distribution (Eq. 5) with parameters determined following section "Technical reliability."

To assess whether the inherent evacuation criteria is fulfilled, the percentage of simulated positive event forecasts for the next day is compared to the specified percentage. The number of available event forecasts on each day depends on the remaining number of sensors. The forecasts for the group of sensors are modeled by a probability distribution, in which no distinction is made between individual sensors (i.e., they are considered as statistically identical and independent). To obtain probability distributions for daily forecasts between April 1 and May 14, Lognormal distributions are fitted to sensor forecasts calculated for extensometers 1–5 and reflectors 1–5, 8, and 9 (Figure 11b). Figure 12 illustrates the empirical and the fitted cumulative distribution functions of the forecasted days to event calculated on April 18 and May 14. The fitted distributions are applied in the Monte Carlo analysis to randomly generate forecasts for each day. The percentage of sensors that forecast an event for the next day is calculated and compared to the percentage defined by the inherent evacuation criterion.

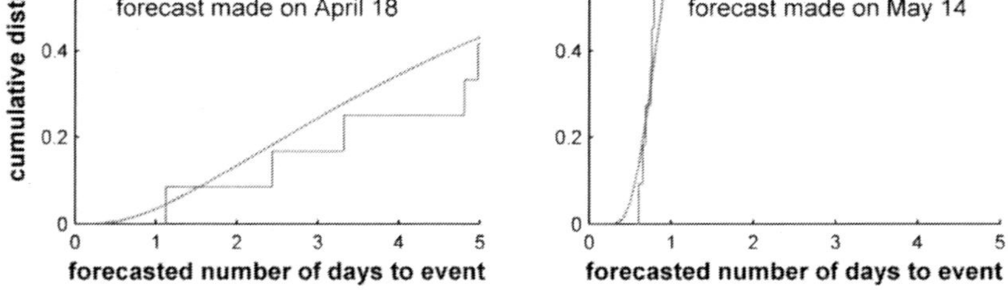

Figure 12. The empirical (observed) and the fitted cumulative probability distribution function of the forecasted days to the event for the group of sensors. The observed sensor forecasts are those shown in Figure 11b. **a** On April 18, the probability of an event on the next day is estimated as 0.034 and for an event in 5 days as 0.43; **b** on May 14, the probability of an event to occur within the next day is estimated as 0.65 and for the second day 0.99

Intervention costs arise whenever an evacuation is initiated. This cost is primarily estimated from interruptions of business processes in the subjacent factories at 100,000 CHF/ day based on the information of local experts. If the technical evacuation criterion is fulfilled, a 5-day evacuation is necessary to install a replacement system. In this time, a temporary monitoring system (e.g., interferometric radar) must be installed to decide if the access for installation to the area is safe. Total intervention costs are estimated at 800,000 CHF, with 500,000 CHF at this due to the 5 days of evacuation, and the remaining 300,000 CHF are the investment cost for the replacement system, including cost for temporarily monitoring measures. Separate costs occurring to different stakeholders are not considered. If the inherent evacuation criterion is fulfilled, a 2-day evacuation is initiated, and the corresponding cost is 200,000 CHF.

In this analysis, the probability of detection is the probability that the evacuation and road closure are in place on May 15, the day of the actual event.

Figure 13a displays the probability of detection calculated for the different risk types as a function of the number of initially installed sensors. For the more risk-averse decision maker, the probability of detection is close to one, only slightly depending on the number of sensors. More risk-tolerant decision makers achieve a probability of detection between 0.65 and 0.85 which reaches its minimum at 11 sensors. Figure 13bdisplays the expected intervention cost calculated for both risk types as a function of the sensors. As expected, the risk-averse decision maker creates higher expected intervention cost, especially with a small number of sensors. In this case, large costs for evacuation and replacement systems are generated. With a highly redundant sensor unit (around 30 sensors), the expected cost for the risk-averse decision maker reaches their minimum at 400,000 CHF. The expected cost for intervention incurred by the risk-tolerant decision maker reaches minimal costs (215,000 CHF) at 20. For larger number of sensors, the expected cost increases slightly up to a maximum of 236,000 CHF. This sum does not include the cost for installing the initial sensors. A comprehensive cost analysis should also consider the acquisition cost of the warning system, which increases with a rising number of sensors, to determine the optimal number of sensors.

The intervention on the day of the event, May 15, can be because the event was correctly forecasted (inherent reliability) or because the failure of multiple sensors triggered an intervention (technical failures). In Figs. 14 and 15, the probabilities of detection and expected intervention costs obtained with the inherent evacuation criterion and the technical evacuation criterion are shown individually for both risk tolerance profiles. To this end, both evacuation criteria are checked daily, independent of whether or not the other criterion has been fulfilled and an evacuation may already be in place. For this reason, the sum of the two individual probabilities of detection may be greater than one. The quality of the inherent forecasts during the time of reinstalling the system after technical failures may be poor, since only few sensors are left during that period. If no more sensors are available 1 day before the event occurred, zero forecasts can be made and the probability of detection associated with the inherent evacuation criteria becomes zero.

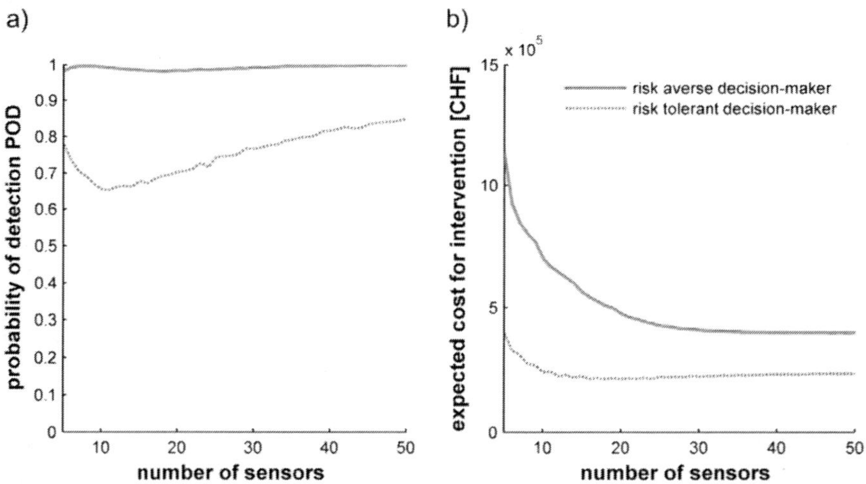

Figure 13. a The probability of detection as a function of the decision maker's risk-taking profile and the number of initially installed sensors. The probability of detection is the probability of an evacuation being in place on the day of the event; **b** the expected cost for intervention (evacuation, road closure) as a function of the decision maker and the number of initially installed sensors

The probability of detection achieved due to technical failures decreases with the increase of sensors (Figure 14). Evacuations due to an insufficient number of sensors are less likely to occur when the decision maker has a high-risk tolerance. For the risk-averse decision maker, an interesting peak is formed around 7 sensors. For minimal numbers of sensors (5–6), an evacuation because of technical failures is low, because the warning system got already substituted before the event. For 7–9 sensors, the probability that the system fails during the event is maximal and so is the probability of detection achieved with the technical evacuation criterion. The probability of detection because of an accurate event forecast increases with increasing number of sensors and is close to 1 for systems with at least 20 sensors for the risk-averse decision maker. The probability of detection achieved by the more risk tolerant person never exceeds the probability of detection reached by the risk-averse decision maker and never exceeds 0.84.

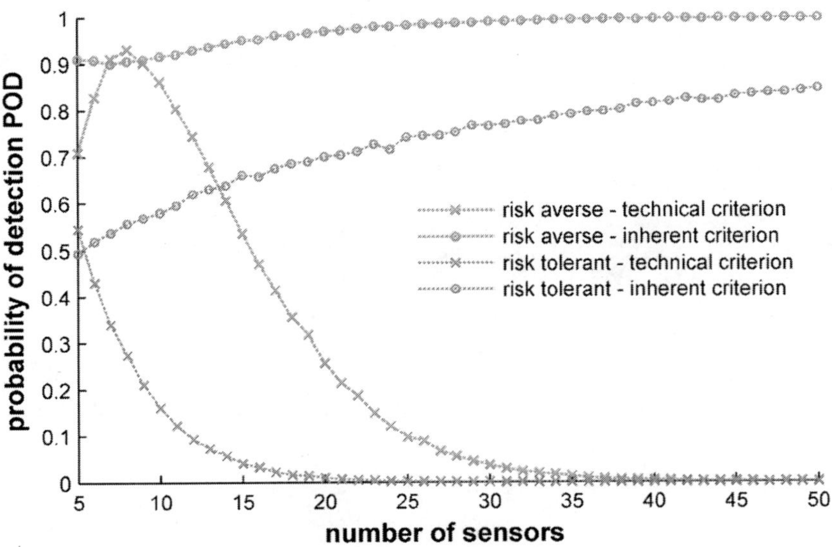

Figure 14. The probabilities of detection achieved with the inherent and technical evacuation criteria are individually shown for both risk-averse and risk-tolerant decision makers

In Fig 15, the expected intervention cost created through the evacuation criteria are separately illustrated for both risk tolerance profiles. The expected cost due to the technical evacuation criterion decreases, when the number of initial sensors increases, because system failures become less likely. The risk-averse type generates immense cost from fewer sensors. The expected intervention cost generated by event forecasts becomes constant for more than 10 sensors and are higher for the risk-averse decision maker.

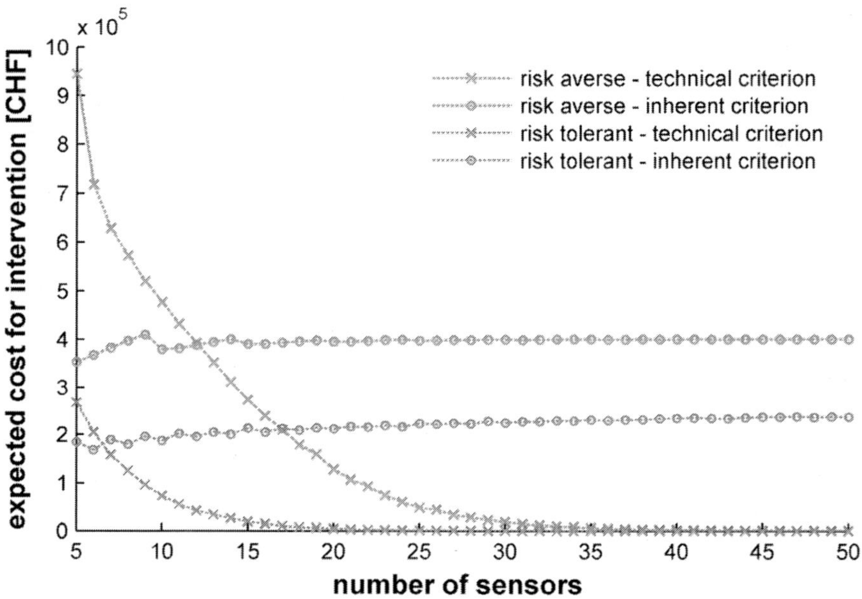

Figure 15. Expected intervention cost achieved with the inherent and technical evacuation criteria are individually shown for both risk-averse and risk-tolerant decision makers

To understand how the intervention costs arise, the probability of an evacuation being mandated due to sensor forecasts (inherent evacuation criterion) is illustrated in Figure 16 for the last 44 days before the event, with 10 sensors (a) and 30 sensors (b). The risk-tolerant person would evacuate on 3 days with a significant probability, namely on days 8, 9, and 1 before the event. When comparing Figure 16a with Figure 16b, it is clear that the forecast accuracy increases with more sensors. The probability of the risk-tolerant person proposing an evacuation due to sensor signals on the day of the event is 0.59 with 10 sensors and 0.76 with 30 sensors. The probability of wrongly initiating an evacuation on days 8 and 9 stays between 0.37 and 0.35 when increasing the number of sensors from 10 to 30. The same tendencies are observed for the risk-averse decision makers, whose evacuation probabilities are significantly higher. With 30 sensors instead of 10, the probability of detection increases from 0.91 to 0.99.

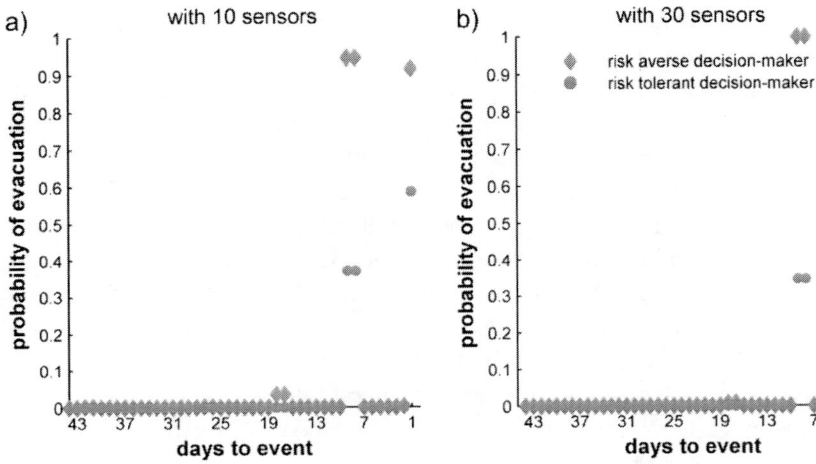

Figure 16. Probability of evacuation due to sensor forecasts (inherent evacuation criteria) calculated for 10 (**a**) and 30 (**b**) initial sensors and different risk-taking profile types for 44 days before the event

4. DISCUSSION

The effectiveness of warning systems was quantified as a function of the achieved risk reduction. Rockfall warning systems for mid- and high-magnitude events reduce the exposure probability of people to a hazardous event if the event is detected in time for intervention measures to be implemented. This effect can be calculated directly from the system reliability, which can be quantified by the probability of detection and the probability of false alarms. For the quantification of warning systems, the effect of false alarms can be neglected, because the events are rare and the warning information is sent to system operators. In other cases, especially when the information is issued directly to endangered people, the effect of false alarms can decrease the compliance to the alarm (Sättele et al. 2015).

Within the Preonzo case study, we conducted a comprehensive reliability analysis, which demonstrated the importance of assessing the technical and the inherent reliability for the displacement and the acceleration phase individually. During the detachment phase, the technical reliability depends on failure probabilities of components and their configuration in the system. In a similar case study, the failure probabilities of individual technical system components of a debris flow alarm system were analyzed in detail (Sättele et al. 2015). In that study, internal failure rates λ_{IF} were derived from specified mean time to failure and for repairable parts for mean time between failure values, with $\lambda_{IF} \sim 5.0 \times 10^{-5}$/day. The rates of external failure λ_{IE} were obtained from repair records and expert estimates as $\lambda_{EF} \sim 4.5 \times 10^{-4}$/day. The mean repair time to detect and to repair was estimated as $E[T_r] = 1$ day for systems equipped with a control system. For the current study, we adopted these values to simplify the reliability analysis. We could show that a high reliability can be achieved with redundancies and is limited through dependencies, e.g., data transmission is not redundant. To obtain more accurate reliabilities, failure probabilities should be determined for each component and field-site individually. The inherent reliability depends on the position of the sensors in the field and warning thresholds. Sensors located in the southern section showed lower displacements (extensometers 1–2) or no movement (reflectors 10–14) and did not exceed the thresholds.

The reliability analysis in the acceleration phase is more complex, because failure probabilities of components increase significantly before the event, and because of the data quality, the forecast accuracy of the model and the risk tolerance of decision makers influence the reliability. For Preonzo, forecasts could be made with an increasing accuracy toward the actual event. Nevertheless, some sensors (Ext. 4 and 5) underestimated the acceleration and forecasted the event with a delay. As in the detachment phase, the positioning of the sensors was a crucial factor for the forecast accuracy. The reliability was high because of the multiple and redundant sensor unit. In our simulation, the days with high probability of evacuation coincide with days on which evacuations were actually initiated by the crisis team in Preonzo.

A rockfall warning system can be designed as a fail-safe system. This implies that even if the technical system fails, damage is prevented because an evacuation is initiated. But, such a fail-safe design can create a large intervention cost. We modeled the effect of the initial number of sensors and human decision-making on the effectiveness and cost in the hypothetical analysis. The behavior of the experts was replicated through simple rule-driven decisions, specified in evacuation criteria. Such an approach is novel and facilitates the quantitative assessment of warning systems that are expert-driven systems. The hypothetical analysis demonstrated that the forecast accuracy increases with the number of initial sensors and becomes maximal for risk-averse decision makers. These less risk-tolerant decision makers will create larger cost for intervention with a small number of sensors but achieve acceptable cost with a highly redundant sensor configuration.

Findings from the case study and the hypothetical analysis can be used to develop a method for quantifying the effectiveness of warning systems and will support practitioners in designing and operating cost-efficient systems. The technical and the inherent probability during the detachment phase can be modeled with Bayesian networks, similarly to the reliability of warning systems installed for processes that are triggered rapidly such as debris flows. A cost-effective warning system includes redundant sensor units with multiple measuring techniques. Sensors have a maximal coverage to ensure that precursors are detected early. They should be redundant to maintain high forecast accuracy when the number of failure increases close to the event. To detect system failures, an independently operated control system is in place. The warning thresholds in the detachment phase are low to detect significant changes, but high enough to avoid constant false alarms due to daily fluctuations, etc.

The quantification of rockfall warning systems in the detachment phase remains a major challenge. We showed an approach to model human decision-making by simple evacuation rules, but predictive models describing the forecast accuracy of the 1/v model or similar models are not available. We used the Preonzo event data from May 15 to model the forecast accuracy retrospectively. To assist decision makers in identifying an optimal risk mitigation strategy, the forecast accuracy of 1/v model must be investigated in the future. Such investigations should quantify the model accuracy with respect to the movement patterns, material characteristics, and the positioning of the sensors in the field.

5. CONCLUSION

The reliability of rockfall warning systems is quantifiable and capable of distinguishing between the detachment and acceleration phase of the rockfall process. The reliability analysis confirmed that the Preonzo rockfall warning system as installed detected the rockfall event in May 2012 and was highly reliable. We showed that with Bayesian networks, one can assess the system's ability to generate automated warning during the detachment phase. A highly reliable system includes redundant and multiple sensor technologies, an area-wide sensor positioning, low warning thresholds in combination with a control system. During the acceleration phase, a maximal probability of evacuation on the day of the event is achieved with a sufficient number of sensors and conservative decision makers. Both factors have a considerable effect on the system effectiveness and on the related intervention cost. The effect of human decision-making could be modeled with rule-driven decision criteria. Those and other findings will be used to develop a generic framework approach that allows the quantification of the reliability and the effectiveness achieved with early warning system for natural hazards.

ACKNOWLEDGMENTS

The project is funded by the Swiss Federal Office for Civil Protection FOCP whose support is gratefully acknowledged. We thank Giorgo Valenti and Marco Franzi from Canton Ticino, who provided essential information and data of the Preonzo system and the reviewers for the valuable comments.

REFERENCES

1. Aguilera PA, Fernández A, Fernández R, Rumí R, Salmerón A (2011) Bayesian networks in environmental modelling. Environ Model Softw 26:1376–1388

2. Balbi S, Villa F, Mojtahed V, Giupponi C (2014) Estimating the Benefits of Early Warning Systems in Reducing Urban Flood Risk to People: A Spatially Explicit Bayesian Model. In: Ames DP, Quinn, N.W.T., Rizzoli, A.E. (ed) Proceedings of the 7th International Congress on Environmental Modelling and Software, San Diego

3. Barla G, Paronuzzi P (2013) The 1963 Vajont landslide: 50th anniversary. Rock Mech Rock Eng 46:1267–1270.

4. Bensi M, Kiureghian AD, Straub D (2012) Efficient Bayesian network modeling of systems. Reliab Eng Syst Saf 112:200–213

5. Breznitz S (1989) Cry wolf: the psychology of false alarms. Erlbaum, Hillsdale

6. Bründl M, Romang H, Bischof N, Rheinberger C (2009) The risk concept and its application in natural hazard risk management in Switzerland. Nat Hazards Earth Syst Sci 9:801–813

7. Busslinger M (2009) Landslide time-forecast methods, a literature review towards reliable prediction of time to failure HSR University of Applied Science, Rapperswil

8. Evans S (2006) The formation and failure of landslide dams: an approach to risk assessment Italian. J Eng Geol and Environ 1:15–20

9. Fell R, Ho K, Lacasse S, Leroi E (2005) State of the Art Paper 1-A framework for landslide risk assessment and management. In: Proceedings of the International Conference on Landslide Risk Management, Vancouver, Canada

10. Fuchs S (2006) Cost-Benefit Analysis of Natural Hazard Mitigation. In: Encyclopedia of Natural Hazards. Springer, pp 121–125

11. Fukuzono T (1990) Recent studies on time prediction of slope failure Landslide News 4

12. Ganerød GV et al (2008) Geological model of the Åknes rockslide, western Norway. Eng Geol 102:1–18

13. Hashimoto T, Stedinger JR, Loucks DP (1982) Reliability, resiliency, and vulnerability criteria for water resource system performance evaluation. Water Resour Res 18:14–20

14. Hattenberger D, Wöllik A (2008) (Naturgefahren-) Mess-und Frühwarnsysteme: Einzelne rechtliche Aspekte. Baurechtliche Blätter: bbl 11:89–101

15. Hewitt K, Clague JJ, Orwin JF (2008) Legacies of catastrophic rock slope failures in mountain landscapes. Earth-Sci Rev 87:1–38

16. Huggel C, Clague JJ, Korup O (2012) Is climate change responsible for changing landslide activity in high mountains? Earth Surf Process Landf 37:77–91

17. Hungr O, Evans S (2004) The occurrence and classification of massive rock slope failure. Felsbau 22:16–23

18. IEEE (2010) IEEE standard framework for reliability prediction of hardware, IEEE Std 1413. Institute of Electrical and Electronics Engineers, Inc, NY

19. Jensen FV, Nielsen TD (2007) Bayesian networks and decision graphs. Information Science and Statistics, 2nd edn. Springer Science + Business Media, New York

20. Kilburn CR, Petley DN (2003) Forecasting giant, catastrophic slope collapse: lessons from Vajont. Northern Italy Geomorphology 54:21–32

21. Krähenbühl R (2006) Der Felssturz, der sich auf die Stunde genau ankündigte. Bull Angew Geol 11:49–63

22. Krautblatter M, Funk D, Günzel FK (2013) Why permafrost rocks become unstable: a rock–ice-mechanical model in time and space. Earth Surf Process Landf 38:876–887

23. Lacasse S, Eidsvik U, Nadim F, Hoeg K, Blikra LH (2008) Event tree analysis of Åknes rock slide hazard. In: 4th Canadian conference on geohazards, Université Laval, pp 20–24

24. Loew S, Gischig V, Moore J, Keller-Signer A (2012) Monitoring of potentially catastrophic rockslides. In: Proc. of 11th Int. and 2nd North Am. Symp. on Landslides and Engineered Slopes, Banff, Canada, pp 101–116

25. Margreth S, Romang H (2010) Effectiveness of mitigation measures against natural hazards. Cold Reg Sci Technol 64:199–207

26. Paté-Cornell ME (1986) Warning systems in risk management. Risk Anal 6:223–234.

27. Penning-Rowsell EJC, Tunstall S, Tapsell S, Morris J, Chatterton J, Green C (2005) The benefits of flood and coastal risk management: a handbook of assessment techniques. Middlesex University Press, London

28. Petley D, Petley D (2006) On the initiation of large rockslides: perspectives from a new analysis of the Vaiont movement record. In: Landslides from Massive Rock Slope Failure. Springer, pp 77–84

29. Petley D, Higuchi T, Dunning S, Rosser N, Petley D, Bulmer M, Carey JA (2005a) A new model for the development of movement in progressive landslides. In: International conference on landslide risk management. Taylor & Francis Ltd., Vancouver, pp 350–358

30. Petley DN, Higuchi T, Petley DJ, Bulmer MH, Carey J (2005b) Development of progressive landslide failure in cohesive materials. Geology 33:201–204

31. Ravanel L, Deline P (2011) Climate influence on rockfalls in high-Alpine steep rockwalls: the north side of the Aiguilles de Chamonix (Mont Blanc massif) since the end of the little ice age. The Holocene 21:357–365

32. Romang H et al (2011) IFKIS-Hydro: an early warning and information system for floods and debris flows. Nat Hazards 56:509–527

33. Rose ND, Hungr O (2007) Forecasting potential rock slope failure in open pit mines using the inverse-velocity method International. J Rock Mechanics and Mining Sci 44:308–320

34. SafeLand (2012) Quantitative risk-cost-benefit analysis of selected mitigation options for two case studies. Deliverable 5.3. SafeLand Project - Living with landslide risk in Europe: Seventh Framework Programme for research and technological development (FP7) of the European Commission

35. Saito M (1969) Forecasting time of slope failure by tertiary creep. In: Proc. 7th Int. Conf on Soil Mechanics and Foundation Engineering, Mexico City, pp 677–683

36. Sättele M, Bründl M, Straub D (2012) A classification of warning system for natural hazards. In: Moormann C, Huber M, Proske D (eds) 10th international probabilistic workshop. Institut für Geotechnik der Universität Stuttgart, Stuttgart, pp 257–270

37. Sättele M, Bründl M, Straub D (2015) Reliability and effectiveness of warning systems for natural hazards: concept and application to debris flow warning Reliability Engineering and System Safety 142:192–202

38. Schröter K, Velasco C, Nachtnebel HP, Bianca H, Beyene M, Rubin C, Gocht M (2008) Effectiveness and Efficiency of Early Warning Systems for Flash-Floods (EWASE).

39. Špačková O, Straub D (2015) Cost-benefit analysis for optimization of risk protection under budget constraints, Risk Analysis 35(5):941–959

40. Stewart MG, Melchers RE (1997) Probabilistic risk assessment of engineering systems. Chapman & Hall, London

41. Straub D (2005) Natural hazards risk assessment using Bayesian networks. In: Augusti G, Schuëller GI, Ciampoli M (eds) 9th International conference on structural safety and reliability ICOSSAR'05. Millpress, Rome

42. Straub D (2012) Lecture notes engineering risk analysis: Part E – Probabilistic models. Technische Universität München

43. Sturny RA, Bründl M (2014) Bayesian Networks for Assessing the Reliability of a Glacier Lake Warning System in Switzerland. In: Fujita et al (eds) Interpraevent 2014 in the Pacific Rim, Nara, Japan, pp 212–213

44. UNEP (2012) Early warning systems: a state of the art analysis and future directions. United Nations Environment Programme (UNEP), Division of Early Warning and Assessment (DEWA), Nairobi

45. UNISDR (2007) Terminology. The United Nations Office for Disaster Risk Reduction. http://www.unisdr.org/we/inform/terminology, last retrieved in October 2014, Geneva

46. Wegmann M, Funk M, Flotron A, Keusen H (2003) Movement studies to forecast the time of breaking off of ice and rock masses. In: Early Warning Systems for Natural Disaster Reduction. Springer, pp 565–568

47. Willenberg H, Eberhardt E, Loew S, McDougall S, Hungr O (2009) Hazard assessment and runout analysis for an unstable rock slope above an industrial site in the Riviera valley. Switzerland Landslides 6:111–119

48. Zimmer VL, Collins BD, Stock GM, Sitar N (2012) Rock fall dynamics and deposition: an integrated analysis of the 2009 Ahwiyah Point rock fall, Yosemite National Park, USA Earth Surface Processes and Landforms 37:680–691

Chapter 3

New insights into the temporal prediction of landslides by a terrestrial SAR interferometry monitoring case study

Paolo Mazzanti[1,2] **, Francesca Bozzano**[1,2]**, Ivan Cipriani**[3]**,**
Alberto Prestininzi[3]

[1]Dipartimento di Scienze della Terra, "Sapienza" Università di Roma, P.le Aldo Moro 5, 00185 Rome, Italy

[2]spin-off "Sapienza" Università di Roma, Via Cori snc, 00177 Rome, Italy

[3]CERI, Research Centre on Prevention, Prediction and Control of Geological Risks, P.zza U. Pilozzi 9, 00038 Valmontone, Italy

ABSTRACT

Ten small rock slides (with a volume ranging from 10^1 to 10^3 m^3) on a slope affected by working activities were detected, located, and timed using pictures collected by an automatic camera during 40 months of continuous monitoring with terrestrial SAR interferometry (TInSAR). These landslides were analyzed in detail by examining their pre-failure time series of displacement inferred from high-sampling frequency (approximately 5 min) TInSAR monitoring. In most of these cases, a typical creep behavior was observed with the displacement starting 370 to 12 h before the collapse. Additionally, an evident acceleration decrease of the displacement a few hours before the failure was observed in some rock/debris slides, thus suggesting the possibility of a mechanical feature of the slope that differs from the classical creep theory. The efficacy of the linear Fukuzono approach for the prediction of time of failure was tested by back-analyzing the ten landslides. Furthermore, a modified Fukuzono approach named average data Fukuzono (ADF) was implemented and applied to our dataset. Such an approach is able to improve forecasting effectiveness by reducing the error due to anomalies in the time series of displacement, like the acceleration decrease before failure. A prediction with a temporal accuracy of at least 2 h was obtained for all the analyzed rock/debris slides.

KEYWORDS

Terrestrial SAR interferometry; Landslide forecasting; Displacement time series; Average data Fukuzono; Failure

1. INTRODUCTION

Landslide prediction is a major step toward reducing the impact of natural disasters and risks related to human activities in mountainous and hilly areas. Landslides are complex natural phenomena involving volumes ranging from some cubic meters up to some hundreds of millions of cubic meters. Landslides do not behave as a perfectly rigid and brittle body; hence, they are always affected by deformation before the failure (Siddle et al. 2007). The trend of deformation vs. time is considered one key factor in allowing the prediction of time of failure. The amount of pre-failure deformation spans several orders of magnitude (from a few millimeters to several meters) depending on the type of material involved in the landslide, the slope geometry, and the landslide trigger. However, landslides are often characterized by a complex geometry and a combination of heterogeneous materials with different features, thus leading to a nonlinear behavior before the failure. Temporal variation of preconditioning and triggering factors such as climate, river activity, tectonics, and human activities further complicate the evolution of landslides over time.

Since the 1960s, several studies have been carried out with the aim to define rules and procedures to estimate the time of failure of landslides (i.e., Saito 1965; Fukuzono1990; Voight 1988; Crosta and Agliardi 2002). Semi-empirical approaches based on the analysis of displacement (or derived values) time series have been developed and tested on both laboratory-scale and real events, sometimes obtaining good results in terms of landslide forecasting. All these approaches require the collection of displacement data before the failure by a suitable displacement monitoring technique. Back-analysis of large landslides, e.g., the 1963 Vajont landslide (Kilburn and Petley2003), has demonstrated the efficacy of these approaches in predicting past landslides, but it also highlights the difficulties in predicting the failure time of present landslides (Crosta and Agliardi 2003). All these past studies suffer from the great complexity and infinite variety of landslides from the low quality of displacement data available in terms of temporal resolution in data collection, accuracy, and spatial resolution. In fact, the most recent studies still employ point-based techniques (e.g., inclinometers, extensometers) or monitoring systems with a low temporal frequency in data collection (days to hours) or low accuracy in displacement measurement, etc. In other words, little attention has been given to the importance of monitoring data and to related monitoring methods (Dunnicliff 1988) to improve the forecasting efficacy, even if over the last few years innovative and advanced monitoring techniques have been developed, especially in the remote sensing field (Mazzanti 2012).

Terrestrial SAR interferometry (TInSAR) (Mazzanti 2011; Luzi 2010), also known as ground-based SAR interferometry (GBInSAR), is one of the most recent and most powerful of these techniques for monitoring landslides. This technique has the following main advantages: (1) fully remote monitoring (no installation in dangerous areas is required); (2) widespread monitoring instead of the monitoring of single points; (3) simultaneous monitoring of a high number of points (up to hundreds of thousands); (4) high accuracy in surface displacement monitoring (up to decimal millimeters); and (5) a high data sampling rate (up to a few seconds). These improved capabilities play a fundamental role in the attempt to predict the time of landslide failures, thus allowing, for example, identification of the precise area affected by the movement (widespread view), cross-validation of the displacement time series by the enormous number of monitoring points, and the acquisition of dense and accurate time series.

A field experiment was carried out to evaluate the new opportunities offered by the TInSAR monitoring data for time of failure prediction purposes. Four years of continuous monitoring of an unstable slope by TInSAR with a data sampling rate of 5 min allowed the study of the pre-failure behavior of ten small-scale landslides identified on optical images.

The relevant improvements in the forecasting methods for landslide time of failure are also investigated in detail.

2. LANDSLIDE FAILURE PREDICTION BY DISPLACEMENT TIME SERIES

The first attempts to predict the time of failure of unstable slopes on the basis of displacement time evolution date back to the early 1960s (Saito and Uezawa 1961; Saito 1965, 1969). By analyzing the rupture of 80 samples on triaxial compression lab tests, Saito observed that displacement was the most useful parameter to predict the time of failure. Hence, Saito (1965) developed a method based on the "slope creep" theory (Terzaghi 1950; Haefeli 1953) to obtain the time before failure using slope displacement data. The creep process is a time-dependent deformation of materials under the action of stresses (e.g., gravity) and is usually divided into three main phases: primary, secondary, and tertiary creep. The failure of materials and slopes usually occurs in the tertiary creep. Therefore, Saito (1969) modified the method to fit nonsteady behavior of landslides in the tertiary creep prior to failure. In the 1980s, improvements to Saito's prediction model were made by Fukuzono (1985) and Voight (1988, 1989a). Fukuzono showed that the logarithm of velocity of the surface displacement is proportional to the logarithm of the acceleration. In other words, under invariant loading conditions, pre-failure behavior can be described by a power law equation in the following form:

$$\frac{d^2 t}{dt^2} = a\left(\frac{dx}{dt}\right)^{\alpha} \quad (1)$$

where *x* is the downward surface displacement along the slope, *t* is the time, and *a* and*α* are dimensionless constant parameters.

According to the results of the experiments by Fukuzono (1985, 1989) and the studies of Varnes (1983) and Yoshida and Yachi (1984), *α* commonly varies within a range of 1.5–2.2. Therefore, the following equation has been suggested to predict the time of failure:

$$\frac{1}{v} = [a \cdot (\alpha - 1)]^{\frac{1}{\alpha-1}} \cdot (t_f - t)^{\alpha \frac{1}{\alpha-1}} \quad (2)$$

where v is the surface displacement velocity and t_f is the failure time. For $\alpha = 2$, the resulting curve is linear; for $\alpha > 2$, it is convex; and for $1 > \alpha < 2$, it is concave (Fukuzono 1985). Hence, in the case of $\alpha = 2$, the failure time can be simply computed by the following equation:

$$t_f = \frac{t_2 \cdot \left(\frac{1}{v}\right)_1 - t_1 \left(\frac{1}{v}\right)_2}{\left(\frac{1}{v}\right)_1 - \left(\frac{1}{v}\right)_2} \tag{3}$$

and the time of failure corresponds to the interception of the x-axis of the interpolated straight line in the diagram's inverse velocity vs. time (Fukuzono 1985).

A more generalized relationship to describe rate-dependent material failure was given by Voight (1988):

$$\ddot{\Omega} = A\dot{\Omega}^{\alpha} \tag{4}$$

where and $\ddot{\Omega}\ddot{\Omega}$ and $\dot{\Omega}\dot{\Omega}$ are the acceleration and velocity of the displacement, respectively.

In terms of time of failure prediction, the following equation can be obtained by assuming $\dot{\Omega}\dot{\Omega}_f$ (velocity at t_f) = infinite:

$$t_f - t = \left[\frac{\dot{\Omega}^{(1-\alpha)} - \dot{\Omega}_f^{(1-\alpha)}}{A(\alpha - 1)}\right] \tag{5}$$

This assumption leads to nonconservative forecasts, but the error is often small and may not be important to decision makers (Voight 1988).

The efficacy of this method in predicting landslide time of failure has been demonstrated by several authors, including Voight and Kennedy (1979), Voight (1989b), Cornelius and Voight (1990, 1994, 1995), Voight and Cornelius (1990, 1991), Rose and Hungr (2007), and Gigli et al. (2011).

3. LANDSLIDE MONITORING BY TERRESTRIAL SAR INTERFEROMETRY

Continuous monitoring is a key requirement for the prediction of a landslide. Failure prediction can be based on the monitoring of landslide-triggering factors (e.g., rainfall, groundwater level) or of landslide effects (e.g., ground displacement). Over the years, it has been demonstrated that the monitoring of landslide effects is the most effective solution for the control and prediction of landslides. Hence, several technical solutions for the monitoring of ground displacement are currently available. These techniques can be divided into two main categories:

- Geotechnical contact techniques, including extensometers, inclinometers, and strainmeters (see Dunnicliff (1988) for an extensive review)

- Remote techniques, including topographic systems, photogrammetry, laser scanning and SAR Interferometry (see Mazzanti (2012) for an extensive review)

Of these techniques, differential SAR interferometry (DInSAR) is among the newest and most powerful, and its features fit well with landslide monitoring requirements. DInSAR was developed for satellite applications in the early 1990s and was originally used to measure ground displacements at a regional scale (Curlander and McDonough 1991; Massonet and Fiegl 1998; Hanssen 2001). In the late 1990s and early 2000s, several innovation occurred in the SAR field such as the development of innovative approaches for satellite data processing based on data stacking (Ferretti et al. 2001) and the development of the first ground-based SAR equipment prototypes. Over the last 10 years, both satellite and terrestrial InSAR have been extensively used for landslide monitoring, thus demonstrating their efficacy (Pieraccini et al. 2002; Leva et al. 2003; Tarchi et al. 2003; Antonello et al. 2004; Hilley et al. 2004; Strozzi et al. 2005; Noferini et al. 2005; Casagli et al. 2006; Colesanti and Wasowski 2006; Farina et al. 2006; Herrera et al. 2009; Bozzano et al. 2010; Intrieri et al. 2012).

The TInSAR technique is based on an active radar sensor that emits microwaves and receives the return of scattering objects. In the most common configuration, the sensor, which has two antennas (one emitting and one receiving), moves along a linear rail, thus collecting an enormous number of radar images with a single resolution (range resolution). Then, by combining images collected during a single scan through the synthetic aperture radar principle, bidimensional images are derived, thus introducing the cross-range resolution (Figure 1). The ground cross-range resolution can be approximated as

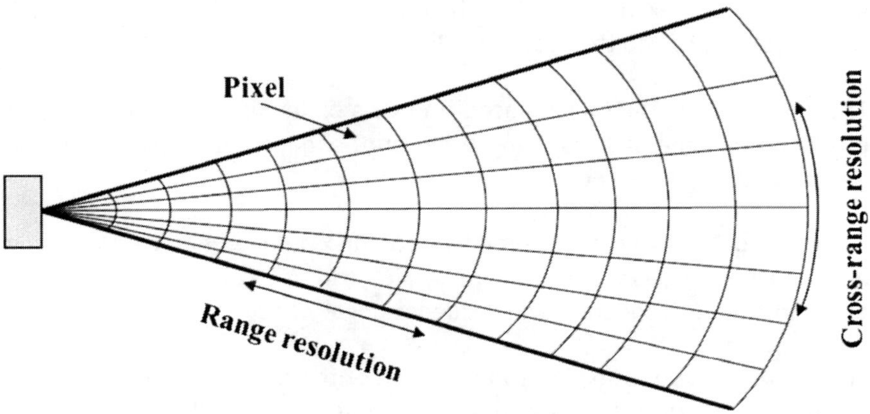

Figure 1. Example of grid resolution

$$\Delta a = \frac{r\lambda}{2L} \tag{6}$$

where Δa is the cross-range resolution, r is the range distance (i.e., the distance between the GB-SAR system and the measured point), λ is the central wavelength, and L is the length of the rail.

This means that considering a wavelength of approximately 17 mm and a rail length of 2 m (the features of the equipment used in this paper), the cross-range resolution is on the order of 0.4 and 4 m at a distance of 100 and 1,000 m, respectively.

For the range resolution, the equipment used in this study is based on the stepped frequency-continuous wave (SF-CW) principle, in which a band of continuous electromagnetic waves having different frequencies are emitted. The range resolution is equal to:

$$\Delta r = \frac{c}{2B} \tag{7}$$

where c is the light velocity and B is the radiofrequency bandwidth. By Eq. 7, we can observe that the range resolution is constant (i.e., it is independent of the range distance) and that it only depends on the bandwidth. For example, with a bandwidth of 200 MHz, the maximum range resolution is on the order of 0.75 m.

The final result of TInSAR data acquisition is a bidimensional image made of pixels (up to some millions), whose footprint size increases with increasing instrument target distance.

Each pixel of the image is featured by a complex number, namely amplitude and phase values that are specific for a certain polarization, electromagnetic frequency, and incidence angle (Ulaby et al. 1982).

The signal to noise ratio (SNR) value is one of the parameters for assessing the backscattering features of a specific target (i.e., localized area) inside the investigated scenario. Both mean values or variability over time of this value on data stack are used.

The phase at each pixel can be used for the estimation of displacement through the interferometric technique. In other words, the phase difference of each pixel of two or more SAR images collected at different times is computed and, in the case of long stacks of images, cumulated over time. In terrestrial SAR interferometry, the computed phase difference ($\Delta\varphi$) is mainly controlled by the following factors:

$$\Delta\varphi = \Delta\varphi_{\text{displ}} + \Delta\varphi_{\text{atmos}} + \Delta\varphi_{\text{noise}} \tag{8}$$

where $\Delta\varphi_{\text{displ}}$ is the phase shift contribution due to the ground displacement, $\Delta\varphi_{\text{atmos}}$ is the phase shift contribution due to the atmospheric changes, and $\Delta\varphi_{\text{noise}}$ is the phase shift contribution due to the instrumental noise. By assuming that the $\Delta\varphi_{\text{noise}}$ is random and not relevant, once the $\Delta\varphi_{\text{atmos}}$ is computed, the $\Delta\varphi_{\text{displ}}$ can be derived. Therefore, the displacement is computed by the following equation:

$$d = \frac{\lambda}{4\pi}\Delta\varphi \tag{9}$$

where d is the displacement computed along the line of sight (LOS), λ is the wavelength of the radar signal, and $\Delta\varphi$ is the phase difference between the two acquisitions.

Hence, the accuracy in the displacement measurement depends on the signal's wavelength, the atmospheric conditions, and the sensing distance. Specifically, the lesser the distance and the more stable the atmospheric conditions, the higher the accuracy of the displacement measurements (from 0.01 mm in the lab to a few millimeters in the field).

It is worth noting that the interferometric analysis implies some limitations concerning the displacement measurement due to the cyclic behavior of the phase. As a matter of fact, the phase unwrapping (e.g., Goldstein et al. 1988) is a key part of the interferometric analysis for displacement measurement. Considering the wavelength of the herein used GBSAR equipment, the phase ambiguity is on the order of 4.5 mm (i.e., the $\lambda/4$ value) that can be assumed as 9 mm if we assume that inversion of displacement direction along the line of sight cannot occur.

The coherence value, which evaluates the correlation among the phases at each pixel, is a good estimator of the phase stability and, therefore, can be used for a preliminary evaluation of the expected displacement accuracy. This value ranges between 0 and 1, where 0 is the complete decorrelation of the phase and 1 is the complete correlation (Hanssen 2001). Hence, multitemporal analysis allows LOS displacement time histories of each pixel of the image to be obtained (Bozzano et al. 2011).

In summary, the following operational features make the TInSAR technique particularly suitable for the monitoring of landslides and for failure forecasting: (1) the ability to yield data and answers within a brief time (a few minutes); (2) efficacy under any weather and lighting conditions; (3) completely remote operability (it does not require the installation of sensors or targets on the monitored slope); (4) continuous distributed monitoring of the entire slope with a high pixel resolution; and (5) long-range monitoring (up to some kilometers).

Notwithstanding the numerous TInSAR successfully applications cited above, little attention has been devoted to the opportunities offered by this technique for landslide prediction purposes and still less for small-scale landslides.

4. THE STUDY AREA

The field experiment presented herein concerns a slope affected by a major deep-seated landslide characterized by a very complex geological and geomorphological setting (Bozzano et al. 2008, 2011). Due to the development of a major communication road, the slope has been extensively investigated since 2007 and controlled for more than 4 years by an integrated monitoring platform composed of traditional sensors (inclinometers, piezometers, load cells, and topographic measures) and a remote station equipped with an IBIS-L terrestrial SAR interferometer, by IDS S.p.A (Table 1), a weather station, and an automatic camera. These equipment was housed in a specially designed box and installed on the opposite slope to the one involved in the project, at a distance ranging from 700 to 900 m (Figure 2). The remote monitoring network has been active 24/7 from November 2007 until now. Collecting TInSAR data at a sampling rate of approximately 5 min, the network has thus far collected more than 20,000 pictures, 400,000 measurements of weather data, and 400,000 SAR images (Bozzano et al. 2011). The digital camera and the weather station, featuring an automatic data acquisition and transfer system, have played a key role in the monitoring activity. The images continuously supplied by the camera have made it possible to continuously monitor the construction activities and the engineering works on the investigated slope, thereby optimizing data collection and processing. This functionality also facilitates the interpretation of the results and the optical first identification and timing of landslides. In fact, during the experimental period, the slope was frequently affected by shallow translational landslides (with a volume ranging from 10^1 to 10^4 m^3), which represent surface evidence of a deep landslide (Bozzano et al.2011). These landslides were identified and located by optical photos (Figs. 3 and 4) and then detailed in terms of location over the slope and time of occurrence by TInSAR displacement images. Detailed and accurate SAR displacement images and cumulative time series of each pixel inside the landslide area were derived for each landslide.

These landslides were investigated in detail by examining their size and geometry, geological material, and location (Table 2). The landslide volumes range from 10^1 to 5×10^2 m^3, with an estimated maximum thickness of approximately 3 m. The landslides were classified as translational and roto-translational movements involving weathered and fractured gneiss widely outcropping over the slope. These landslides were mainly triggered by rainfall. Some of these landslides occurred in portions of the slope covered by spritz beton. It is worth noting that the sizes of some of the landslides reported in Table 2 are similar to those of the experimental landslides described in the papers by Fukuzono (1990) and Moriwaki et al. (2004).

Table 1. Instrumental configuration of the TInSAR instrument

Rail length	2 m
Central frequency	16.75 GHz
Bandwidth	300 MHz
Polarization	VV
Antenna gain	20 dB
Number of scans in the SAR image	401
Range resolution	0.50 m
Cross-range resolution	4.5 mrad (3.6 m at a distance of 800 m)
Max distance	1,200 m
Inter-scan delay (waiting time between the end of one scan and the start of the next)	6 s
Measurement time interval	~5 min

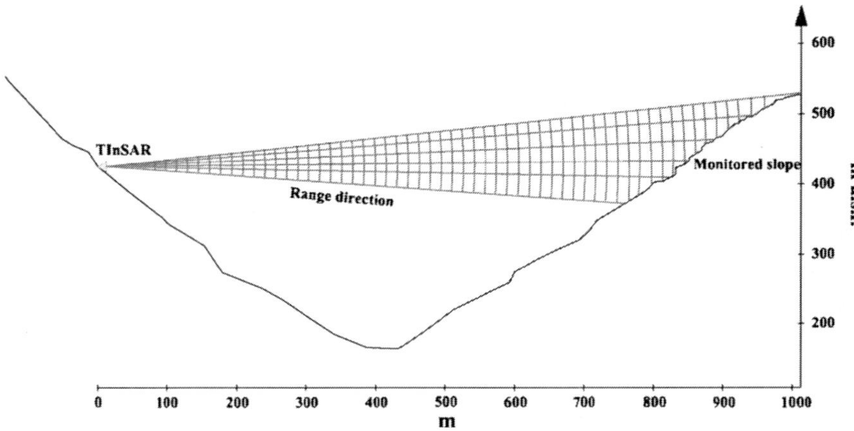

Figure 2. Sketch showing the monitoring geometry

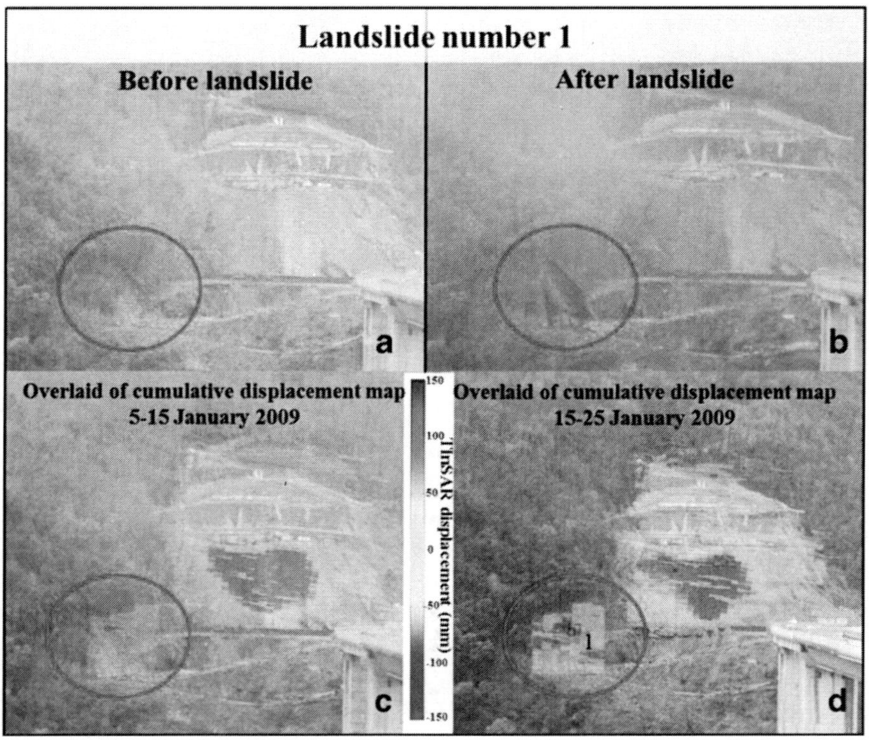

Figure 3. **a**, **b** Pictures before and after landslide 1 (*red circle*); **c**, **d** TInSAR displacement images before and after the landslide overlain on the picture (**a**, **b**)

Figure 4. Picture of the slope showing the location of the ten shallow landslides described in Table 2

5. PRE-FAILURE BEHAVIOR OF SMALL-SCALE LANDSLIDES DETECTED BY CONTINUOUS TINSAR MONITORING

The displacement behavior of the ten shallow landslides described above, especially at their pre-failure stage, were analyzed in detail to infer information about the total amount of displacement, the duration of the entire process, the velocity, the acceleration, and other factors (Table 3).

Table 2. Synoptic table showing the physical features of the investigated landslides

Landslide ID	Type	Material	Width (m)	Length (m)	Area (m^2)	Thickness (m)	Failure time	15 days cumulative displacement before landslide failure
1	Rotational slide	Mobilized gneiss	7	15	105	3	24 January 2009; h 14:32–15:32	238
2	Flow	Gneiss colluvium	2	7	14	1	18 February 2009; h 12:53–13:53	160
3	Translational slide	Gneiss colluvium + spritz beton	10	8	80	2	20 December 2009; h 16:46–17:46	73
4	Rotational slide	Gneiss colluvium + spritz beton	10	13	130	1.5	Night between 15 and 16 January 2010	38
5	Translational slide	Gneiss colluvium + spritz beton	8	7	56	1.5	03 February 2009; h 16:58–17:58	60
6	Translational slide	Gneiss colluvium + spritz beton	20	13	260	2	Night between 10 and 11 February 2010	76
7	Flow	Mobilized and altered gneiss	15	13	195	1	12 February 2010; h 11:58–12:58	66
8	Translational slide	Gneiss colluvium + spritz beton	10	15	150	2	12 February 2010; h 11:58–13:58	75
9	Translational slide	Mobilized and altered gneiss	13	4	52	1.5	17 February 2010; h 12:57–13:57	73
10	Translational slide	Mobilized and altered gneiss	15	5	75	2	Night between 9 and 10 March 2010	35

Table 3. Synoptic table showing the pre-failure behavior of the investigated landslides

ID	Time span deformation (min)	Total displacement (mm)	Maximum velocity (mm/h)	Maximum acceleration (mm/h^2)	Delay between the acceleration peak and the collapse (min)	Delay between the last significant rainfall and the acceleration peak (min)
1	6,585	144.52	17.07	2.85	203	4,686
2	13,080	769.42	13.23	0.83	2,225	7,236
3	455	20.29	9.81	6.74	77	1,236
4	416	23.26	33.75	82.11	42	654
5	1,255	21.89	8.08	3.87	137	2,646
6	2,460	63.98	10.63	2.25	132	138
7	3,920	105.14	19.56	5.87	52	798
8	753	99.22	27.21	16.18	132	84
9	22,255	764	44.37	5.61	223	6,420
10	1,340	148.18	64.06	41.92	47	2,658

For each landslide, we computed the time series of displacement of those pixels that were considered the most representative of the overall landslide behavior. Specifically, pixels were chosen on the basis of the following criteria: (1) the highest quality TInSAR data available on the basis of SNR and coherence analysis; (2) pixels clearly located within the main landslide body (and not in the surrounding parts affected by tensional release); and (3) pixels located preferably in the topographic upper part of the landslides, as they are assumed to be less influenced by internal deformation.

Thanks to the high temporal information density (one displacement value every about 5 min), it was possible to process the identified time series using a digital filter to remove the noise. A direct-form FIR equiripple lowpass filter, implemented in the "filtfilt" algorithm (Matlab 7), was used for the filtering, thus significantly improving the quality of the time series (Figure 5). Data filtering is very important for our purposes, as the used models for landslide prediction are based on landslide velocity, i.e., the displacement noise is commonly amplified.

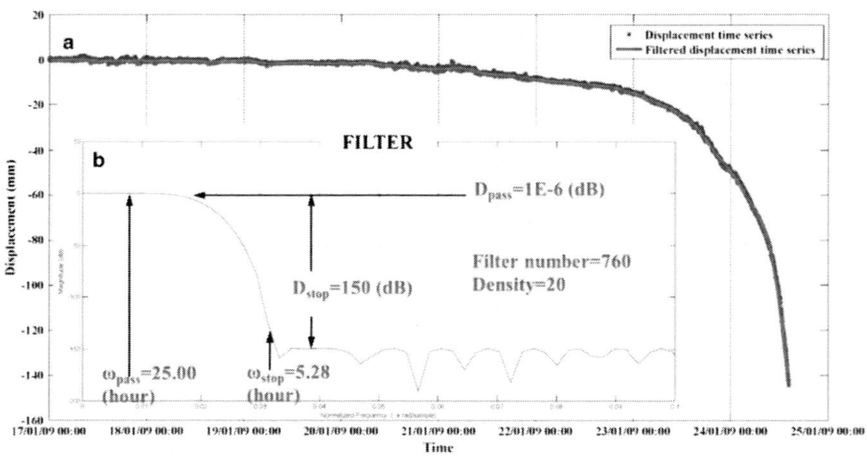

Figure 5. *a* Displacement time series and filtered displacement time series; *b* applied direct-form FIR Equiripple lowpass filter: D_{pass} and D_{stop} = passband and stopband ripples; ω_{pass} and ω_{stop} = transition width

Once achieved, the final time series landslides were analyzed, thus deriving the following:

1 The temporal evolution of displacement, i.e., the initiation of the movement (manually identified on the time series as the intersection point between horizontal trend lines and a clearly detectable inclined trend lines when a cumulative displacement of 2 mm is achieved), and the final collapse of each landslide were exactly dated and timed with a precision of a few minutes, thus obtaining values ranging from a couple of weeks to a few hours (Figure 6).

2 The total landslide displacement from its onset to its collapse, thus achieving values ranging from a couple of centimeters up to 1 m (Figure 7).

3 Maximum velocity, thus achieving values ranging from 8 to 64 mm/h (Figure 8).

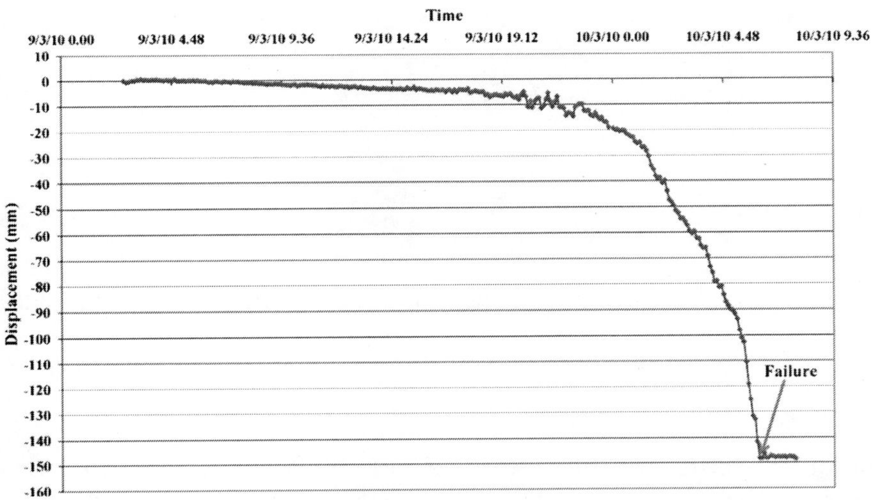

Figure 6. Example of time series of displacement derived from TInSAR monitoring data (landslide n.10). The failure time can be clearly identified with a precision of a few minutes thanks to the sudden stop in the displacement

It is worth noting that all the landslides, apart from 2, 4, and 8, were characterized by the maximum velocity just before the collapse, thus showing a trend of velocity continuously increasing since the landslide start until the failure. The peak of acceleration was registered in all the landslides some hours before the collapse instead (Figure 8).

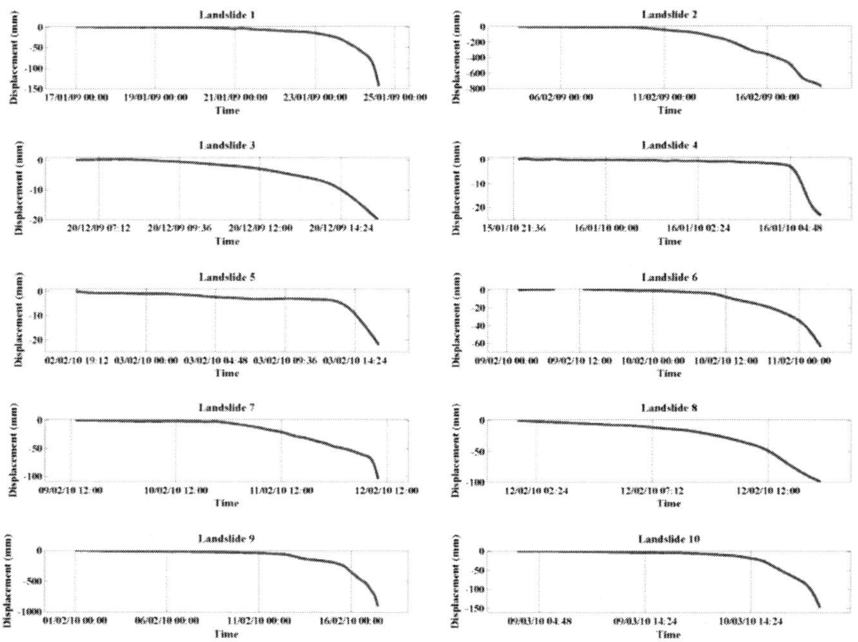

Figure 7. Time series of displacement of the ten landslides of Figure 5 and Table 2

In other words, the final stage of the landslide displacement was characterized by an increasing velocity but a decreasing acceleration rate. The peak acceleration of the landslides ranged from 1 to 82 mm/h^2, depending on the landslide. The value of acceleration decrease (computed as the difference between the peak acceleration and the last value available before the failure) ranged from 2 to 154 mm/h^2. Furthermore, the value of acceleration decrease is closely proportional to the peak acceleration value of the landslides. Specifically, the ratio between the value of acceleration decrease and the acceleration peak ranges from 0.8 to 2.1. It is worth noting that the achieved values of total displacement, velocity, and acceleration were very similar to those measured in the experiments by Fukuzono (1985) and Moriwaki et al. (2004), which is a further confirmation of the quality of the dataset presented herein, which is comparable to the data derived from a controlled experiment.

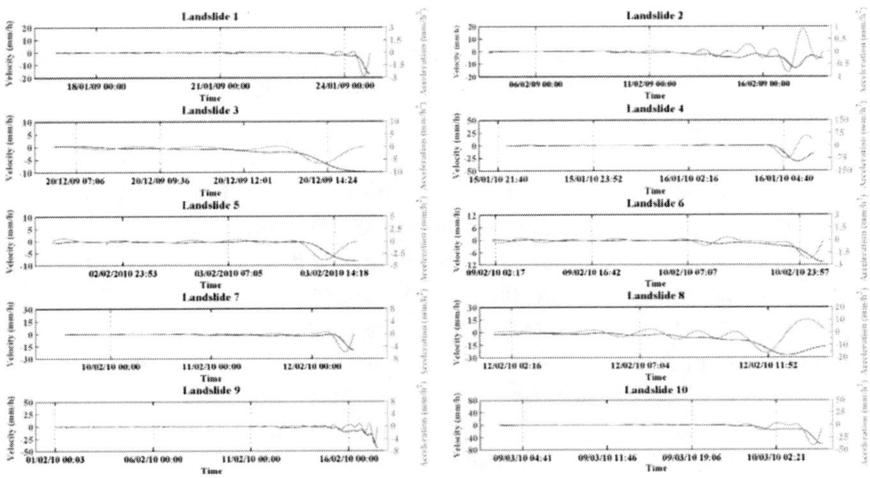

Figure 8. Time series of velocity and acceleration of the ten landslides of Figure 5 and Table 2

Nevertheless, it is worth noting that pre-failure acceleration decrease is something never observed before, whether in field or experimental landslides. This evidence, which could play a relevant role in time of failure prediction, was obtained thanks to the availability of accurate and high-sampling rate data.

6. TIME OF FAILURE ANALYSIS OF DETECTED LANDSLIDES

The large dataset of events occurred on the same slope (which means similar conditions and features of the landslides), and the detailed displacement data available represent a rare occasion to test the efficacy of semi-empirical approaches based on time series of displacements (or derived quantities such as velocity and acceleration). Furthermore, to the authors' knowledge, semi-empirical approaches have previously been tested mainly on large-scale real landslides, except for the large-scale experiments by Fukuzono (1985) and Moriwaki et al. (2004). Hence, the presented dataset is quite new in this field of research.

In this paper, analyses based on the simple linear Fukuzono ($\alpha = 2$) approach are presented (as it is the most commonly used for time of failure prediction purposes) to infer their efficacy in predicting the time of failure of the landslides investigated herein. For each landslide, the predicted time of failure was computed iteratively since the beginning of the displacement phase (looking at the tertiary creep phase) by increasing the number of displacement data step by step. Hence, the real prediction of the time of failure based on the newly collected data over time was computed. In other words, the dataset used for the computation of the time of failure is progressively enlarging with time, and consequently, the prediction is updated. For each landslide, the obtained results were plotted in a diagram showing the predicted time of failure in the *x*-axis and the time of elaboration before failure (in hours) in the *y*-right axis (blue line in Figure 9). For example, 1.6 h before failure, the predicted time of failure of landslide 3 computed by using the time series of displacement allowable until this time was 14:42 instead of 15:36 (red line in Figure 9) as what really occurred. In contrast, 0.1 h before failure, the time of failure of landslide 3 was exactly predicted. Starting elaboration time (yellow line in Figure 9) was roughly identified on the time series of displacement by using a graphical method similar to that one used in soil mechanics to identify yield stress on the stress-strain odometric curve.

Diagrams showing the temporal discrepancy between the real and predicted time of failure over time (e.g., the quality of prediction over time) were obtained for the ten landslides (Figure 10).

Three hours is the maximum absolute error in the landslide time of failure prediction except for landslides 2 and 9 (Figs. 4 and 11 and Tables 2 and 3). In these two cases, the maximum error was on the order of 600 to 1,000 h, i.e., longer than the entire timespan of the landslide. Landslides 2 and 9 were characterized by longer time series of displacement and by a spurious creep behavior characterized by several acceleration increase and decrease phases before the failure. Such a pattern of displacement can be considered the main reason for the low performances of the linear Fukuzono analysis.

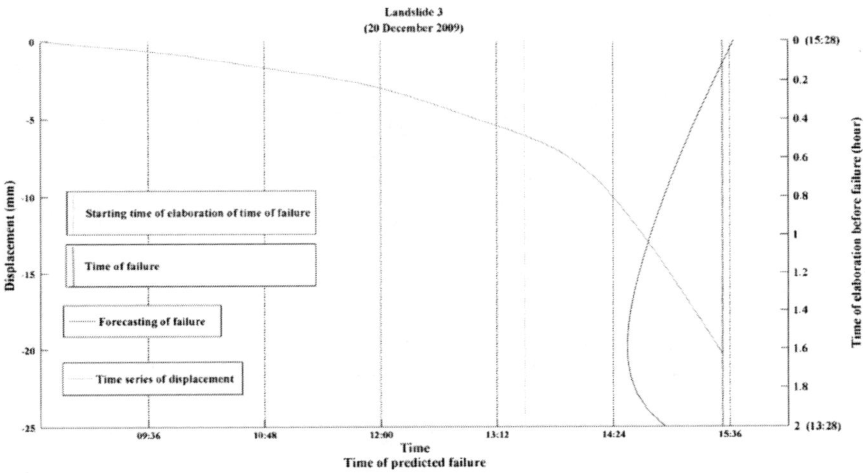

Figure 9. Diagram showing the variation of the time of failure prediction using an incremental dataset length for landslide 3. *Green line* is the "time series of displacement" (*left y*-axis vs. time in abscissa); *blue line* is the updating of the prediction during the 2 h before the failure (*right y*-axis vs. time of failure in abscissa). Recorded time of failure is marked by the *red line*; *yellow line* is the starting time for the computation

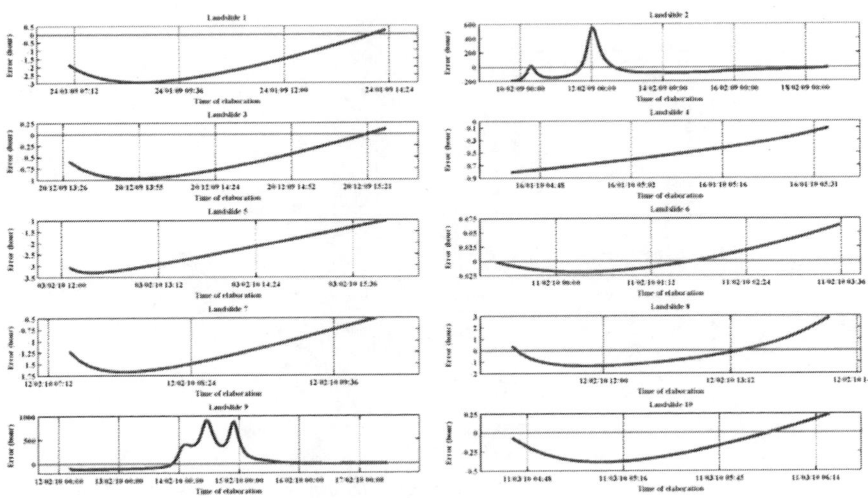

Figure 10. Temporal discrepancy between the real and predicted time of failure over time for the ten landslides of Figure 5and Table 2

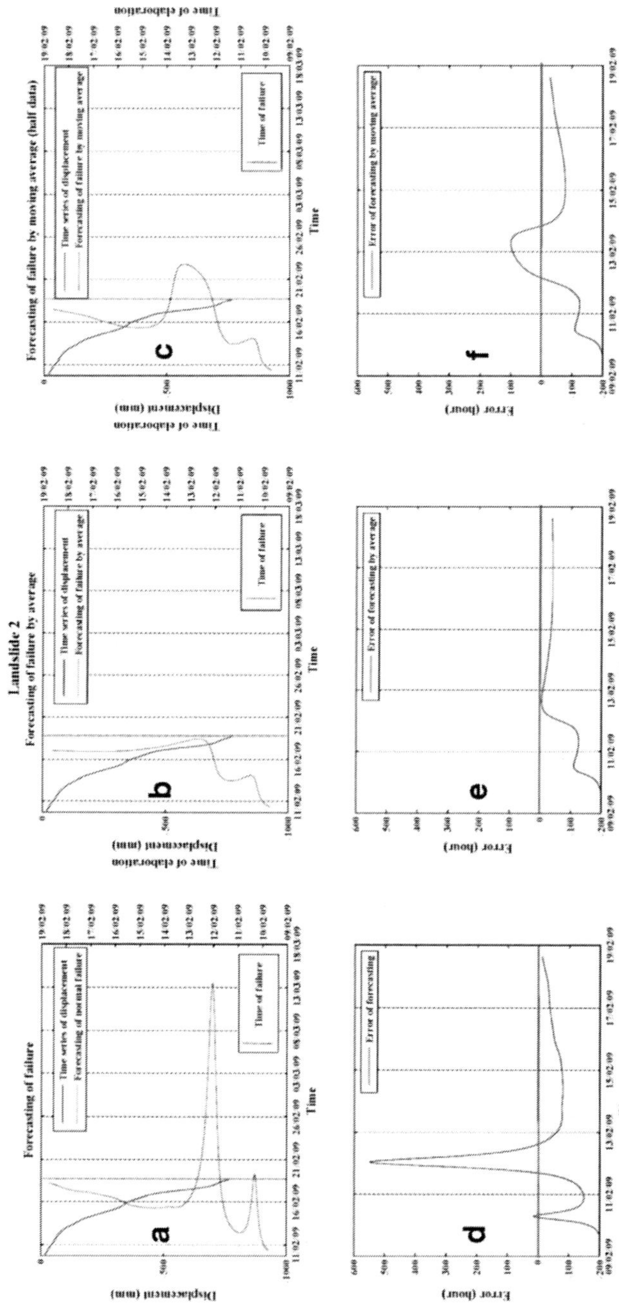

Figure 11. a–f From *left* to *right* (*upper* and *lower diagrams*), the results of the Fukuzono analysis using the conventional approach, the average approach, and the moving average approach for landslide 2

For the other eight remaining landslides, the error in the predicted time of failure is very small. Specifically, for five landslides (1, 3, 6, 8, and 10), positive prediction error was found (i.e., predicted time of failure postponed with respect to the real landslide occurrence), while for two landslides (4 and 7), negative error (i.e., predicted time of failure anticipated respect to the real one) was found. Our attention was mainly concentrated on the positive errors as they are contrary to general precautionary principles.

To ameliorate these problems, a new approach named average data Fukuzono (ADF) was developed. ADF is a modified version of the Fukuzono method consisting of the average and moving average velocity computed from temporal consecutive data. In the first case, the data were averaged iteratively, starting from the first data collected. In the case of the moving average, the data were averaged by using half of the dataset moved iteratively by one single step until the last half before the failure.

Figure 11 shows the results achieved using the ADF approach for landslide 2. As can be seen, both the average and the moving average approaches (Figure 11b, c, e, f) significantly reduce the prediction error achieved by using the standard Fukuzono method and reported in Figure 11a, d.

Figure 12 shows the results achieved by using the ADF approach for landslide 10. In this case, the error obtained by the standard Fukuzono analysis (Figure 12a, d) is quite small; hence, the quality of the prediction is only slightly improved by the ADF approach (Figure 12b, c, e, f). However, the moving average significantly reduces the error in the final phase of displacement, thus avoiding the negative effect due to the decelerating phase.

The efficacy of the ADF approach in reducing the positive prediction errors is clearly visualized in Figure 13, where the mean value of the prediction error and the related standard deviation of the dataset of the five landslides are shown by normalizing the time.

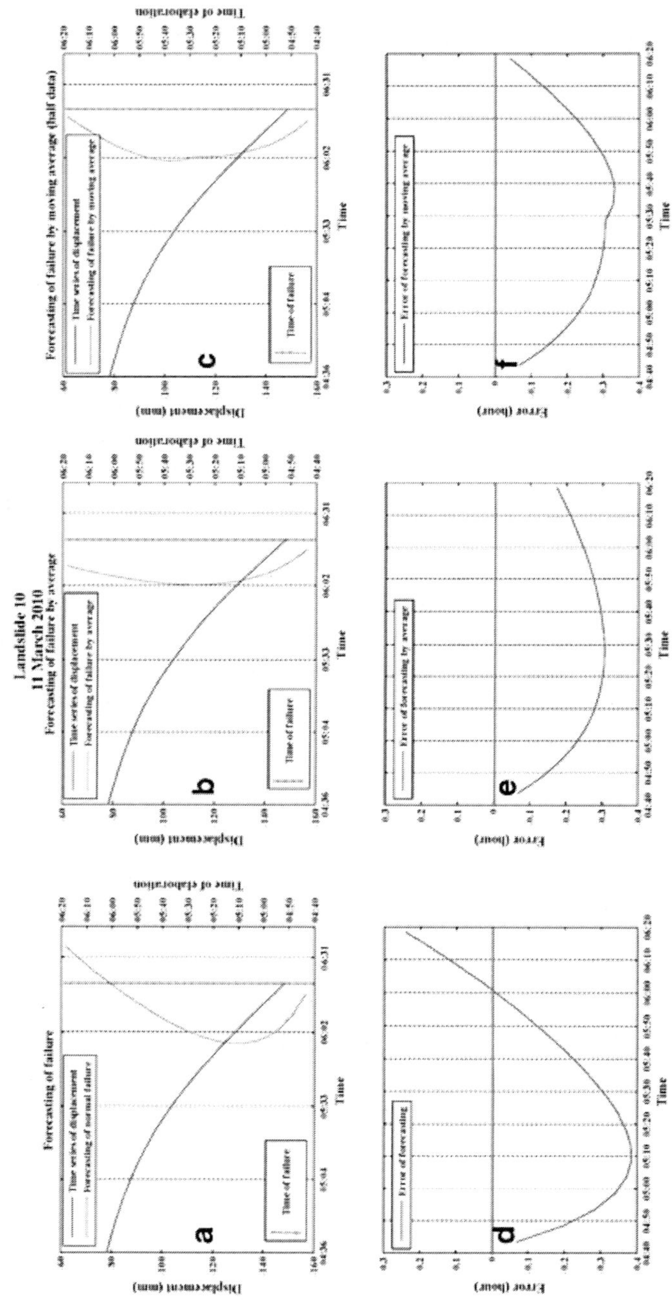

Figure 12. a–f From *left* to *right* (*upper* and *lower diagrams*), the results of the Fukuzono analysis using the conventional approach, the average approach, and the moving average approach for landslide 10

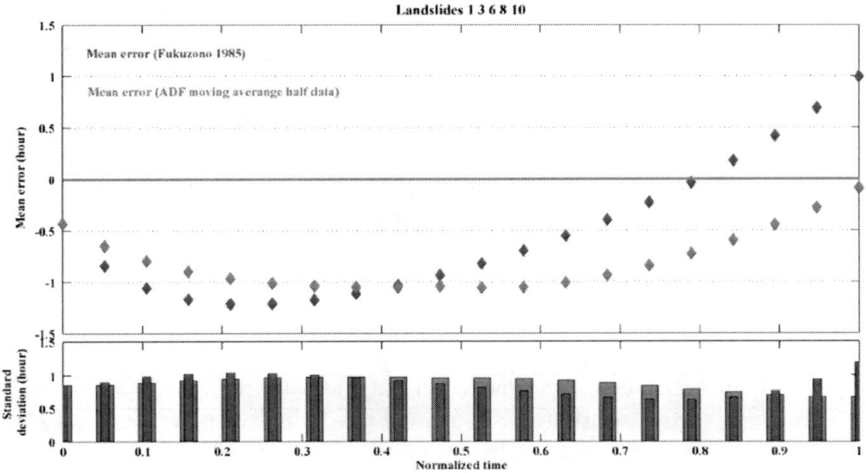

Figure 13. Comparison between the error of the prediction of the failure time for landslides 1, 3, 6, 8, and 10 by using the linear Fukuzono method and the moving average ADF method. Time in abscissa is normalized with respect to the time of failure of each five landslides and error at each normalized time step is the mean error value for the five landslides; *bars* are referred to the standard deviation associated to the mean value at each time step

7. DISCUSSION

A long-term continuous monitoring of the displacement of an unstable slope (whose size is on the order of 0.2 km^2) by TInSAR allowed us to build a spatio-temporally high-resolution dataset describing the pre-failure behavior of minor small and shallow landslides (a few tenths of cubic meters) with an accuracy of 1 mm. Some of these landslides were similar in size to the experimental ones proposed by Fukuzono (1985) and Moriwaki et al. (2004) and were characterized by similar displacement behavior before failure (in terms of total displacement, velocity, and acceleration values). However, while the abovementioned experiments were performed on a single "ideal" landslide (whose features were previously defined), our dataset was made of ten landslides that occurred in a field environment. Continuous monitoring by terrestrial SAR interferometry provided an unquestionable contribution to this study, thus allowing us to achieve data as of high quality as can be achieved on a lab experiment. Specifically,

1 The high temporal resolution (about 5 min) on data collection allowed us to characterize also the displacement of landslides that occurred in a short time span (few hours) with a high data sampling rate, thus allowing us to identify the last steps of the creep process before failure.

2 The spatially continuous monitoring (i.e., displacement maps instead of single points) with a high ground resolution (few meters) allowed us to measure the displacement also of small landslides (few tenths of square meters) and, in some cases, to cross-check the inferred displacement from adjacent pixels.

3 The high accuracy in the displacement measurement allowed us both to characterize landslides with a small pre-failure displacement (even a couple of centimeters) and to well define their acceleration behavior.

4 The capability of TInSAR to collect data under any weather and lighting conditions allowed us to get information about events that occurred on complex environmental conditions and during the night.

Furthermore, it is worth stressing that the opportunity of monitoring an entire slope with a high spatial continuity was a key feature that allowed us to detect and characterize landslides that occurred also in sectors that were not originally planned.

Nevertheless, it is important to account also for some of the limitations of terrestrial InSAR monitoring and on their effects on the results achieved in this study:

1 First of all, as TInSAR provides only surface displacement data, no information was available on the deep behavior of the investigated landslides. However, considering the shallow feature of the herein discussed landslides, such a limitation is not considered relevant for this study.

2 Phase ambiguity problem, especially in the final evolution phases before the failure when the displacement velocities are higher. As a matter of fact, considering the usual small size of the landslides, spatial unwrapping algorithms demonstrated to be not very effective and only a temporal phase unwrapping was used, supported by the authors' expertise in the TInSAR data processing. In this regard, it is worth noting that, for most of the landslides, the velocity in the final stage is well below the phase ambiguity threshold (4.5 or 9 mm in 5 min, if we assume that inversion of direction of movement is not reasonable for a

landslide), and it is more than reasonable to assume that the phase ambiguity problems have not happened. However, in some cases (such as numbers 4, 9, and 10), this occurrence cannot be excluded. Furthermore, no other data derived from different sensors have been collected.

Anyway, instead of what was stated above, all these landslides were characterized by a similar displacement pattern in terms of peak values of velocity and acceleration and time-dependent evolution. However, a peculiar acceleration decrease before failure was observed in several landslides in our dataset. Such a behavior represents an anomaly with respect to available data from the literature and, at present, cannot be explained by existing theoretical creep models that assume a continuous increasing of acceleration until failure. This pre-failure acceleration decrease has several implications in the application of the already-existing semi-empirical models based on the creep theory, as demonstrated by the analyses performed by the simple linear Fukuzono approach (Figure 9). The pre-failure acceleration decrease leads to a delay in the failure time with respect to that previously predicted by using the linear Fukuzono approach. Such behavior cannot be fully understood by the available data, and it requires further study. Nevertheless, we cannot neglect that the creep theory assumes a deformation under the condition of constant stress. Such a condition is not considered in our experiments because rainfall occurred during the experiments; hence, we cannot exclude the possibility that this condition could be one of the reasons for the pre-failure acceleration decrease. However, it is worth noting that the constant strain condition is also not considered in actual large landslides, such as those studied by Crosta and Agliardi (2003). The effect of rainfall was observed in landslides 2 and 9, where multiple acceleration and acceleration decrease phases occurred before the failure. In fact, these landslides are characterized by a high prediction error using the linear Fukuzono approach. The other landslides are always predicted with an error lower than 3.5 h. A drastic reduction in the forecasting errors was achieved by the herein proposed average data Fukuzono (ADF) approach (Figure 10). This method, which consists of the average and moving average of the displacement data over time, is very effective at reducing the forecasting error, especially for unsteady cases characterized by important decelerating phases (Figure 11). Furthermore, the ADF approach seems to be very effective in reducing the effect of the final decelerating phase (Figure 13).

However, it is worth noting that the ADF approach is very sensitive to the number of available data, i.e., detailed time series significantly increase its efficacy, thus confirming the importance of high-resolution monitoring data for landslide prediction.

8. CONCLUSIONS

With the aim of deepening our understanding of landslide pre-failure behavior and, therefore, landslide prediction capabilities, a long-term field experiment using TInSAR was carried out. By analyzing the pre-failure behavior of ten landslides that occurred during the experiment, we showed that high-accuracy and high-sampling rate (a few minutes) of surface displacement data can be successfully used to obtain a good prediction of a landslide's time of failure. Furthermore, the widespread and high-resolution information retrieved by TInSAR displacement images allows investigation from a remote position and the assessment of the pre-failure behavior of small landslides. In fact, traditional contact monitoring methods may fail due to a low data sampling rate, low accuracy and, particularly, punctual and localized information that may lead to a lack of information or to misinterpretations, especially for small-sized landslides.

A landslide is a very complex process characterized by an infinite range of variable features. The ability to obtain detailed and accurate information about the behavior of landslides is crucial to predict their future evolution (i.e., failure). Existing semi-empirical models based on the creep theory are very effective at predicting time of failure in laboratory experiments, but they are effective for real landslides only in a few cases. However, we demonstrated that by using accurate and high-sampling rate data, these models may be effectively used in real practice. Furthermore, a high data sampling rate allows the effective use of the ADF approach, which is able to drastically improve the efficacy of creep-based prediction methods in the case of "unsteady landslides." Hence, it can be said that by using TInSAR continuous monitoring and advanced data processing solution such as the ADF approach, we can achieve the same results previously achieved for "ideal" laboratory landslides for real field landslides.

ACKNOWLEDGMENTS

This research was partially funded by MIUR grants—Prin 2009, Research title "Relationships between human activities and geologic instabilities by integrating monitoring data and geological models related to already studied case histories". Principal investigator: Francesca Bozzano.

REFERENCES

1. Antonello G, Casagli N, Farina P, Leva D, Nico G, Siebar AJ, Tarchi D (2004) Ground-based SAR interferometry for monitoring mass movements. Landslides 1:21–28

2. Bozzano F, Mazzanti P, Prestininzi A (2008) A radar platform for continuous monitoring of a landslide interacting with an under-construction infrastructure. Ital J Eng Geol Environ 2:35–50

3. Bozzano F, Mazzanti P, Prestininzi A, Scarascia Mugnozza G (2010) Research and development of advanced technologies for landslide hazard analysis in Italy. Landslides 7(3):381–385

4. Bozzano F, Cipriani I, Mazzanti P, Prestininzi A (2011) Displacement patterns of a landslide affected by human activities: insights from ground-based InSAR monitoring. Nat Hazards 59(3):1377–1396.

5. Casagli N, Farina P, Leva D, Tarchi D (2006) Application of ground based radar interferometry to monitor an active rockslide and implication for emergency management. In: Evans SG et al (eds) Landslide from massive rock slope failure. Springer, Dordrecht, pp 157–173

6. Colesanti C, Wasowski J (2006) Investigating landslides with space-borne Synthetic Aperture Radar (SAR) interferometry. Eng Geol 88:173–199

7. Cornelius RR, Voight B (1990) Feasibility of material failure approach to eruption prediction for Mount St. Helens, 1985 and 1986. EOS Trans Am Geophys Union 71:1693

8. Cornelius RR, Voight B (1994) Seismological aspects of the 1989–1990 eruption at Redoubt Volcano, Alaska: the materials failure forecast method (FFM) with RSAM and SSAM seismic data. J Volcanol Geotherm Res 62:469–498

9. Cornelius RR, Voight B (1995) Graphical and PC-software analysis of volcano eruption precursors according to the materials failure forecast method (FFM). J Volcanol Geotherm Res 64:295–320

10. Crosta GB, Agliardi F (2002) How to obtain alert velocity thresholds for large rockslides. Phys Chem Earth 27(36):1557–1565

11. Crosta GB, Agliardi F (2003) Failure forecast for large rock slides by surface displacement measurements. Can Geotech J 40:176–191

12. Curlander JC, McDonough RN (1991) Synthetic aperture radar systems and signal processing. Wiley, New York

13. Dunnicliff CJ (1988) Geotechnical instrumentation for monitoring field performance. Wiley, New York

14. Farina P, Colombo D, Fumagalli A, Marks F, Moretti S (2006) Permanent scatterers for landslide investigations: outcomes from the ESA-SLAM project. Eng Geol 88:200–217

15. Ferretti A, Prati C, Rocca F (2001) Permanent scatterers in SAR interferometry. IEEE Trans Geosci Remote Sens 39:8–20

16. Fukuzono T (1985) A new method for predicting the failure time of a slope. Proceedings of the fourth international conference and field workshop on landslides, Tokyo, Japan. Landslide Soc 145–150

17. Fukuzono T (1989) A simple method for predicting the failure time of slope—using reciprocal of velocity, Technology for Disaster Prevention, Science and Technology Agency, Japan. Int Coop Agency, Japan 1(13):111–128

18. Fukuzono T (1990) Recent studies on time prediction of slope failure. Landslide News 4:9–12

19. Gigli G, Fanti R, Canuti P, Casagli N (2011) Integration of advanced monitoring and numerical modeling techniques for the complete risk scenario analysis of rockslides: the case of Mt. Beni (Florence, Italy). Eng Geol 120:48–59.

20. Goldstein RM, Zebker HA, Werner C (1988) Satellite radar interferometry: two-dimensional phase unwrapping. Radio Sci 23:713–720

21. Haefeli R (1953) Creep problems in soils, snow and ice. Proc 3rd Int Conf Soil Mech Found Eng 3:238–251

22. Hanssen R (2001) Radar interferometry. Kluwer, Dordrecht

23. Herrera G, Fernández-Merodo JA, Mulas J, Pastor M, Luzi G, Monserrat O (2009) A landslide forecasting model using ground based SAR data: the Portalet case study. Eng Geol 105(3):220–230

24. Hilley GE, Burgmann R, Ferretti A, Novali F, Rocca F (2004) Dynamics of slow-moving landslides from permanent scatterer analysis. Science 304:1952–1955

25. Intrieri E, Gigli G, Mugnai F, Fanti R, Casagli N (2012) Design and implementation of a landslide early warning system. Eng Geol 147:124–136.

26. Kilburn CJ, Petley DN (2003) Forecasting giant, catastrophic slope collapse: lessons from Vajont, Northern Italy. Geomorphology 54(1–2):21–32

27. Leva D, Nico G, Tarchi D, Fortuny J, Siebar AJ (2003) Temporal analysis of a landslide by means of a ground-based SAR interferometer. IEEE Trans Geosci Remote Sens 41(4):745–752

28. Luzi G (2010) Ground based SAR interferometry: a novel tool for geoscience. Geosci Remote Sens New Achiev 1–26

29. Massonet D, Fiegl KL (1998) Radar interferometry and its application to changes in the earth's surface. Rev Geophys 36(4):441–500

30. Mazzanti P (2011) Displacement monitoring by terrestrial SAR interferometry for geotechnical purposes. Geotech Instrum News 29(2):25–28

31. Mazzanti P (2012) Remote monitoring of deformation. An overview of the seven methods described in previous GINs. Geotech Instrum News 30(4):24–29

32. Moriwaki H, Inokuchi T, Hattanji T, Sassa K, Ochiai H, Wang G (2004) Failure processes in a full-scale landslide experiment using a rainfall simulator. Landslides 1:277–288

33. Noferini L, Pieraccini M, Mecatti D, Macaluso G, Atzeni C (2005) Long term landslide monitoring by ground based SAR interferometer. Int J Remote Sens 27:1893–1905

34. Pieraccini M, Casagli N, Luzi G, Tarchi D, Mecatti D, Noferini L, Atzeni C (2002) Landslide monitoring by ground-based radar interferometry: a field test in Valdarno (Italy). Int J Remote Sens 24(6):1385–1391

35. Rose ND, Hungr O (2007) Forecasting potential rock slope failure in open pit mines using the inverse-velocity method. Int J Rock Mech Min Sci 44:308–320

36. Saito M (1965) Forecasting the time of occurrence of a slope failure. Proc 6th ICSMFE 2:537–539

37. Saito M (1969) Forecasting time of slope failure by Tertiary creep. Proc 7th ICSMFE 2:677–683

38. Saito M, Uezawa H (1961) Failure of soil due to creep. Proc 6th ICSMFE 1:315–318

39. Siddle HJ, Moore R, Carey JM, Petley DN (2007) Pre-failure behaviour of slope materials and their significance in the progressive failure of landslides. In: Mathie E, McInnes R, Fairbank H, Jakeways J (eds) Landslides and climate change: challenges and solutions. Proceedings of the International Conference on Landslides and Climate Change, Ventnor, Isle of Wight. Taylor & Francis, UK, pp 21–24.

40. Strozzi T, Farina P, Corsini A, Ambrosi C, Thuering M, Zilger J, Wiesmann A, Wegmuller U, Werner C (2005) Survey and monitoring of landslide displacements by means of L-band satellite SAR interferometry. Landslides 2(3):193–201

41. Tarchi D, Casagli N, Fanti R, Leva D, Luzi G, Pasuto A, Pieraccini M, Silvano M (2003) Landside monitoring by using ground-based SAR interferometry: an example of application to the Tessina landslide in Italy. Eng Geol 68:15–30

42. Terzaghi K (1950) Mechanisms of landslides. In: Paige S (ed) Applications of geology to engineering practice. Geol. Soc. Amer. Spec. Pub, Berkey, pp 83–123

43. Ulaby FT, Moore RK, Fung AK (1982) Microwave remote sensing: active and passive, vol II. Addison-Wesley, Advanced Book Program, Reading

44. Varnes DJ (1983) Time-deformation relations in creep to failure of earth materials. In: Proc. of 7th Southeast Asian Geotechnical Conference 2:107–130

45. Voight B (1988) A method for prediction of volcanic eruptions. Nature 332:125–130

46. Voight B (1989a) A relation to describe rate-dependent material failure. Science 24(3):200–203

47. Voight B (1989b) Materials science law applies to time forecasts of slope failure. Landslide News, Tokyo, Japan Landslide Soc 3:8–10

48. Voight B, Cornelius RR (1990) Application of material failure approach to eruption prediction with RSAM at Redoubt, 1989–1990. EOS Trans Am Geophys Union 71:1701

49. Voight B, Cornelius RR (1991) Prospects for eruption prediction in near real-time. Nature 350:695–698

50. Voight B, Kennedy BA (1979) Slope failure of 1967–1969, Chuquicamata mine, Chile. In: Voight B (ed) Rockslides and avalanches, vol 2. Elsevier, Amsterdam, pp 595–632

51. Yoshida T, Yachi M (1984) On the velocity of landslide (in Japanese). In: Proceeding of 23rd Meeting of Japan Landslide Society 136–139

Chapter 4

Formation, failure, and consequences of the Xiaolin landslide dam, triggered by extreme rainfall from Typhoon Morakot, Taiwan

Chun-Hung Wu, Su-Chin Chen, Zheng-Yi Feng

ABSTRACT

An extreme rainfall event on August 9, 2009, which was close to setting a world record for 48-h accumulated rainfall, induced the Xiaolin deep-seated landslide, which was located in southwestern Taiwan and had volume of 27.6×10^6 m^3, and caused the formation of a landslide dam. The landslide dam burst in a very short time, and little information remained afterward. We reconstructed the process of formation and failure of the Xiaolin landslide dam and also inferred the area of the impoundment and topographic changes. A 5 × 5-m digital elevation model, the recorded water stage of the Qishan River, and data from field investigation were used for analysis. The spectral magnitude of the seismic signals induced by the Xiaolin landslide and flooding due to failure of the landslide dam were analyzed to estimate the timing of the dam breach and the peak discharge of the subsequent flood. The Xiaolin landslide dam failure resulted from overtopping. We verified the longevity of the Xiaolin landslide dam at about

2 h relying on seismic signals and water level records. In addition, the inundated area, volume of the impoundment behind the Xiaolin landslide dam, and peak discharge of the flood were estimated at 92.3 ha, $19.5 \times 10^6\ m^3$, and $17 \times 10^3\ m^3/s$, respectively. The mean velocity of the flood-recession wave front due to the dam blockage was estimated at 28 km/h, and the peak flooding velocity after failure of the dam was estimated at 23 km/h. The Xiaolin landslide provides an invaluable opportunity for understanding the mechanism of deep-seated landslides and flooding processes following a landslide dam failure.

KEYWORDS

Deep-seated landslide; Landslide dam; Overtopping; Typhoon Morakot; Xiaolin

1. INTRODUCTION

The formation of a landslide dam results from the natural blockage of the river channels by hillslope-derived mass movements (Costa and Schuster 1988). Most landslide dam disasters due to excessively rainfall or earthquake events (Schuster1993) occur in landslide-prone countries, including New Zealand (Korup 2004, Korup2005; Korup and Tweed 2007), Japan (Tabata et al. 2002), China (Chen et al. 1992; Dai et al. 2005; Liu et al. 2010), and especially Taiwan with its frequent torrential rainfall and earthquake events (Chen 1999; Shou and Wang 2003; Hsu and Hsu 2009). The floods resulting from the bursting of a landslide dam can result in serious bank erosion and inundation of populated areas and farmland near the river. The longevity of a landslide dam depends on the discharge into the impoundment, geomorphologic factors, and the composition of the dam (Costa and Schuster 1988). The existence time of a landslide dam directly determines the seriousness of the subsequent flooding. Due to the seriousness of the event and insufficient reaction time, the failure of a landslide dam and the subsequent flooding have often resulted in numerous deaths (Costa and Schuster 1988; Shroder 1998; Chen et al. 1992; Dai et al. 2005; Duman2009).

There have been many huge landslide dam failure events worldwide, including the failure of landslide dams in Southwestern China in 1786

(Chanson 2005; Dai et al.2005; Liu et al. 2010; Xu et al. 2009). It has been claimed that at least 5,500 deaths have occurred due to landslide dam failure events in China since 1900. Why the landslide dam events often lead to catastrophic disasters can be explained from two angles. First, some landslide dam events have short longevity, especially in extreme rainfall-triggered cases. Individuals who live along rivers cannot easily detect the formation of a landslide dam during extreme rainfall and, therefore, do not have sufficient time to evacuate. Second, the volume of the impoundment behind an earthquake-triggered landslide dam with long longevity is often too large to be discharged safely. Two very large landslide dams, Tsaoling (Hsu and Hsu 2009) and Chiufenershan (Shou and Wang 2003) and 15 smaller ones (Chen 1999) occurred in Taiwan due to the 1999 Chichi earthquake. Most of these dams were of short longevity, except the Chiufenershan landslide dam, which still exists with remedial engineering measures more than 12 years later.

Detailed information concerning the geomorphologic characteristics, time of formation, and time of failure of rainfall-triggered landslide dams is difficult to collect because of their short longevity. The Xiaolin landslide dam caused by Typhoon Morakot in 2009 in Taiwan was similarly a short-longevity one with little remaining information. This study reconstructs the process of formation and failure of the Xiaolin landslide dam by using the few pieces of remaining information, which include seismic signals, water stage records, and data from field investigations. The 5 × 5 m digital elevation model (DEM) and geomorphological data from field surveys were employed to infer the location and the original deposited landform of the Xiaolin landslide dam. The size of the landslide dam and the volume of the impoundment lake were also estimated. The seismic signals caused by the Xiaolin landslide and the subsequent flooding were recorded by a regional seismic network 14 km distant from Xiaolin and used to confirm the time of formation and failure of the Xiaolin landslide dam. This study interpreted the detailed geomorphologic information from stereoscopic aerial photo and field investigations to explain the process of the Xiaolin landslide and the formation and failure of the Xiaolin landslide dam. We verified the existence of the Xiaolin landslide dam and explained the flooding process after the failure of the landslide dam. Records from a downstream gauge station were used to check the existence and failure timing of the Xiaolin landslide dam and estimate the possible discharge during the flooding process.

2. OVERVIEW OF THE XIAOLIN LANDSLIDE

2.1 Regional setting

The Xiaolin landslide dam was located to the west of Xiaolin Village in the watershed of the Qishan River, which is a tributary of the Kaoping River in Southwestern Taiwan (Figure 1). The area of the sub-watershed upstream of the Xiaolin landslide dam is about 353 km^2. The terrain along the river reach near Xiaolin Village consists of river terraces; this area is known as the Xiaolin Gorge featuring an active river down cutting (Kung and Chyi 2004). Before the landslide, Xiaolin Village was located on a fluvial terrace on the left hand side of the river; this terrace had a drop height of 5–10 m to the riverbed and a gentle slope of less than 15°.

The geological setting of the Xiaolin landslide (Tsou et al. 2011; Wu et al. 2011) is shown in Figure 2. The upslope portions of the Xiaolin landslide are chiefly located in Miocene/Pliocene Tangenshan sandstone. The downslope portions are mainly located in Pliocene Yenshuikeng shale. The Tangenshan sandstone consists of thick inter-bedded sandstone and clayey sandstone and has a thickness of 450–500 m. The Yenshuikeng shale consists of massive shale with a thickness of 200–250 m. The fluvial terraces and alluvial formations in the vicinity of Xiaolin Village chiefly consist of gravel and sand. Three active faults, including the Qishan fault, Wulipu fault, and a newly discovered fault (Lee et al. 2009), are close to or in the landslide. Three folds also pass through the landslide area, namely the Xiaolin syncline, the Jiaxian anticline, and the Shihchangeli syncline. The flanks of the three folds dip gently with angles of 10°–20°; however, the flanks occasionally have high dipping angles of 30°–40° (Keng1965).

According to 15-year records from four nearby rainfall stations, including Minsheng (the station number is C1V160), Biaohu (C0V150), Jiaxian (01P660), and Xinfa stations (C1V240), the mean annual rainfall at Xiaolin is about 2,440–3,270 mm (Figure 3a). The average rainfall during the rainy season is about 2,620 mm (approximately 90 % of the annual rainfall); frequent typhoon events, typically occurring from May to October, contribute to the rainfall amount. In contrast, average rainfall during the dry season is normally about 280–300 mm.

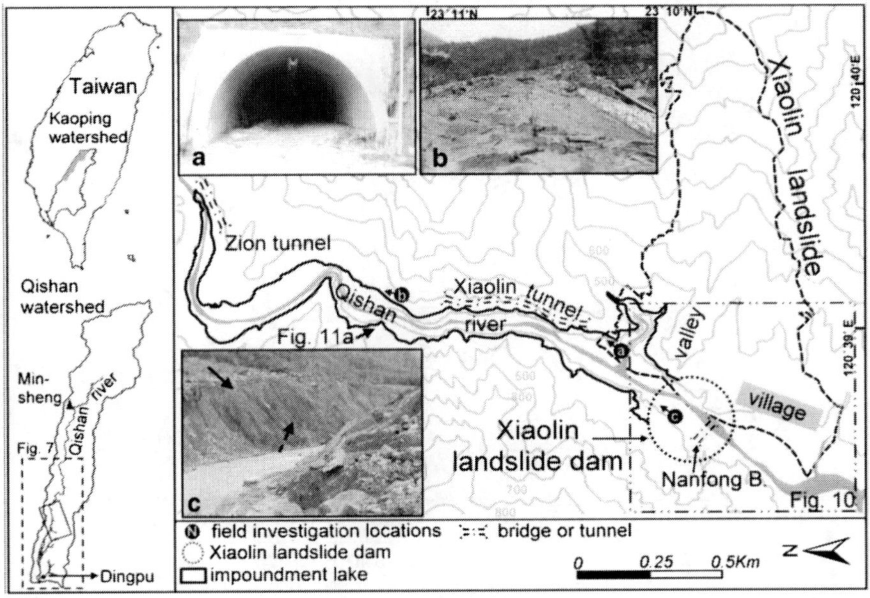

Figure 1. Location of the Xiaolin landslide dam on a contour base map of the area before the Xiaolin landslide. Field survey pictures taken after the failure of the Xiaolin landslide dam and impoundment lake with an estimated depth of 52.4 m are shown in Figure 1a, b, and c. **a** Large amounts of fine sediments were deposited to a thickness of 1.3 m on the road, and grass was found on the lamps hanging in the Xiaolin tunnel. It is conjectured that the Xiaolin tunnel was completely submerged during the impoundment after the landslide and before failure of the landslide dam. **b** Large amounts of fine sediment were found on the road along the upstream Qishan River, but no obvious fine sediment was found near the Zion tunnel. It is inferred that the impoundment lake, with an estimated elevation of about 422 m, submerged the area from the Zion tunnel (upstream) to the location of the Xiaolin landslide dam (downstream). The submerged area and volume of the impoundment behind the Xiaolin landslide dam are estimated to have been 92.3 ha and 19.5×10^6 m^3. **c** The burst section of the Xiaolin landslide dam eroded by the overtopping floodwaters. Some of the collapsed materials, the yellow surface soil (*black arrow*) were even deposited on the original bank hillslope and the black surface soil (*dashed arrow*) opposite to that of the Xiaolin landslide

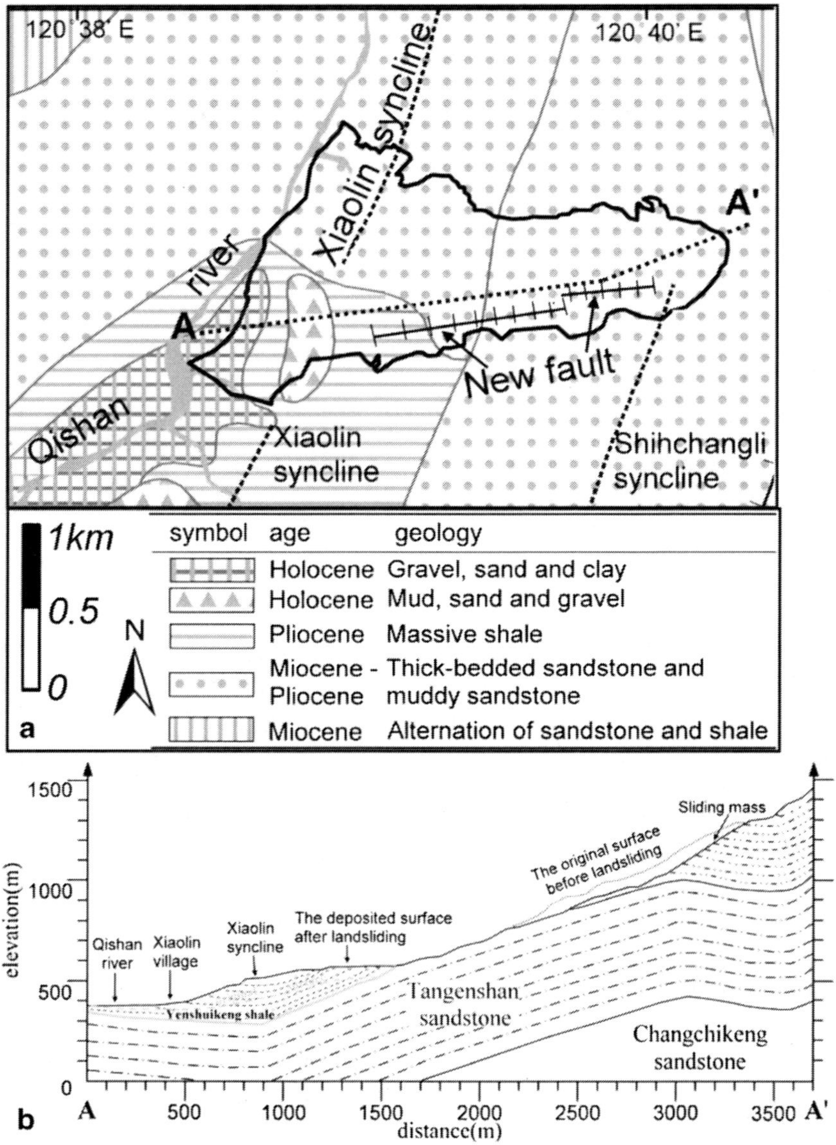

Figure 2. **a** The geological setting (Wu et al. 2011) and **b** the geological section map (Lee et al. 2009) of the Xiaolin landslide. The new fault within the Xiaolin landslide area was found by Lee et al. (2009)

2.2 Rainfall distribution

The torrential rainfall brought by Typhoon Morakot was concentrated in Southern Taiwan and caused 17 landslide dams in mountainous areas (Chen et al. 2011a, b). Torrential rainfall during Typhoon Morakot (August 6–10, 2009) was equivalent to roughly 75 % and 72 % of average annual rainfall at the Jiaxian Station (2,839 mm, Figure 3) and Xinfa Station (3,305 mm, Figure 3). The accumulated rainfall at Xiaolin Village during Typhoon Morakot was about 1,998 mm, which was estimated based on rainfall records at 18 nearby rainfall stations, and was equivalent to roughly 85 % of the total rainfall in 2009.

Based on the four rainfall stations closest to Xiaolin, the heaviest rainfall period, which had total rainfall of 578–787 mm with and a mean intensity of 48–66 mm/h, occurred from 12:00 to 24:00 on August 8, 2009 (Figure 3). The accumulated rainfalls for 24-h, 48-h, and 72-h periods at the Jiaxian Station during Typhoon Morakot were 1,040, 1,614, and 1,915 mm, respectively. The corresponding rainfall amounts for the three durations over 200-year return periods are 967, 1,436, and 1,761 mm (Wu et al. 2011), respectively. The high-accumulated rainfall records during Typhoon Morakot were 1,931 mm in 24 h, 2,361 mm in 48 h, and 2,748 mm in 72 h at Weiliaoshan station, also in Southwestern Taiwan. Here, the 2,361 mm rainfall in a 48-h period is only 139 mm less than the world record of 2,500 mm during the same duration (Quetelard et al.2009; The Ohio Weather Library 2012; Hydrometeorological Design Studies Center2012; see the black dashed circle in Figure 3b). Heavy rainfall exceeding the 200-year return period forecast and close to the world record rainfall at Xiaolin triggered the serious disasters during Typhoon Morakot. This rainfall was characterized by high intensity and a long duration, while torrential rainfall during the 1996 Typhoon Herb was high-intensity and short-duration (Lin and Jeng 2000; see the black dashed rectangle in Figure 3b). This change of the rainfall characteristics in Taiwan during the past two decades is a worthwhile subject for further study.

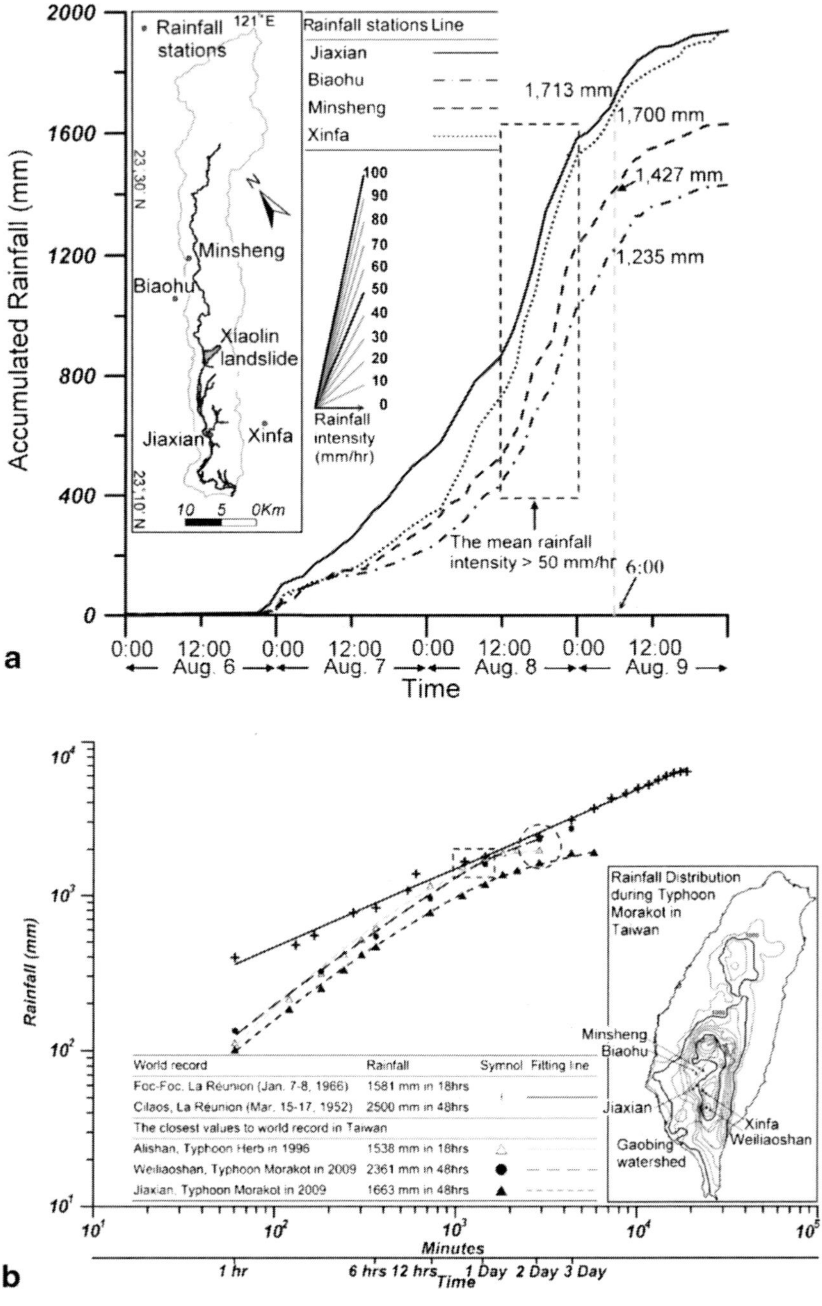

Figure 3. The characteristics of heavy rainfall events in Taiwan, including Typhoon Herb in 1996 and Typhoon Morakot in 2009. **a** Hourly rainfall intensity near Xiaolin landslide during Typhoon Morakot. The heaviest rainfall intensity, shown in the *dashed rectangle* in (**a**), was in the afternoon

and night of August 8, 2009. **b** Comparison of cumulative rainfall records in Taiwan with world's cumulative rainfall records (Quetelard et al. 2009; The Ohio Weather Library 2012; Hydrometeorological Design Studies Center 2012). The heavy rainfall during Typhoon Herb in 1996, shown in the *dashed rectangle* in (**b**), was characterized by high intensity and short duration. Rainfall during Typhoon Morakot in 2009, shown in the *dashed circle* in (**b**), was characterized by high intensity and long duration

2.3 The Xiaolin deep-seated landslide

The Xiaolin landslide resulted in 465 deaths in Xiaolin Village and the Xiaolin landslide dam blocking the Qishan River. The Xiaolin landslide source area was separated by a ridge (shown in Figure 4; Wu et al. 2011) into two zones, including Zone A and Zone B (shown in Figure 4). Because the displaced volume in Zone B did not contribute to the formation of the Xiaolin landslide dam, this study chiefly discusses the incision and deposition in Zone A with the landslide area and volume of 1.9 km^2 and 26×10^6 m^3.

This study estimated the landslide depth along the D profile lines in the depletion area with an elevation above 800 m (Wu et al. 2011) and the deposition depth along the E profile lines in the deposition area with an elevation under 600 m (Wu et al. 2011) by employing the 5 × 5 m DEMs before and after Typhoon Morakot (Figs. 4, 5, and 6). The deepest incised depth along D profile lines is 73 m in profile D3, and the deepest incised area is located at the areas with original elevation of 880–980 m in profile D3 (the dashed rectangle in Figure 5). The landslide mass was mainly deposited along profiles E1 to E4 (Figure 6), i.e., the valley uphill from Xiaolin Village (Figure 1), and the maximum deposited depth was roughly 50 m in profile E3 (shown in Figure 6). The landslide mass from this deep-seated landslide heavily modified the original topography in the downslope area (the gray dotted rectangles from profiles E6 to E9 in Figure 6). The slopes from profiles E1 to E4 near the Qishan River (the gray dotted circles in Figure 6) were very steep after Typhoon Morakot, but those in the upper deposited area were gentle. This study inferred that the original deposited surface should be gentle both in the upper deposited area and in the downslope area from profiles E1 to E4. The landslide mass of Zone A was transported along the original valley and deposited at the intersection of the valley and the Qishan River, forming the Xiaolin landslide dam. The steep slope from profiles E1 to E4 near the Qishan River is an indication of the erosion of the dam body by the overtopping flow. The flood also washed away much of the deposited landslide mass further down the river.

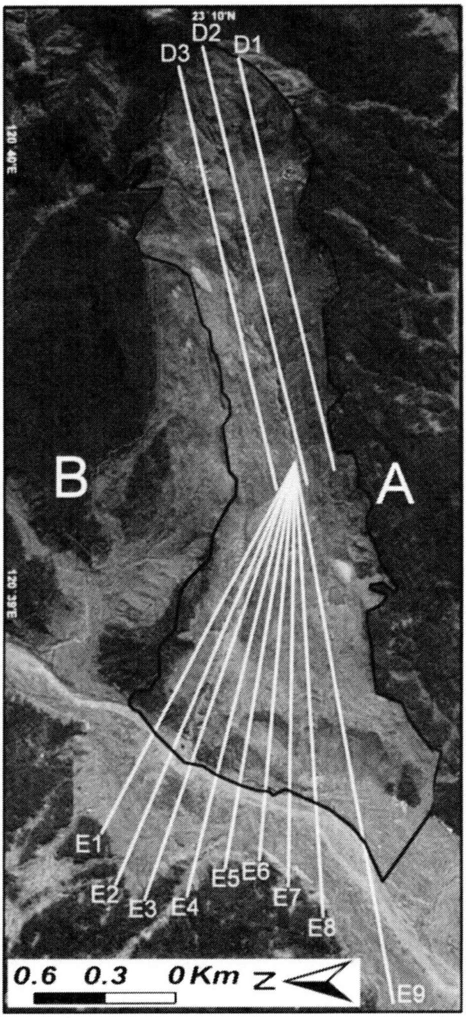

Figure 4. Survey profiles of the Xiaolin landslide, which was separated by a ridge into landslide A and landslide B. The landslide B is not studied in the study because the formation of Xiaolin landslide dam mainly resulted from the landslide volume of Landslide A. The *yellow lines* are the profile lines in Figure 5 (*D profile lines*) and Figure 6 (*E profile lines*)

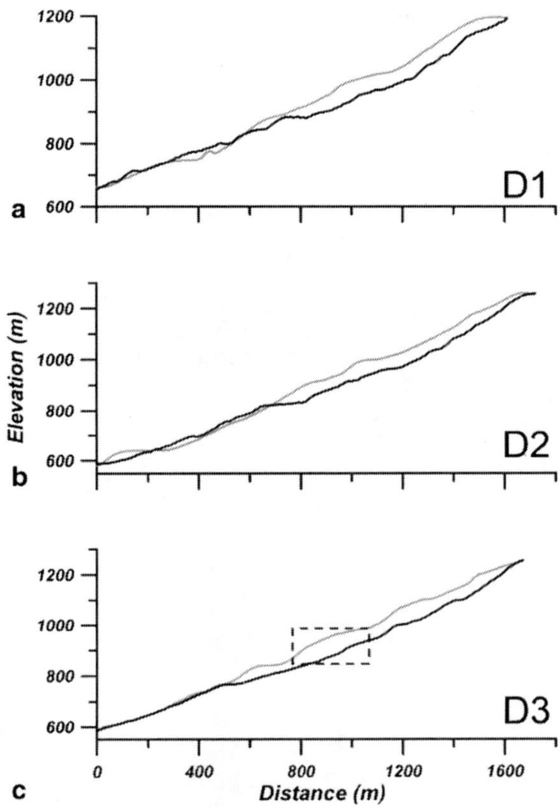

Figure 5. Elevations along the longitudinal direction (*D1 to D3 profile lines*, shown in Fig 4) of the Xiaolin landslide before (*gray line*) and after (*black line*) Typhoon Morakot. The *dashed rectangle* in Figure 5c is the deepest incised area

3. LANDSLIDE DAM AND RESERVOIR

3.1 Water stage records

The water stage records at Shanlin Bridge with the subscript S and Nanfong Bridge with the subscript N in Figure 7 were used to estimate the occurrence time of the Xiaolin landslide and the failure time of the Xiaolin landslide dam. The stream lengths from Nanfong Bridge to Jiaxian and from Jiaxian to Shanlin Bridge are 10.7 and 17.1 km, respectively. Water stage records and the rainfall intensity near the Xiaolin landslide are shown in

Figure 7. Three apparent peaks (B_N and B_S, D_N and D_S, and I_S in Figure 7) can be observed in the water stage records.

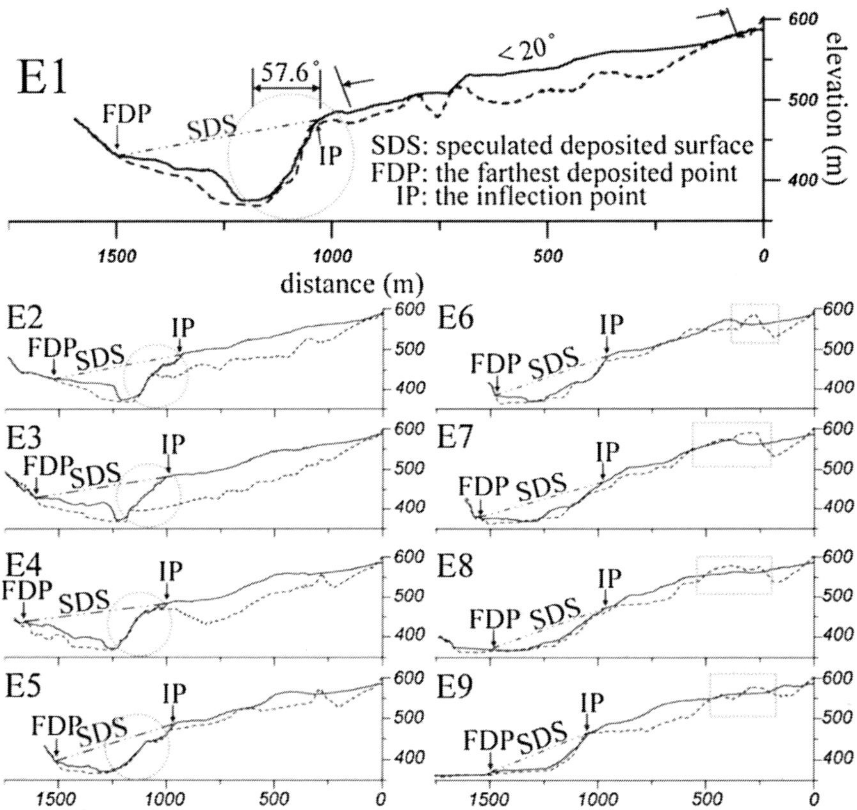

Figure 6. Elevations of sediment deposits from the Xiaolin landslide before (*dashed line*) and after (*solid line*) Typhoon Morakot along the *E profile lines* (shown in Figure 4)

The water elevations at Nanfong Bridge and Shanlin Bridge were 368.8 and 116.1 m, respectively, before Typhoon Morakot. The first peak (B_S and B_N in Figure 7) had an average water stage increasing rate of 0.61 m/h and was caused by the heavy rainfall, which had a mean intensity of 26–37 mm/h from 10:00 to 16:00 on August 8. The second peak (D_N at 23:20 on August 8 and D_S at 0:40 on August 9, Figure 7) had an average water stage rising rate of 0.77–0.83 m/h and was caused by the torrential rainfall, which had a mean intensity of 60 mm/h from 20:00 to 23:00 on August 8. The accumulated rainfall of 261 mm within 3 h was observed at Minsheng station. The flood, with an estimated water depth of 7 m at 0:30, destroyed the gauge station at Nanfong bridge on August 9. Based on the official investigation report of the Water Resources Agency (2009), the water stage diminished rapidly at a rate of 0.92 m/h at Shanlin Bridge after the second peak, and this resulted from decreasing rainfall intensity and the broken levee at Dingpu (Figure 7) from 0:00 to 0:30 on August 9.

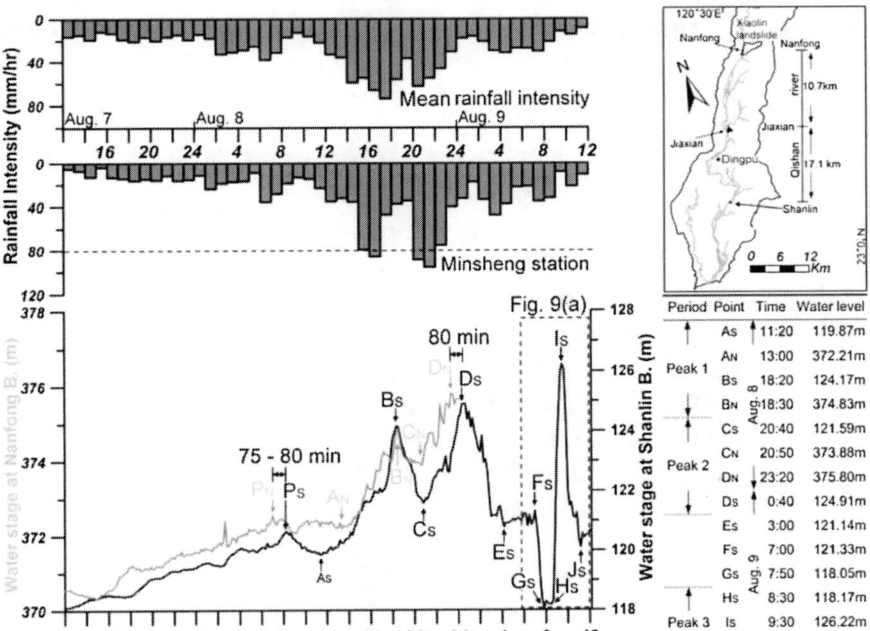

Period	Point	Time	Water level
Peak 1	As	11:20 (Aug. 8)	119.87m
	AN	13:00	372.21m
	Bs	18:20	124.17m
	BN	18:30	374.83m
Peak 2	Cs	20:40	121.59m
	CN	20:50	373.88m
	DN	23:20	375.80m
	Ds	0:40 (Aug. 9)	124.91m
	Es	3:00	121.14m
	Fs	7:00	121.33m
	Gs	7:50	118.05m
Peak 3	Hs	8:30	118.17m
	Is	9:30	126.22m
	Js	11:20	120.11m

Figure 7. Water stage records and rainfall intensity (*top*) in the Qishan watershed during Typhoon Morakot. The mean rainfall intensity was calculated from the rainfall records of the Minsheng, Jiaxian, and Yumei stations

The water stage at Shanlin Bridge station should have decreased gradually in view of the accumulated rainfall of 150–197 mm at the four rainfall stations (Figure 3) from 0:00 to 6:00 on August 9, but it instead dropped at a rate of 3.94 m/h in 50 min and rose at a rate of 8.05 m/h in 1 h. We inferred that the unexpected drop and rise of the water stage were due to the formation and failure of the Xiaolin landslide dam. The changes in the gauge were also consistent with the survivors' accounts.

3.2 The formation of the Xiaolin landslide dam

This study reconstructed the original deposited landform of the landslide dam before its failure. The inflection point (IP in Figure 6) is the intersection point of the gentle deposition slope and the steep slope in the deposition landform. The farthest deposited point (FDP in Figure 6) is the point on the hillslope opposite to the Xiaolin landslide and closest to the Qishan River without change before and after Typhoon Morakot. The line which links the IP and FDP is termed the speculative deposited surface (SDP in Figure 6). The landform composed by the gentle deposition surface and SDP in each profile line is considered to be the original deposition surface before the failure of the landslide dam. This study estimated the depositional area and volume of the Xiaolin landslide dam to be 0.18 km^2 and 10.5×10^6 m^3. The maximum deposited height from the riverbed before the landslide to the SDP and width of the dam are around 107 and 348 m, respectively, on Profile E3. We classified the Xiaolin landslide dam as a rock-avalanche dam and deduced that the dam was formed around 6:16–6:17 on August 9, 2009 (Tsou et al.2011; Feng 2012).

4. DAM FAILURE AND RIVER DISCHARGE

4.1 Water level variation

This study used the water stage records at Shanlin Bridge (Figure 8a) and the time-frequency spectra of the Jiaxian broadband seismic signals from 6:00 to 10:00 on August 9, 2009 (Feng 2012) to estimate and verify the formation and failure time of the Xiaolin landslide and the resulting dam. The time-frequency spectral magnitudes of the Jiaxian broadband seismic signals were calculated using the Hilbert-Huang transform (HHT; Huang, et al. 1998). The occurrence of the Xiaolin landslide was indicated around 6:16 (Lin et al. 2010, 2010b; Feng 2012) and can be observed in the spectral magnitude at 2.94 Hz (the spike at roughly 6:16 with the maximum averaged value of 0.0157 during 6:10–25 in Figure 8b). In addition, we can see that the spectral magnitudes gradually decrease from 6:40 to 7:00, remained almost constant from 7:00 to 8:00, increased rapidly from 8:00 to 8:45 and decreased again from 8:45 to 9:30.

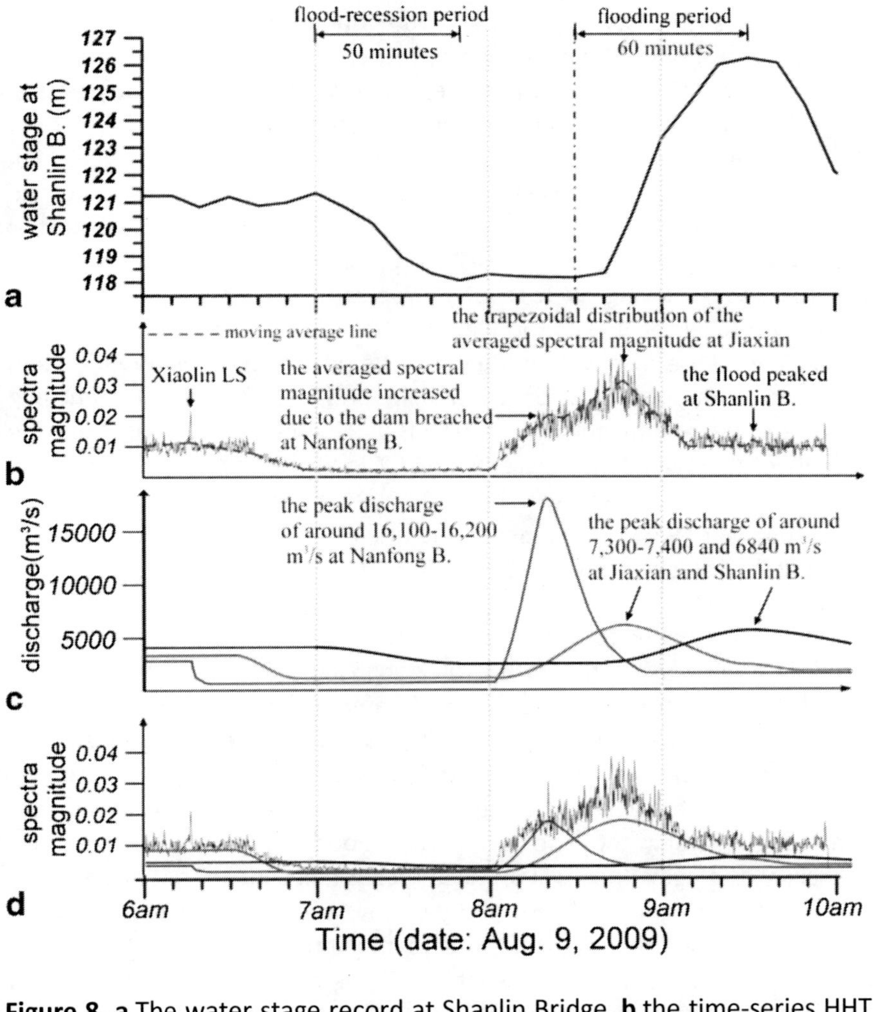

Figure 8. a The water stage record at Shanlin Bridge, **b** the time-series HHT spectra graph, **c** the inferred time-series river discharge at Nanfong B (*blue line*), Jiaxian station (*red line*), and Shanlin Bridge (*black line*), and **d** a comparison of the spectral magnitude and the decayed/scaled river discharge at Nanfong Bridge (*blue line*), Jiaxian station (*red line*), and Shanlin Bridge (*black line*) during Typhoon Morakot

The flood-recession period caused by the blockage of the river by the Xiaolin landslide dam began at 6:16 at Nanfong Bridge (Figure 8b) and between 7:00 and 7:50 at Shanlin Bridge (Figure 8a). The flood-recession period at Jiaxian began at 6:33 and ended at 6:52 (Figure 9), which can be seen in Figure 8b around 6:30–7:00. The velocity of the flood-recession wave front at the start and end of the flood-recession period can be estimated at about 38 and 17.7 km/h, respectively, and the mean velocity of those was 28 km/h. Furthermore, we hypothesize that the spectral magnitude can serve as an analog signal indicating the seismic signature of the landslide and flood, and we use the moving average values (Figure 8b) of the seismic signals to indicate the formation and failure processes of the Xiaolin landslide dam event. The increase and decrease of the spectral moving average magnitude reveals the passing of the flood in Figure 8b. The maximum moving average of the spectral magnitude, which had a value over 0.030 at 8:43–47 in Figure 8b, can be inferred to be the peak flood after failure of the landslide dam. The peak flood took about 45 min to travel from Jiaxian (8:45) to Shanlin Bridge (9:30 with the moving averaged spectral magnitude of 0.011 in Figure 8a), implying that the velocity of the peak flood was 22.8 km/h. The time at which the peak flood reached Nanfong Bridge can be estimated at 8:20, which also has a maximum averaged spectral magnitude of 0.020 from 8:10–25 in Figure 8b, by using the velocity of the peak from Jiaxian to Shanlin.

The "convolution effect" of the seismic signals induced by the whole flooding process after the failure of Xiaolin landslide dam. A large dam breach flow will transmit a considerable amount of outgoing energy that shall register seismic signals on a sensitive broadband seismic device in the Jiaxian broadband station; even the flow source is far away. Therefore, when a dam breaches, the seismic waves will be transmitted out continuously along the river while the flood travels down the river, especially, the peak flood will produce most significant seismic signals. The seismic levels of any position will be accumulated and contributed by the flood flows continuously from upstream to downstream. This is the convolution effect concept we point out. To better explain this convolution effect concept and how the spectral magnitude curve shall be interpreted and corresponding to discharge, we choose three representative discharge curves with three respective peaks depicted in Figure 8c, d.

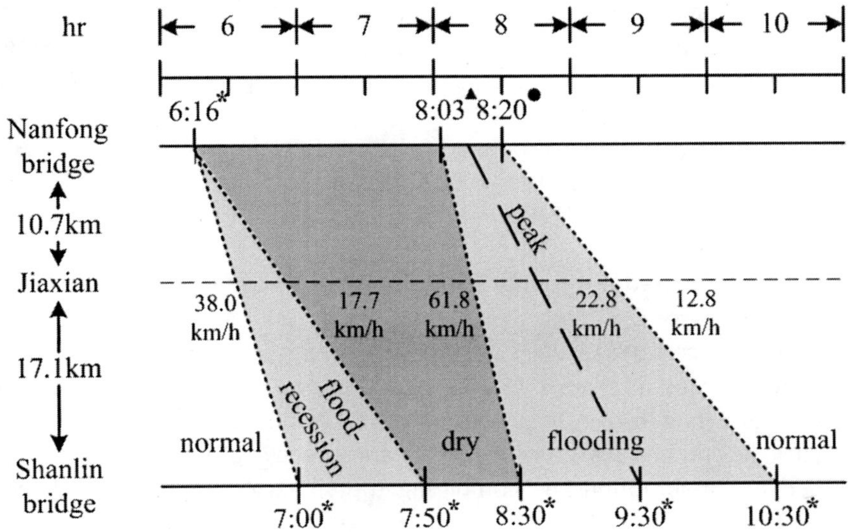

Figure 9. A schematic diagram of the estimated times of the starting and ending of waves of flood-recession at Jiaxian due to the formation and failure of the Xiaolin landslide dam. The times indicated with a *superscript asterisk* or*filled triangle* have been determined from water stage records and seismic signal records, while times indicated with a *superscript filled circle* have been estimated from water stage records and the seismic signal records

The seismic signals decayed at a rate of 1/d√d (Lin et al. 2010), where *d* is the shortest distance from the location of the landslide or flood to the Jiaxian broadband station. Feng (2012) suggested that the seismic signals around 2.75 and 2.94 Hz of the Jiaxian broadband station are the most obvious signals to show the seismic signals caused by the flooding due to failure of Xiaolin landslide dam. The study also used the amplitudes of the 2.5–5.5 Hz bandpass-filtered seismic signals of the Jiaxian broadband station to correlate the river discharges. The shortest distances from Nanfong Bridge and Shanlin Bridge to the Jiaxian broadband station are 10.4 and 10.5 km, respectively. Averaged from the spectral magnitudes of 2.75 and 2.94 Hz, the spectral magnitude contributed by the peak flood at Nanfong Bridge can be estimated as 0.1811 based on the distance of 10.4 km from Nanfong Bridge to Jiaxian, the spectral magnitude of 0.0561 at 8:20 (Figure 8b), and the decay rate. Similarly, the spectral magnitude contributed by the peak flood at Shanlin Bridge can be estimated as 0.0772 from the spectral magnitude of 0.0238 at 9:30 (Figure 8b).

We correlate and scale the spectral magnitudes with the peak discharge for the dam break flooding process. We therefore estimate the peak discharge at the Jiaxian station is set as Q_P; the peak discharge at Nanfong Bridge and Shanlin Bridge can be estimated as 2.19 Q_P and 0.93 Q_P, respectively, based on the averaged spectral magnitudes 0.1811 from 2.75 and 2.94 Hz around 8:20 at Nanfong station, 0.0827 around 8:45 at the Jiaxian station, and 0.0772 around 9:30 at Shanlin Bridge.

Q_P can be estimated at around 7,300–7,400 m^3/s based on the measured peak discharge of 6,840 m^3/s at Shanlin Bridge during Typhoon Morakot (Water Resources Agency 2009). The peak discharge at Nanfong Bridge can be also estimated at around 16,100–16,200 m^3/s. The river discharge time series at Nanfong Bridge, Jiaxian station, and Shanlin Bridge are illustrated in Figure 8c and are scaled with the distance-decay rate in Figure 8d. We verified our assumptions by comparing the spectral magnitude and the decayed/scaled river discharges.

Furthermore, we adopt the derived discharge at Nanfong Bridge to estimate the flood flow velocity when the Xiaolin landslide dam breached to validate the estimated discharge. The flood flow velocity at E2 profile is estimated at 15 to 17 m/s as the failure of the Xiaolin landslide dam, while that at E7 profile is estimated at around 10 m/s. The study also infers that the failure process of Xiaolin landslide dam, which is explained in following paragraph, is pertaining to gradual failure resulting from scouring the boundary of the Xiaolin landslide dam.

We cannot directly validate the estimated river discharge at Nanfong Bridge as the failure of Xiaolin landslide dam at 8 to 10 PM on August 9, 2009 because the gauge station was destroyed around 1 AM on August 8, 2009. Although, some literatures (Newman and Bennell 2002; Feng 2012) have shown that the proportionality between the vibration magnitude and river flow condition exists, this relationship between the flood discharge and the spectral magnitude presented in the study is only suitable for Qishan River at Jiaxian.

4.2 Morphological changes in the river and the disaster caused by the Xiaolin landslide dam failure

This study found vestiges of the dam and pieced together the dam failure process (Figure 10a) based on stereoscopic aerial photos and field investigations. The yellow arrows in Figure 10a indicate the Xiaolin landslide and the ensuing deposition, while the white lines and arrows in Figure 10a indicate the subsequent sediment transport caused by surface runoff originating from the source area of the watershed after the landslide. The blue lines in Figure 10a indicate the Qishan River, while black arrows indicate scouring (Figure 10b, c, d, and e) of the downslope side of the sediment deposited near the Qishan River. This study inferred that the landslide mass flowed in two directions: One portion of the landslide mass buried the original Xiaolin Village (Figure 1), and the other portion flowed along the original valley (Figure 1).

When the water depth of the impoundment behind the landslide dam was higher than the minimum deposit thickness of around 52 m in Profile E3 (Figure 6), i.e., equals to the elevation of 442 m, the floodwaters would begin to flow around the boundary of the deposit fan and scoured the boundary of the landslide dam. We found several evidences to proof the water depth of the impoundment behind the landslide dam. Firstly, the withered trees hanged in the lamp of the Xiaolin tunnel with the estimated elevation of 428 m (Figure 1a) stands for the Xiaolin tunnel had been totally submerged under the impoundment behind the landslide dam. Secondary, a mass of the withered trees deposited on the top of the hillslope with the elevation of 439 m (Figure 11a) also indicates that the water stage of the impoundment behind the landslide dam should be over 439 m. Furthermore, the key information used to reconstruct the failure process of the Xiaolin landslide dam is the concave surface of the original downslope sediment deposit area (Figure 10e), which was where Xiaolin Village was located, and the vestiges of the bank erosion in the opposite bank of Xiaolin landslide (Figure 11b). This study infers that a concave surface should have been formed by scouring from the flood, and we also found that a key ridge (Figure 10a) on the hillslope opposite from that of the Xiaolin landslide would have served to divert the flood during the failure of the dam. As the water from the impoundment behind the landslide dam gradually scoured the boundary of the landslide dam, the flow direction was diverted by the ridge. The diverted flood gradually scoured the toe of the ridge (Figure 11b) and the deposit fan of the landslide dam (Figure 10b, c, and d), and the route of scouring became the outlet for the about-to-fail impoundment lake. The landslide mass deposited in the streambed of the Qishan River by the Xiaolin landslide was ultimately washed away, and the diverted flood also scoured the downslope side of the original downslope sediment deposited by the landslide, forming a concave surface (Figs. 10e and 11c). The maximum hollow depths in the two concaves were estimated at around 15 m by using the 5 m Dem after Xiaolin landslide.

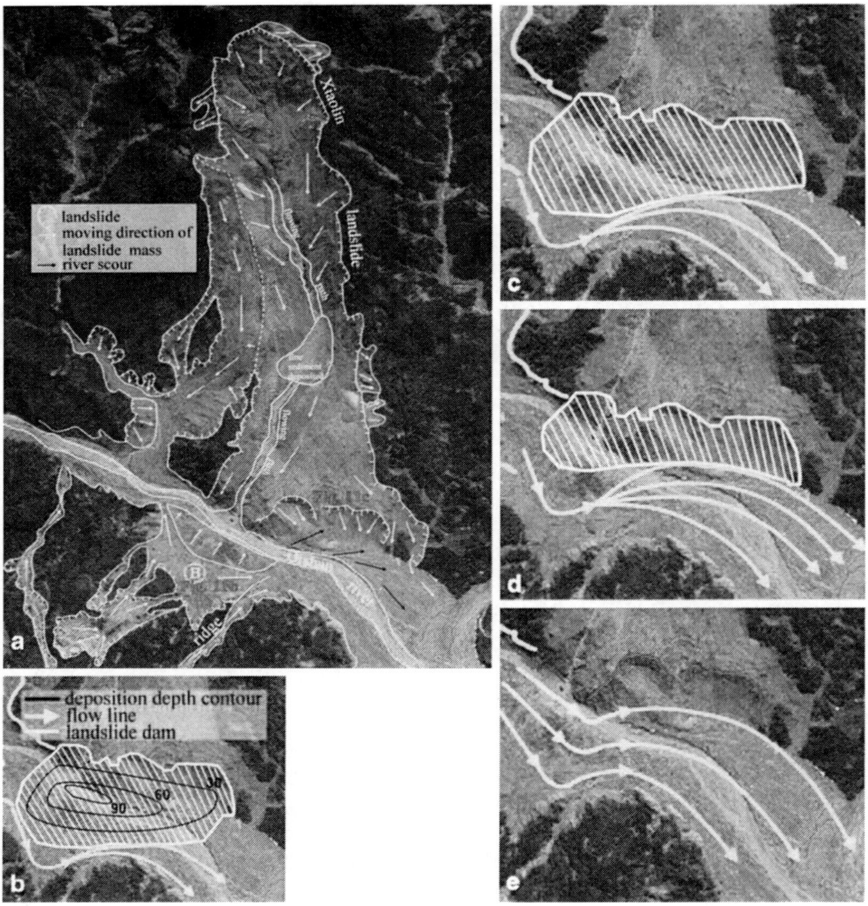

Figure 10. Interpretation of aerial photo of the landslide scar after the Xiaolin landslide (**a**) and reconstructed flood flow paths after failure of the Xiaolin landslide dam (**b**, **c**, **d**, **e**). When the Xiaolin landslide dam started to fail due to overtopping from the impoundment, the overtopping flow started to run around the dam and scoured the dam boundary (**b**). As the flow gradually scoured the boundary of the landslide dam, the volume of the landslide dam was gradually reduced, and the outburst water gradually increased (**c** and **d**). The outburst water also scoured the downslope side of the deposited sediment and formed two concave surfaces (**e**) due to the diversion of water by the ridge. The overtopping flow ultimately destroyed the landslide dam, and all the impounded water surged downstream. The local geomorphology after the end of flooding, including vestiges of the dam, are shown in Figure 1c

Figure 11. The vestiges of the impoundment behind the Xiaolin landslide dam. **a** A mass of withered trees (*red dashed circle*) deposited on the top of the hillslope with the elevation of 439 m. The picture-taken location is shown in Figure 1. **b** The vestiges of bank erosion (*black dashed line*) by the scour due to the diverted flood in the hillslope of the ridge (Figure 10). **c** The concave surface (*red dashed line*) with the maximum hollow depth of around 15 m located in the downslope side of the original downslope sediment deposited by the Xiaolin landslide. The picture-taken location is shown in Figure 10

This unpredictable catastrophic deep-seated landslide was caused by the extreme rainfall with the high intensity of >50 mm/h and lasting over 12 h, from 13:00 to 24:00 on August 8, 2009. It happened so fast that villagers had no time to respond and evacuate. The loss of life, 465 deaths, is what made this event a disaster. The volume and speed of the Xiaolin landslide made it too large and fast for people to survive. Deep-seated landslides have been reported since 70 years ago (Stini 1941), but the definition and understanding of these landslides remain difficult to clarify. Although typical characteristics and quantifiable definition of deep-seated landslides have been proposed, such as by Agliardi et al. (2001) and Uchida et al. (2011), it is still extremely difficult to predict where and when a deep-seated landslide will occur. The characteristics of rainfall in Taiwan have been gradually changing to heavy rainfall with high intensity or long duration during recent years (Shiu et al 2009). The maximum hourly rainfall intensity of 181.5 mm during Typhoon Megi in 2010 is a good example of the changing characteristics of rainfall in Taiwan. Due to its fragile geological conditions, steep slopes (Chen et al. 2011a, b) and the increasing intensified-rainfall events with long duration (Chiang and Chang 2011; Wu et al. 2011), Taiwan is especially prone to deep-seated landslides.

5. CONCLUSIONS

We reconstructed and verified the existence of the failed Xiaolin landslide dam using topographic change data, water stages, seismic signals caused by the landslide and flood, and the results of a detailed field investigation. We also explain the new concept of "convolution effect" for the seismic signals induced by the landslide dam breach. Also, new information such as the timing and discharge of the dam breach flooding process is provided.

The Xiaolin landslide was triggered by a rainfall event with a 200-year recurrence period and formed the Xiaolin landslide dam on August 9, 2009. This study used the seismic signals caused by Xiaolin landslide and the subsequent breach flooding to estimate the timing of the Xiaolin landslide, the peak discharge, and the flooding process resulting from the failure of the dam. The Xiaolin landslide dam with the deposition area and volume of 0.18 km^2 and 10.5×10^6 m^3 was resulted from the deep-seated Xiaolin landslide with the landslide area and volume of 1.9 km^2 and 26×10^6 m^3. The Xiaolin landslide dam with the maximum deposition height of 50 m was a rock-avalanche dam, formed at 6:16 AM on August 9, 2009 and started to fail from 8:03 AMdue to overtopping by the impounded water. The landslide dam claimed 465 deaths in Xiaolin village and existed for approximately 107 min. As the water depth of the impoundment behind the landslide dam exceeded the minimum deposited depth, the flood began to scour the boundary of the deposit fan. A ridge on the hillslope opposite side to Xiaolin landslide and the concave surface of the original sediment deposit area where Xiaolin Village was located are the two key information allowing the reconstruction of the dam failure process in this study. The concave surface was formed due to scouring by the flood, had been diverted by the ridge after the flood gradually washed out the landslide mass of the dam, and the impoundment lake finally burst. We reconstructed the flooding process of the Xiaolin landslide dam based on information of the seismic signals, the water stage records, and the remaining landforms after the failure of the landslide dam. In addition, the result of the study not only contributes to develop a better warning system for the deep-seated landslide and predict the possible location of landslide dams, but also draw attention to deep-seated landslide events caused by extreme events, especially in mountainous countries, in view of the changing rainfall characteristics observed in recent decades.

ACKNOWLEDGMENT

The authors would like to thank the National Science Council of the Republic of China, Taiwan, for supporting this research under Contract No. NSC 99-2625-M-005-002-MY3 and NSC 99-2625-M-005-004-MY3.

REFERENCES

1. Agliardi FG, Crosta G, Zanchi A (2001) Structural constraints on deep-seated slope deformations kinematics. Eng Geol 59:83–102

2. Chanson H (2005) The 1786 earthquake-triggered landslide dam and subsequent dam-break flood on the Dadu River, southwestern China. Geomorphology 71:437–440

3. Chen SC (1999) Failure mechanism and disaster mitigation on landslide-dammed lakes. J Chin Soil Water Conserv 30(4):299–311

4. Chen YJ, Zhou F, Feng Y, Xia YC (1992) Breach of a naturally embanked dam on Yalong River. Can J Civ Eng 19:811–818

5. Chen H, Lin GW, Lu MH, Shih TY, Horng MJ, Wu SJ, Chuang B (2011a) Effects of topography, lithology, rainfall and earthquake on landslide and sediment discharge in mountain catchments of southeastern Taiwan. Geomorphology 133:132–142

6. Chen SC, Hsu CL, Wu TY, Chou HT, Cui P (2011b) Landslide dams induced by typhoon Morakot and risk assessment. 5th International Conference on Debris-Flow Hazards Mitigation: Mechanics, Prediction and Assessment, Padua, Italy, pp. 653-660

7. Chiang SH, Chang KT (2011) The potential impact of climate change on typhoon-triggered landslides in Taiwan, 2010–2099. Geomorphology 133:143–151

8. Costa JE, Schuster RL (1988) The formation and failure of natural dams. Geol Soc Am Bull 100:1054–1068

9. Dai FC, Lee CF, Deng JH, Tham LG (2005) The 1786 earthquake-triggered landslide dam and subsequent dam-break flood on the Dadu River, southwestern China. Geomorphology 65:205–221

10. Duman TY (2009) The largest landslide dam in Turkey: Tortum landslide. Eng Geol 104:66–79

11. Feng ZY (2012) The seismic signatures of the surge waves of the 2009 Xiaolin landslide-dam breach in Taiwan. Hydrol Processes 26:1342–1351

12. Hsu YS, Hsu YH (2009) Impact of earthquake-induced dammed lakes on channel evolution and bed mobility: Case study of the Tsaoling landslide dammed lake. J Hydrol 374:43–55

13. Huang NE, Shen Z, Long SR, Wu MC (1998) The empirical mode decomposition and the Hilbert spectrum for nonlinear and non-stationary time series analysis. Proc R Soc Lond A: Math Phys Eng Sci 454:903–995

14. Hydrometeorological Design Studies Center (2012). "World record point precipitation measurements." NOAA's Weather Service, <http://www.nws.noaa.gov/oh/hdsc/index.html> (Apr. 9, 2012)

15. Keng WP (1965) Geology of the Minsheng-Chiahsien area, Southern Taiwan. Bull Geol Surv Taiwan 16:3–15

16. Korup O (2004) Geomorphometric characteristics of New Zealand landslide dams. Eng Geol 73:13–35

17. Korup O (2005) Geomorphic hazard assessment of landslide dams in South Westland, New Zealand: fundamental problems and approaches. Geomorphology 66:167–188

18. Korup O, Tweed F (2007) Ice, moraine, and landslide dams in mountainous terrain. Quat Sci Rev 26:3406–3422

19. Kung CL, Chyi SJ (2004) River terraces and landscape evolution in the Nantzuhsien drainage basin. J Geogr Sci 38:47–62

20. Lee CT, Dong JJ, Lin ML (2009) Geological investigation on the catastrophic landslides in Xiaolin village, southwestern Taiwan. Sino-Geotech 122:87–94

21. Lin ML, Jeng FS (2000) Characteristics of hazards induced by extremely heavy rainfall in Central Taiwan—Typhoon Herb. Eng Geol 58:191–207

22. Lin CH, Kumagai H, Ando M, Shin TC (2010) Detection of landslides and submarine slumps using broadband seismic networks. Geophys Res Lett 37:L22309–L22314

23. Liu N, Zhang JX, Lin W, Cheng WY, Chen ZY (2010) Draining the Tangjiashan Barrier Lake. J Hydraul Eng 136:914–923

24. Newman JD, Bennell JD (2002) A mountain stream flowmeter. Flow Meas Instrum 13:223–229

25. Quetelard H, Bessemoulin P, Cerveny RS, Peterson TC, Burton A, Boodhoo Y (2009) World-record rainfalls during tropical cyclone gamede. Am Meteorol Soc 90:603–607

26. Schuster RL (1993) Landslide dams—a worldwide phenomenon. Proceedings Annual Symposium of The Japanese Landslide Society, Kansai Branch, 27 April, Osaka, 1–23

27. Shiu CJ, Liu SC, Chen JP (2009) Diurnally asymmetric trends of temperature, humidity and precipitation in Taiwan. J Clim 22:5635–5649

28. Shou KJ, Wang CF (2003) Analysis of the Chiufengershan landslide triggered by the 1999 Chi-Chi earthquake in Taiwan. Eng Geol 68:237–250

29. Shroder JF Jr (1998) Slope failure and denudation in the western Himalaya. Geomorphology 26:81–105

30. Stini J (1941) Unsere Täler wachsen zu. Geol U Bauw, Wien 13(3):71–79

31. Tabata S, Mizuyama T, Inoue K (2002) The disasters of natural dam. Kokon Shoin, Tokyo

32. The Ohio Weather Library (2012) World point rainfall records. Ohio Weather Library Inc, <http://weatherrecords.owlinc.org/RainfallRecords1.html> (Accessed Apr. 9, 2012)

33. Tsou CY, Feng ZY, Chigira M (2011) Catastrophic landslide induced by Typhoon Morakot, Shiaolin, Taiwan. Geomorphology 127:166–178

34. Water Resources Agency (2009) The analysis of the rainfall and river discharge during Typhoon Morakot. Water Resources Agency, Ministry of Economic Affairs, Taipei, Taiwan, pp 24–25

35. Wu CH, Chen SC, Chou HT (2011) Geomorphologic characteristics of catastrophic landslides during typhoon Morakot in the Gaobing Watershed, Taiwan. Eng Geol 123:13–21

36. Uchida T, Yokoyama O, Suzuki R, Tamura K, Ishizuka T (2011) A new method for deep catastrophic landslide susceptibility. Int J Erosion Cont Eng 4:32–42

37. Xu Q, Fan XM, Huang RQ, Westen CV (2009) Landslide dams triggered by the Wenchuan Earthquake, Sichuan Province, south west China. Bull Eng Geol Environ 68:373–386

Chapter 5

Terrestrial laser scanning for rockfall stability analysis in the cultural heritage site of Pitigliano (Italy)

Riccardo Fanti, Giovanni Gigli, Luca Lombardi, Deodato Tapete, Paolo Canuti

ABSTRACT

Traditional surveying methods are often not sufficient to achieve a complete geomechanical characterization of the rock mass, to analyze the instability mechanisms threatening the cultural heritage of hilltop historic towns. In Pitigliano (Tuscany, Central Italy), terrestrial laser scanning was employed complementarily to conventional geomechanical techniques. The overall 3D survey of the exposed surfaces was combined with scanlines of the inner walls of the subterranean cavities running underneath the historic centre. The rock mass discontinuities geometry was extracted, and the most critical instability mechanisms were mapped, with particular interest in the potential impacts on the ancient buildings located along the cliff edge. The geomechanical analysis of the surveyed joint sets confirmed a structural control on the cliff morphology by two main joint sets. Thanks to the laser scanner-based kinematic analysis, flexural toppling and wedge failure were found as the main hazardous instability mechanisms in Pitigliano. Finally, the conservation criticalities were identified and a pilot monitoring system was installed in a sector highly susceptible to block detachment.

KEYWORDS

Terrestrial laser scanning; Kinematic analysis; Rockfall; Hazard assessment; Discontinuity extraction; Pitigliano (Italy)

1. INTRODUCTION

The impact of landslide hazard on the immovable cultural heritage represents a multi-disciplinary theme, which requires several different approaches (Canuti et al. 2009). A complete analysis involves geotechnical, structural and engineering issues and can lead to design adequate countermeasures. In Italy, ancient buildings, archaeological sites and even entire historic towns are frequently threatened by slope instability phenomena (Canuti et al. 2000; Fanti 2005; Lollino and Audisio 2006). In Tuscany, Umbria and Latium (Central Italy), where geological and geomorphological backgrounds determined the formation of hilltop-sites occupied since Etruscan time, well-known dynamics of instability characterize these places. The case of Orvieto (Tommasi et al. 2006) is the most famous, but in the same geological province there are many towns built on the top of volcanic slabs overlying normal- or over-consolidated clays. In these conditions, some landslide mechanisms can develop: slow movements in the clays, toppling and rockfalls of the marginal zones of the slab, deep-seated gravitational slope deformations (Ribacchi et al. 1988).

Pitigliano, situated in Southern Tuscany (Figs. 1 and 2), summarizes several of these aspects and represents a very interesting case in the field of relationship between slope instability and cultural heritage conservation. However, the significance is not only associated with the historical relevance of the site and of its monuments but also with the implementation of an integration approach of different techniques of geomechanical investigation and data collection, whose main results are presented in this paper. In particular, in addition to traditional methods, the rock mass was characterized by means of terrestrial laser scanner along with an extensive survey of the very complex system of chambers and caves excavated underneath the town through the centuries.

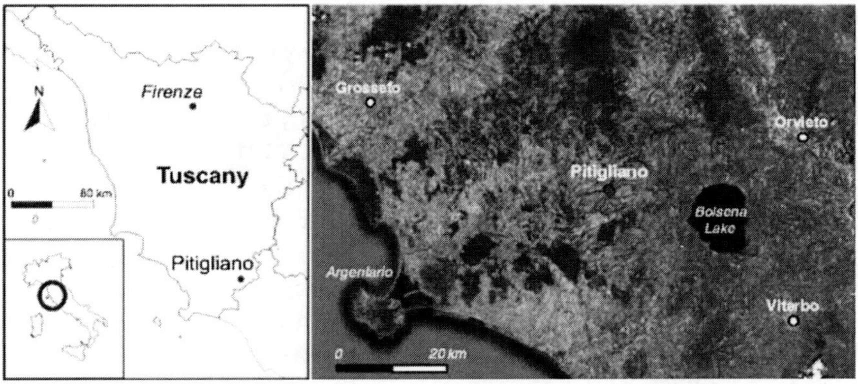

Figure 1. Location map of Pitigliano (aerial view modified from Google Earth)

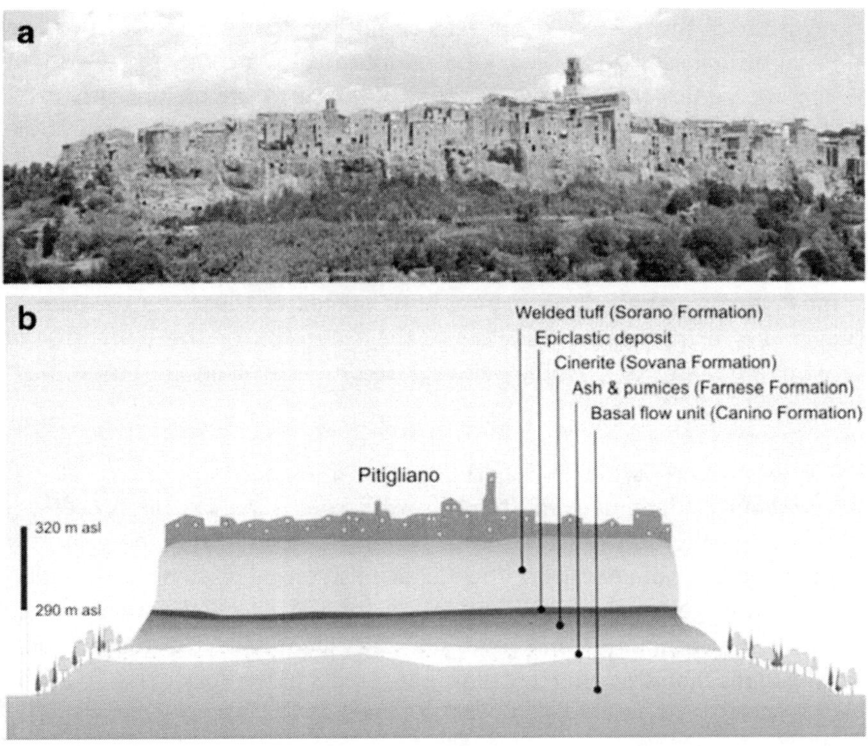

Figure 2. Overall view of the southern side of Pitigliano (**a**); sketch of the rock formations constituting the geological sequence (**b**)

Terrestrial laser scanning (TLS) supplies high resolution point clouds of the surveyed surfaces, and new TLS devices with growing range and resolution are increasingly becoming a remarkable tool in landslide characterization. TLS is employed in different task of slope instability analysis: for monitoring of deformations and displacements (Gordon et al. 2001; Abellan et al. 2009; Oppikofer et al. 2009), for the characterization of discontinuities on rock cuts (Feng and Roshoff 2004; Slob et al. 2005; Lato et al.2009; Sturzenegger and Stead 2009a, b; Gigli and Casagli 2011; Gigli et al. 2011), for geomorphological analyses of coastal cliffs, vertical slopes and freestanding boulders (Lim et al. 2005; Rosser et al. 2005; Armesto et al. 2009). This technique is also used for the documentation and 3D modelling of building and cultural heritage sites (Arayici2007; Yastikli 2007), and these products can be integrated with geomechanical analysis in the characterization of rock masses (Gigli et al. 2009). For its geometrical and geological features, the tuff rock mass of Pitigliano represents a good site for testing the potentials of TLS for geomechanics.

In regard with this case study, it is to be noted that the rock mass has an unusual distinguishing element, i.e. the presence of a complex network of manmade cavities which allows a survey of the slab from an inner point-of-view, providing significant geostructural data, otherwise not easily collectable. The habit of cutting corridors and rooms under the houses was relatively frequent in the past, especially when the foundation soil was cohesive but not very hard, as in the case of tuff or other volcanic rocks. In Italy, there are several historical cases, including also important cities as Rome (Crescenzi et al. 1995) and Naples (Evangelista 1991), but the case of Pitigliano is extraordinary because of the extension of the underground network and for its possible active role in controlling the rock mass stability.

The present work is mainly focused on the integrated analysis of all the geomechanical data derived from the different surveys and processing phases performed on the case study of Pitigliano. The usefulness of new digital products obtained from the TLS acquisitions is discussed in relation to extracting geomechanical information for the exposed rock surfaces, detecting discontinuities and joint sets and identifying the most critical instability mechanisms affecting different sectors of the rock mass.

1.1 Site description

The historic centre of Pitigliano is placed above a flat slab, which provided an ideal location for defence purposes, being a 25-m thick plateau of Pleistocene tuffs (about 310 m a.s.l.).

The tuff cliff, oriented in E–W direction, is the result of extensional tectonics occurred since Miocene to Quaternary and an intense volcanic activity predominantly explosive since the middle-Pleistocene (Varekamp 1980; Vezzoli et al. 1987), which affected the Northern part of the Roman Comagmatic Province in the area of the Vulsini Volcanic District (VVD). Among the four domains recognized in the VVD on a structural and volcanological basis (Conticelli et al. 1987; Vezzoli et al. 1987; Canuti et al. 2004), the Latera Volcanic Complex (LVC) characterizes the study area.

Volumetrically, the eruptive products are dominantly pyroclastic deposits and, secondarily, pyroclastic fall deposits, which are common at various stratigraphic levels throughout the western area of VVD (Palladino and Agosta 1997 *cum biblio*).

Within the stratigraphic sequence of the Pitigliano cliff, the following pyroclastic formations are found from bottom to the top (Figure 2):

- *Canino formation*—It is the oldest formation of LVC and it lies in direct contact with the Messinian sedimentary basement. This deposit, thick up to 45 m, consists of several pyroclastic flow units, from poorly cohesive up to coherent.

- *Farnese formation*—Thick up to 2–5 m, this non-welded pyroclastic deposit can be locally absent. At its top a basal ash fall, surge levels and a 30-cm thick black paleosoil are detectable (Canuti et al. 2004), as visible along the inner walls of some of the surveyed underground caves.

- *Sovana formation*—This cineritic and pyroclastic deposit, permanently thick of about 10 m, consists of an upper not highly coherent deposit lying above a lower non-welded flow unit.

- *Epiclastic deposit*—Lying between Sovana and Sorano Formations, this alluvial/lacustrine deposit derives from a static phase of the magmatic activity. Very poorly coherent, it reaches up to 4-m thickness along the northern side of the cliff. Sometimes it can be totally absent.

- *Sorano/Grotte di Castro formation*—It is the welded tuff plateau on which Pitigliano is built, thick up to 25 m and lying directly on the Epiclastic deposit. Through the centuries it provided solid rock material for cutting underground caves immediately beneath the historic buildings.

- Aerial view highlights that the present morphology of the Pitigliano cliff has been influenced by fluvial erosion processes. The exposed rock surfaces are actually similarly spaced from the sides of the rivers flowing along the toe of the cliff.

According to the morphologic features typical for this region, the two sides of the cliff are significantly different. The upper portion of the southern side displays very steep cliffs with a slope inclination of about 40°, which abruptly changes and gradually decreases due to both lithological variation and the presence of debris coverage. The northern side, on the contrary, is more regular, with a constant slope of about 35°, only locally interrupted by the presence of few metre-long scarps. Combined actions of lower thermal variations, higher and more permanent humidity content and more intensive chemical alteration and cryoclastism processes justify the higher thickness and the extension of the debris accumulated at the toe of the northern side in respect to the southern side.

A remarkable historical and structural feature of Pitigliano is the huge underground network of chambers and passages excavated during Etruscan and Medieval ages, crossing the rock mass with tunnels long from few metres up to over a hundred metres and characterized by a 1- to 2-m wide rectangular cross-section (Figure 3). Formerly used as local quarries to extract tuff blocks for building construction and volcanic ash for packing cements and mortars, the caves also constituted systems of naturally air-conditioned wine cellars.

In absence of any architectonical or security criteria, most part of the caves were excavated within Sorano/Grotte di Castro Formation due to the higher cementation degree and the rock strength, jointly to the major proximity to the surface. Few caves are found in Canino e Sovana Formations, no one within Farnese Formation, whereas the ones within Epiclastic deposit have been frequently affected by structural collapses.

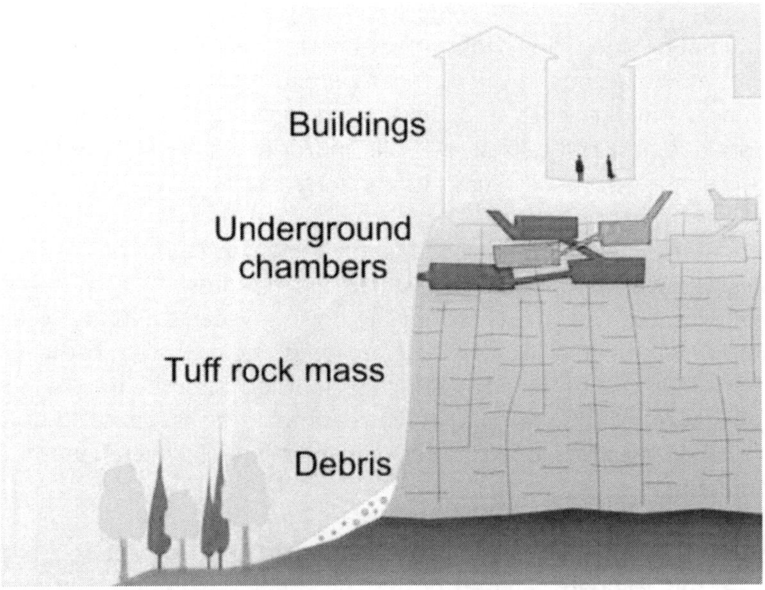

Figure 3. Schematic representation of the spatial relationship between the underground network of chambers and passages and the overlying buildings. Debris produced by weathering, and rock falls is accumulated at the cliff toe

Undoubtedly, a complete and reliable landslide risk assessment for the historic site of Pitigliano should take into account the influence of this huge underground system on the general weakening of the rock mass and its related stability.

The geomorphologic evolution caused by the combination of physical weathering and erosion processes on exposed tuff surfaces, the existing joint systems and the anthropic impact have triggered slope instability phenomena, significantly damaging the historic centre and cultural heritage of Pitigliano.

An overall survey of the main geomorphological and structural features of the exposed rock surfaces suggests flexural toppling and wedge failure represent the main instability mechanisms affecting the cliff. Such instability mechanisms could have affected the exposed surfaces several times in the past, causing a progressive retreat along the edge of the cliffs and, consequently, involving the overlying buildings of the historic centre.

Field observations and comparisons with the ancient graphic documentation and maps confirm that impression. Holes and caves visible at different heights on the cliff surfaces were formerly exterior entrances to the underground cavities and, at the present, they are not accessible anymore due to past localized rock falls and/or erosion processes that have severely damaged pre-existing stairs and terraces.

The comparison with the ancient map known as *Pianta della Terra di Pitigliano* (Giachi XVIII century) testifies the disappearance of a bastion originally located near the centre of the northern sector and the rock spur on which it was built, as well as some buildings at the western corner of the cliff. Some portions of the edge along the southern side seem to have been involved by the cliff retreat. It is reasonable to suppose that such phenomenon severely affected the conservation of several pre-existing constructions.

Although it is only a qualitative estimation of past impacts on historic heritage, this evaluation based on historical documentation confirms the key role of the geomorphological processes and weathering for the triggering of instability phenomena in Pitigliano, and suggests the occurrence of dynamics of progressive evolution of the rock mass strictly related to the most critical instability mechanisms.

2. METHODS

2.1 Geomechanical survey

The volcanic origin and the tectonic history of the formations outcropping in the Pitigliano area represent the geological background of the rock mass joint geometry, that, in turn, controls the observed mechanisms of instability.

The geomechanical characterization of discontinuities has been carried out starting from a traditional geomechanical survey of the rock mass, following prescriptions of ISRM (1978) in a significant number of scanlines along the cliff. A wide set of information on discontinuities (orientation, spacing, persistence, roughness, aperture, filling and seepage) has been collected.

However, in Pitigliano some additional data on the rock mass geometry can be gathered from a geomechanical survey in the cave network, whose presence points out the following issues:

a. The underground network allows an unusual inner look-out of the rock mass, from which some additional data on joints geometry can be obtained;

b. The role of the underground tunnels and chambers in the general behaviour of the rock slab could be assessed, considering these elements as non-persistent discontinuities, in particular if their spatial distribution showed some preferential directions; and

c. The position of the caves could form an obstacle in the realization of stabilisation measures as anchors, which efficiency could be strongly diminished if the tie bar went through an unexpected hollow.

Concerning the spatial distribution of subterranean passages, an extensive survey of the most relevant of such tunnels and chambers has been realized, in order to obtain the exact reconstruction of their dimension and length. The survey has been rationalized on the basis of special speleological criteria: starting from the georeferenced position of the entrance (mostly constituted of some descending steps carved in the tuff at the ground floor of the buildings), the dimensions of rooms and passages and the trend and plunge of their axes have been measured.

The final result of the survey is the complete geometric description of 93 cavities, organized as underground systems formed by one or more corridors (typical section, 2 × 1.5 m) and rooms (up to 5 × 10 × 2 m). In general, the total length of each system is about 15–20 m, and the slope is quite gentle. Nevertheless, there are also intricate structures, crossing several tuff layers through flight of stairs (difference in level up to 8 m), and very long system, as the most famous of them, i.e. a tunnel (106 m along the main direction, with a total linear length of 252 m) excavated underneath the historic centre of Pitigliano along SSE–NNW direction, connecting the two sides of the rock mass.

During the field surveys of the present work, a very interesting historical document has been found in the municipal archive of Pitigliano, i.e. the report of a similar survey work carried out in 1832 (Anonymous 1832), describing the caves used at that time, with several data about their dimension and extent.

The comparison between the two surveys (Anonymous 1832 vs. present time) has led to the conclusion that a relevant number of caves had been modified in the last 180 years, with the construction of parting walls and even the sealing of entire sectors, not accessible nowadays. For this reason, the present survey cannot be considered as completely exhaustive, being possible the presence of some unknown blocked chambers inside the tuff slab.

As regards to the role of the caves as inner point of view for rock mass survey and geomechanical analysis, this could help to minimize the orientation bias of scanlines surveyed on the cliff surfaces, complementarily to the traditional line sampling methodologies (Priest 1993; Brady and Brown 2004), and it could improve the quality and reliability of the datasets concerning the rock mass features. However, the fact that the caves themselves have been probably excavated along some weakness direction of the tuff mass could reduce this potentiality. Spatial distributions of the underground network and the natural joint system could be not independent.

Some considerations in this matter are presented in the next section, together with a synoptic view of the results obtained from laser scanning and speleological surveys, as elements useful for the assessment of the relationship between the external surface of the cliff and the inner manmade architecture.

2.2 Terrestrial laser scanning

The morphology of the Pitigliano cliff is continuously changing due to progressive retreat processes and localized rock falls that have been affecting the northern and southern sides. An updated survey of the entire site, including both the exposed exterior surfaces and underground caves, was considered fundamental in order to assess the current state of the geomorphological evolution of the cliff.

The risk assessment presented in this paper has been carried out, primarily, through the employment of the innovative digital technology of TLS. Survey activities have been undertaken using a Long Range and High Accuracy 3D Terrestrial Laser Scanner system, Riegl LMS-Z420i (RIEGL 2010a), associated operating and processing RiSCAN PRO software package (RIEGL 2010b).

As a time-of-flight laser scanner, this remote sensing ground-based device allows measurements of scanner–object distances, up to about 1,000 m, by calculating the round-trip time a laser pulse (near-infrared wavelength) takes to reach the object surface from the point of emission and come back. The entire field of view is scanned by changing view directions of the laser rangefinder through a system of rotating mirrors, and the related horizontal and vertical angles are measured, with maximum data acquisition rate of 12,000 points/s.

The Cartesian coordinates (X, Y and Z) of each point on the scanned object surface are worked out on the basis of the measured distance and scan angles, allowing the acquisition of point clouds available for the reconstruction of 3D models. These products can also be textured in true colours, thanks to the calibrated and definitely oriented high-resolution digital camera associated to the scanner.

The acquisition of point clouds and subsequent processing of digital products allowed a highly detailed documentation of the present state of the entire rock mass. Developing some efficient tools for a reliable characterization of the rock mass discontinuities, the following issues have been taken into account:

a. The cliff morphology, the geomorphologic difference between the northern and southern sides and the limited accessibility required the designation of point-of-views distributed in order to avoid, as much as possible, 'shadow zones' and

b. The acquisition and registration approach employed in the present case has been properly selected to achieve an optimal quality of the image registration process and related final accuracy of the 3D models suitable for the specific study purposes.

Seven points-of-view have been distributed on stable places located around the Pitigliano cliff, ensuring efficient overlapping between the neighbouring scans as a fundamental requirement for successful matching.

Simultaneously to the TLS survey, a network of some targets was created in order to allow the subsequent registration phase of the scans relative to each other in a single coordinate system. The coordinates of the selected targets have been exactly detected by using a couple of GPS Leica SR530 and they have been georeferenced to a reference point established in correspondence of the local Archaeological Museum, within an area easily accessible and totally covered by the satellites.

Each 3D point cloud has been projected into the corresponding registered 2D image, and true colour values of the image pixels have been assigned to the projected 3D points.

As final digital product, the holistic point cloud representation of the Pitigliano cliff has been realized by merging the seven point clouds acquired from the correspondent points-of-view (Figure 4).

Figure 4. Holistic true coloured point cloud of the Pitigliano cliff and the historic centre

3. RESULTS

3.1 Conventional surveys

Several data on the rock mass have been collected during the geomechanical survey, leading to characterization of the tuff layer of the cliff (Sorano Formation). The geomechanical characterization of discontinuities has been carried out starting from a traditional geomechanical survey of the rock mass, following prescriptions of ISRM (1978) in a significant number of scanlines along the cliff. A wide set of information on discontinuities (orientation, spacing, persistence, roughness, aperture, filling and seepage) has been collected.

As above mentioned (cf. 'Geomechanical survey'), an in situ characterization of rock mass discontinuities was performed according to the indications of ISRM (1978), coupled with point load and tilt tests on laboratory samples.

The obtained data, representative of the geomechanical properties of the rock mass, are summarized in the Table 1, and they were used as input parameters for the kinematic analysis described in 'Kinematic analysis'.

Table 1. Geomechanical properties of the rock mass of Pitigliano obtained from conventional surveys

Set_id	α (deg)	β (deg)	ϕ (deg)	c (MPa)	RMR	Q	RMi
JN1	073	87	48	0.14	62	3.8	1.65
JN2	163	86					

Set_id discontinuity set, α dip direction, β dip, φ average friction angle, c average cohesion, *RMR* rock mass rating according to Bieniawski (1989), *Q* Q-index according to Barton (in Palmström 2009), *RMi* Rock Mass Index according to Palmström (2009)

From these results, in order to accomplish the distinctive featuring of the rock mass, its geomechanical classification has been carried out by applying the most widespread classification systems (Bieniawski 1989; Grimstad and Barton 1993; Palmstrom 1995,2000, 2009; Palmstrom and Broch 2006).

For the definition of geometry of the main joint sets, the discontinuity system of the rock mass has been analyzed starting from the results of the scanline surveys (i.e. those surveyed both on the cliff and along the tunnel walls in the underground cavities).

The synthesis of the collected data is represented in Figure 5 where two sub-vertical perpendicular joint systems, with orientation JN1 = 073/87°, and JN2 = 163/86° are highlighted. At a first glance, the latter seems the less important, but the orientation bias must be considered, as most of the surveyed scanlines result parallel to JN2 and, consequently, the set is under-represented.

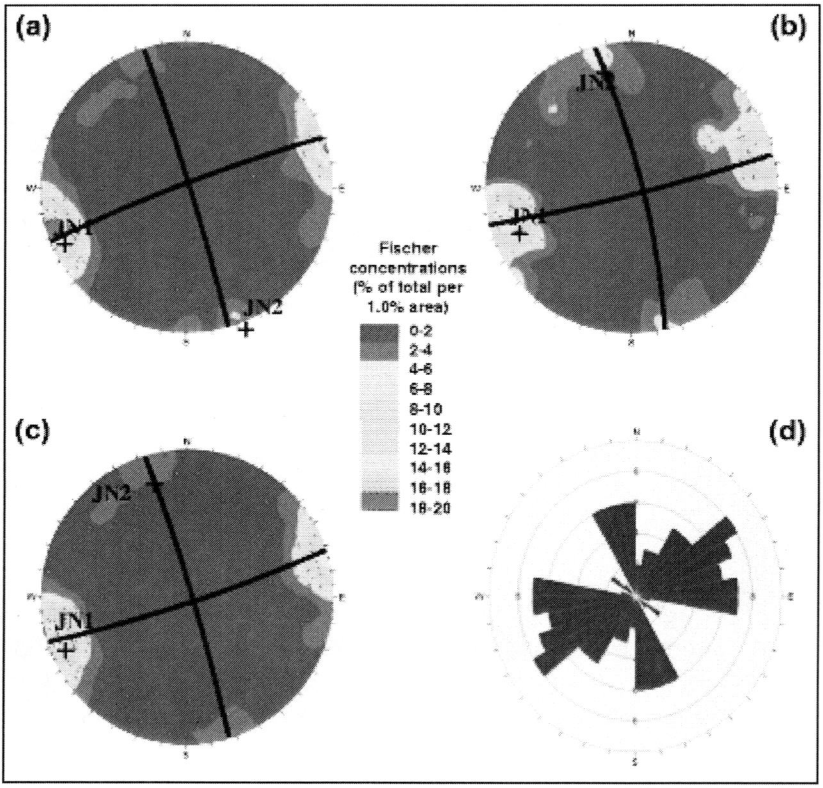

Figure 5. Synthesis of the geomechanical survey data. Stereographic representations of the surveyed discontinuities: *a*underground, *b* on cliff surfaces, *c* total number, and *d* rose diagram of the directions of the underground chambers and corridors

Particular attention has been reserved to the results of the underground survey, starting from its use for the evaluation of the cavities spatial position in comparison with the cliff geometry.

As stated before, these results can be important during the realization of structural works for the stabilization of the rock mass, but a more challenging task is represented by the identification of the more effective approach for the data elaboration and output. A solution has been found in the integration in a CAD environment of the topographic data and the output of laser scanner software that allows the visualization of the exact position of the caves with reference to the cliff and to the buildings. The result is a dynamic 3D model of the town which includes its hidden underground elements that can be surfed from different points of view, as shown in Figure 6.

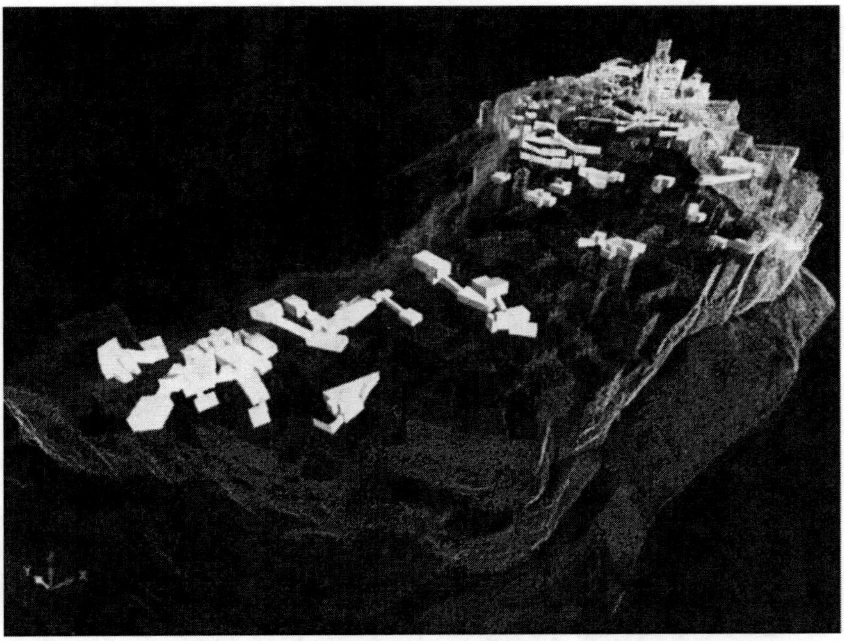

Figure 6. Decimated point cloud of Pitigliano integrated with the 3D reconstruction of the underground network of the surveyed chambers and corridors

Such 3D product is quite easily understandable and provides an efficient and smart tool for localizing the cavities in relation to the cliff surfaces and overlying buildings. At a first glance, it is clearly visible that the historic centre of Pitigliano is actually superimposed to another 'town', excavated underneath the building. This constitutes a relevant part of the cultural heritage in Pitigliano that needs to be preserved and documented, also in regard with its role in the general behaviour of the rock mass.

Such 3D products could be useful in the perspective of stabilization and consolidation works, especially in order to avoid some mishaps, like the injection of a big amount of concrete on unknown caves during the realization of anchors, as what happened in the past.

As regards to the spatial orientation of these tunnels and chambers, the amount of collected data has been considered statistically significant, and the analysis of their spatial distribution has been carried out (Figure 5d). The surveyed caves are mostly E–W oriented, in the same direction of the joint set JN2 and it is possible to infer that this is a direction of weakness of the rock mass, as confirmed by the morphology of the slab (Figure 7).

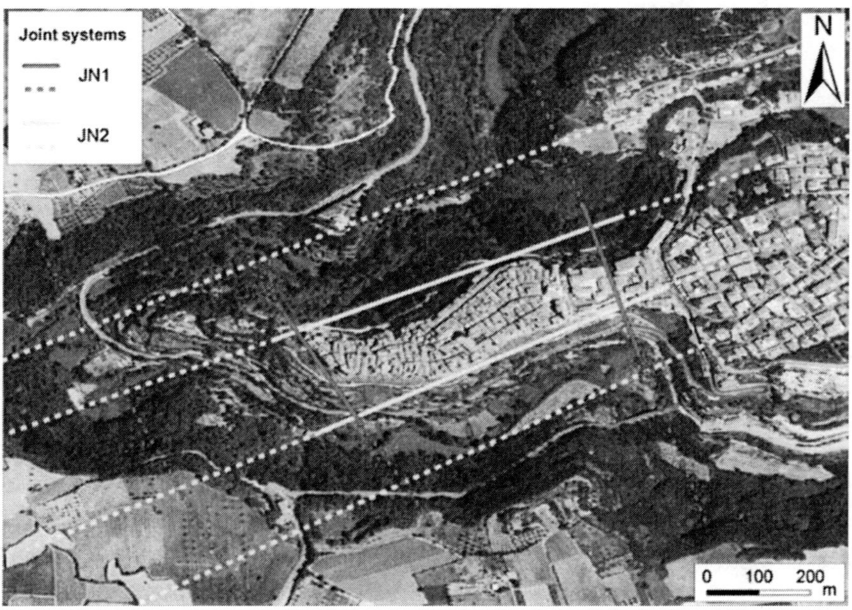

Figure 7. V Synthesis of the structural data and the directions of the two predominant joints sets (JN1 and JN2) overlapped on aerial view of the Pitigliano cliff and surroundings (modified from Google Earth). The joint system geometry (see *continuous lines*) structurally influences the morphology of the Pitigliano cliff

The hypothetical role of the man-made riddling as factor for the rock-mass instability has been finally assessed, coming to the conclusion that the number and the total volume of the underground elements are largely insufficient to act in this sense. In fact, it should be borne in mind that the total volume of surveyed caves is about 9,000 m^3, i.e. almost 1 % of the total volume of the rock slab. Similar considerations can be made on the significance of the set of excavation plans as non-persistent discontinuities, even if single cases (tunnels and chambers very close to the cliff) can be relevant for the stability of some tuff blocks.

3.2 Laser scanner-based survey

Among the results obtained from the TLS survey of the Pitigliano rock mass, the acquired point clouds represent the raw product of laser scanning. Merging all the point clouds in the holistic one (Figure 4) provides an overall and true colour 3D representation of the natural and built environment; this product contains a large amount of data (about 68 million points and related measurements of spatial coordinates and colour parameters) available for further processing.

Thanks to the high density of the 3D points, a detailed modelling of the rock mass morphology has been performed as basic tool to evaluate the most hazardous mechanisms of instability for both sides of the cliff. The overall view of the rock mass surfaces has been used as an effective tool to verify the sequence and contacts of the outcropping geological strata and their correspondent thicknesses, with precise referencing and spatial relationship with the overlying historical buildings located along the edge.

The first outcomes from the laser scanning survey have been an improvement of the geological sketches representing the different sectors of the cliff and the provision of an updated 3D documentation of the present state of the rock mass. The structural relationships between the historic buildings and their rock foundations have also been established and updated at the current state; the sectors more exposed to erosion processes, rock falls and collapses have been identified.

Prior to produce a triangulated mesh necessary for creating a digital terrain model (DTM) of the cliff surfaces from the acquired point clouds, it has been undertaken a pre-processing stage consisting in two main operations:

a. Removal of the vegetation coverage and

b. Distinction, within the point clouds, of the rock outcrops respectively from the built environment and the debris at the toe of the cliff.

These segmentation and decimation steps were realized with both automated and manual procedures, also filtering point clouds on the basis of the different intensity of the return signals coming from the scanned materials.

Such time-consuming preparation of point clouds has been justified by the final aim, i.e. the creation of a reliable DTM suitable for the characterization of the rock mass discontinuities. It is quite well known that vegetation and non-geologic points, jointly to registration errors, can contribute to the noise in point clouds and affect automatic meshing procedures (Buckley et al. 2008).

The resulting points extracted on the rock mass outcrop were re-sampled with the aim of obtaining a regular and uniform spatial distribution.

In order to create a continuous surface from discrete data, the sampled point cloud was finally triangulated, taking care to fill holes and to obtain a congruent surface with correctly orientated triangle normals.

After the triangulation procedure, a total of more than 240,000 triangles were obtained and the following parameters were calculated for each of them: dip, dip direction and outwarding normal.

As first processing stage, the cliff has been virtually divided in six sectors, clockwise from the SE limit, and transverse contour lines have been directly extracted from the continuous surfaces of the point clouds, in order to obtain topographic elevation maps allowing morphologic evaluation for each sector in correspondence of the outcrops.

'Contour line models' were realized by coupling a contour line representation with the true coloured point clouds (Figure 8). By means of the visualization through contour lines, morphologic features of the outcropping surfaces have been highlighted directly on the 3D model, performing an estimation of the elevation values with a centi-decimetre accuracy. Due to the dense vegetation coverage existing at the moment of the TLS survey, the quality of the contour line models for the northern side of the rock mass has not been so high in comparison with the correspondent modelling for the southern side.

Figure 8. True coloured point cloud of the southern side of the Pitigliano cliff; *contour line* representation highlights morphologic features of the outcropping rock mass

As added value of this modelling, such DTMs also represent a suitable archive database for future monitoring campaigns of the temporal evolution of the rock mass morphology. Digital comparisons with new scans can be easily performed. This potential of TLS was taken into account during the selection of the acquisition and registration approach, starting from the localization of the seven point-of-views.

4. ANALYSIS

Thanks to the high resolution of the mesh obtained from the laser scanning data, detailed analyses were performed on the whole rock wall with the aim of extracting the discontinuity sets and of identifying those sectors more prone to instability mechanisms.

On the basis of the orientation of each triangle of the mesh, and of the direction of its outwarding normal, the overhanging sectors of the rock mass have been extracted and directly localized on the 3D point cloud. Distribution of dip and dip direction values has been also obtained for all the outcropping surfaces and the visualization on the 3D point cloud clarifies that most of the rock mass surfaces are sub-vertical and overhanging (Figure 9).

Figure 9. Point cloud of the southern side of the Pitigliano cliff with the distribution of rock mass dip (**a**) and dip direction values (**b**) calculated and extracted by processing 3D laser scanning data

Such analysis has contributed to a first fundamental identification of the hazard levels for the different sectors identified along both sides of the rock slab and constitutes the basis of the 3D kinematic analysis presented in 'Kinematic analysis'.

4.1 Identification of the discontinuity sets

A specific Visual Basic software was developed (Lombardi 2007) for the semi-automatic extraction of the main discontinuity sets within a rock mass, by analyzing a mesh representing the rock mass outcrop.

Based on the dip and dip-direction of each triangle of the mesh, the modal planes of the discontinuity sets are extracted through a traditional stereoplot analysis. The spreading of the triangle poles around each modal plane is also evaluated.

All the triangles whose orientation lies within the area of influence of a modal plane are associated to the correspondent discontinuity set. All unassigned triangles are given an 'other' label and can be re-assigned to each discontinuity set in successive iterations if closeness and angular divergence conditions are satisfied.

Figure 10 reports the contour plot of the poles of each triangle constituting the mesh of the southern sector of the cliff. By observing the areas with highest density, two discontinuity sets can be distinguished: JN1 (054/88°) and JN2 (165/88°).

4.2 Kinematic analysis

As the orientation of the fractures within the rock mass seems not to be random, but related to the tectonic processes that have been acting in the investigated area, a kinematic analysis can be useful to identify the rock-wall sectors which are more prone to instability processes. These are investigated by combining fracture dip and dip directions with local slope orientations.

The rock plate is, in fact, characterized by very steep slopes, sometimes overhanging and, due to its irregular shape (Figure 7), by very different slope aspect (Figure 9).

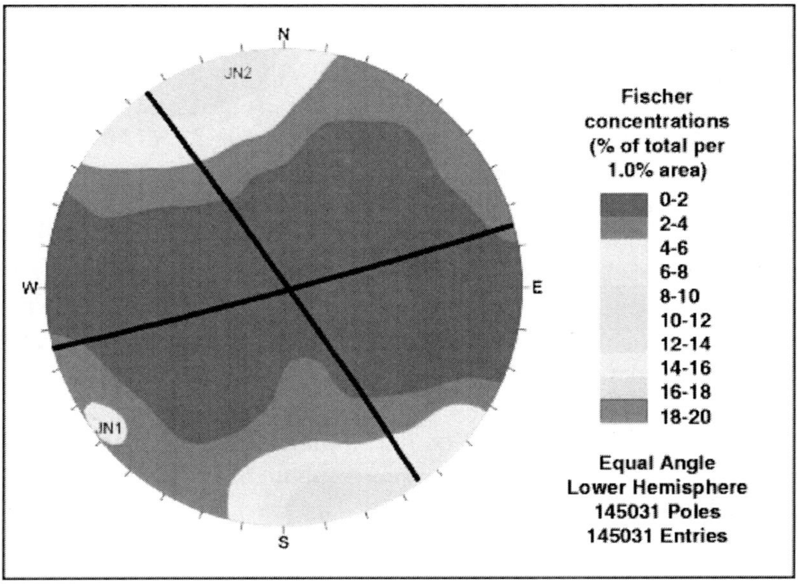

Figure 10. Stereographic projection of the orientation of the *triangles* constituting the 3D mesh obtained from laser scanning data

The results of the analysis after three iterations are presented in Figure 11, where the triangles of the mesh are colored based on the discontinuity set they belong to.

The main instability mechanisms investigated with this approach are:

- Plane failure (Hoek and Bray 1981);
- Wedge failure (Hoek and Bray 1981);
- Block toppling (Goodman and Bray 1976; Matheson 1983);
- Flexural toppling (Goodman and Bray 1976; Hudson and Harrison 1997).

Figure 11. Point cloud of the southern side of the Pitigliano cliff. The main joint sets are distinguished and localized on the basis of the extraction phase from 3D laser scanning data

Casagli and Pini (1993) introduced a Kinematic Hazard Index for each instability mechanism. These values are calculated by counting poles and discontinuities falling in critical areas within the stereographic projection:

$$C_{pf} = 100 \times (N_{pf}/N) \quad \text{for plane failure;}$$

$$C_{wf} = 100 \times (I_{wf}/I) \quad \text{for wedge failure;}$$

$$C_{bt} = 100 \times (N_{bt}/N) \times (I_{bt}/I) \quad \text{for block toppling;}$$

$$C_{ft} = 100 \times (N_{ft}/N) \text{ for flexural toppling,}$$

where

N_{pf} Number of poles satisfying plane failure conditions

I_{wf} Number of intersections satisfying wedge failure conditions

N_{bt} Number of poles satisfying block toppling conditions

I_{bt} Number of intersections satisfying block toppling conditions

N_{ft} Number of poles satisfying flexural toppling conditions.

The analysis input parameters are the slope dip and dip direction (Figure 9), the discontinuity surface orientations (obtained from the field survey and from laser scanning data analysis) and discontinuity friction angle (obtained from the geomechanical survey).

Intersection lines are calculated automatically, together with the equivalent friction angle, based on the intersecting planes friction angles and the shape of the wedge (Casagli and Pini 1993).

A new Visual Basic software developed specifically for true 3D kinematic analyses called Rock Slope Stability (RSS) (Lombardi 2007; Gigli et al. 2012) has been applied in the present study. This tool overcomes many limitations of the traditional kinematic analyses, and even of the more advanced analyses performed on 2.5D surfaces. In fact, it is possible to employ the 3D mesh obtained from laser scanning data, and the kinematic conditions leading to the investigated instability mechanisms have been extended to overhanging slopes.

The resulting analysis has therefore a true 3D character, and can be performed on high resolution surfaces, making it possible to distinguish even small details of the investigated area.

The results from this analysis are summarized in Figure 12, where the kinematic index of the triangles of the mesh is reported for each instability mechanism for a portion of the southern side of the cliff. The probability of occurrence for each mechanism is expressed by means of a colour scale, varying from green to red as the hazard increases. It means that green areas correspond to sectors with very low, up to negligible, hazard with reference to the specific instability mechanism, while orange/red colours mark the sectors highly prone to instability. The maximum values of the kinematic indexes which were calculated range from 25 to 30 %.

Figure 12. Results obtained from the 3D kinematic analysis for the different analyzed instability mechanisms in correspondence of the southern side of the cliff: **a** block toppling, **b** flexural toppling, **c** wedge failure, and **d**plane failure. The probability of occurrence (hazard) is expressed following a common colour scale

In this regard, the highest kinematic index (28 %) is related to the wedge failure (WF; Figure 12c), with a respective hazard which affects several sectors of the cliff surfaces. The spatial distribution of the WF index confirms that conservation criticalities can be raise both for the tuff blocks located immediately underneath the buildings located along the edge and the lower sectors of the exposed cliff. Consequently, a progressive undermining can destabilize huge portions of the cliff at different heights.

Comparing the spatial distribution of the kinematic indexes for the other instability mechanisms, it is quite evident that block toppling represents a negligible deterioration process (kinematic index up to 2 %; Figure 12a) while flexural toppling (FT; Figure 12b) and plane failure (PF; Figure 12d) reach maximum values, respectively, of 22 and 20 %. In particular, hazard patterns of the FT and PF mechanisms show some similarities in their spatial distribution, with values of kinematic index averagely higher for the FT over the same sectors.

5. DISCUSSION AND CONCLUSIONS

The usefulness of the proposed procedure mainly consists in remotely having surveyed rock mass discontinuities, especially in correspondence of impervious areas, gathering an overall characterization of the outcropping surfaces and estimating the probability of occurrence of the different instability mechanisms for each sector.

Looking at Figure 9, we can observe that a considerable portion of the rock wall is overhanging. The analysis of the 3D mesh obtained from the laser scanning data shows that the area of the overhanging sectors is about 33 % of the total area of the rock outcrop.

The advantage of collecting the orientation of rock mass discontinuities by applying different techniques (traditional geomechanical surveys and by analyzing 3D laser scanning data) and from different portions of the rock plate (from chambers inside the rock plate, from scanline surveys at the base of the rock walls, and from the whole rock exposure by analyzing 3D laser scanning data) is useful to reduce the bias introduced due to the employed technique (sampling method) and to the direction of analysis. A clear inversion of concentration values and a slight different orientation of JN1 modal plane are evident by comparing Figs. 5 and 10. Such differences can be explained by considering the bias introduced by the traditional geomechanical survey in favour of discontinuity set JN1. In fact, most of the scanline surveys were performed along the southern portion of the cliff, which is almost parallel to the discontinuity set JN2 modal plane (Figure 7). For the same reason, the analysis of the 3D mesh obtained from laser scanning data tends to overestimate the surfaces belonging to JN2 discontinuity set, as these are parallel to the mean rock slope direction (Figure 11).

These integrated geomechanical analysis is very useful to reduce some of the uncertainties due to partial data collection and provides reliable data for a complete 3D kinematic analysis.

Thanks to georeferencing of all 3D data, it is possible to identify the most critical sectors directly on 3D mapping products and compare their localization with the correspondent on the cliff surfaces, in order to verify the present conditions at the most critical points.

Classification of hazardous areas based on discontinuity extraction and subsequent processing of 3D laser scanner data should find a fundamental validation in the in situ evidence, such as in the present case. Considering the southern side, several sectors have been classified as prone to wedge failure (Figs. 12c and 13a) and some of them are, hazardously, located directly beneath the buildings of the historic centre situated along the edge of the cliff. In particular, Figure 13b shows a block located in an area classified at high wedge failure hazard (Figure 13a). A close view of this block highlights the presence of cracks which undoubtedly define wedge geometry, suggesting a probable future detachment of the block by wedge failure (Figure 13c).

This example is also interesting in terms of evaluation of the direct impacts on the built heritage due to instability mechanisms. The buildings overlying the block show quite recent repair interventions on the masonry structures, with new construction phases made using blocks of the local tuff (Figure 13c). Such repairs suggest the occurrence, in the past, of damages and structural collapses, hinted also by the niche located directly east to the block, where a similar detachment plausibly has taken place.

Identifying and localizing the critical points represents the first step in order to plan and realize adequate countermeasures for the preservation of the built heritage of the historic centre of Pitigliano. Permanent monitoring could be one of the best strategies to tackle instability issues, especially when consolidation works cannot be cost-effective and not easily to carry out, as it is frequent in case of historic contexts.

Figure 13. Example of area on the cliff surface prone to wedge failure (*WF*) as identified on **a** the n mapping of WF instability mechanism (detail of Figure 12c) and **b** correspondent photograph. **c** Detailed view shows the wedge geometry of the block and recent repair works on the overlying masonry

Based on the results of the integrated geomechanical analysis, a test area has been selected on the southern side of the Pitigliano cliff, in correspondence of an area classified as highly prone to the occurrence of wedge failure or flexural toppling, and a monitoring system was installed.

Figure 14 shows the test area where the buildings are built above a tuff block affected by a joint suggesting the presence of a probable plane of failure. Detachment of the block would compromise the stability of the overlying buildings, even causing structural collapses. A further element of instability is here constituted by the presence of the underground cave, whose excavation could have influenced the overall structural setting of the rock mass underneath the buildings. This case clarifies how a geomechanical analysis of instability mechanisms in built-up contexts requires a wider approach which would take into account all different geological and not-geological hazard factors.

In particular, to monitor eventual deformations and displacements along the joint, a 24-h/24 real-time monitoring system has been installed, employing some deformometers with automatic data collection and radio systems to transmit the deformometric data. A local gateway has been positioned in order to receive the data directly from the radios and forward them by GPRS to a remote server, allowing visualization of the data on an easy-to-use website, available for the end-users for constant control activity.

Such monitoring network, specifically adapted for reducing the impact on the buildings and the landscape, has to be considered as a pilot system, potentially extendable to others critical sectors of the Pitigliano cliff or even exportable to other similar contexts.

ACKNOWLEDGEMENTS

The work described in this paper was supported by the World Monuments Fund (World Monument Watch Programme) and by the Municipality of Pitigliano. We gratefully acknowledge the assistance of Dr. Francesco Mugnai and Dr. Massimiliano Nocentini in the field investigation.

Figure 14. Test area selected on the southern side of the Pitigliano cliff. Deformometers were installed to monitor the joint (*dashed line*) which controls the wedge geometry. Spatial relationship with the overlying building suggests probable direct impacts on the preservation of the built heritage

REFERENCES

1. Abellan A, Jaboyedoff M, Oppikofer T, Vilaplana JM (2009) Detection of millimetric deformation using a terrestrial laser scanner: experiment and application to a rockfall event. Nat Hazards Earth Syst Sci 9:365–372

2. Anonymous (1832), Rapporto sopra gli scavi esistenti al di sotto dei fabbricati della terra di Pitigliano. Ms. available at Public Library of Pitigliano

3. Arayici Y (2007) An approach for real world data modelling with the 3D terrestrial laser scanner for built environment. Automat Constr 16(6):816–829

4. Armesto J, Ordóñez C, Alejano L, Arias P (2009) Terrestrial laser scanning used to determine the geometry of a granite boulder for stability analysis purposes. Geomorphology 106(3–4):271–277

5. Bieniawski ZT (1989) Engineering rock mass classifications. Wiley, New York

6. Brady BHG, Brown ET (2004) Rock mechanics for underground mining, 3rd edn. Kluwer, London

7. Buckley SJ, Howell JA, Enge HD, Kurz TH (2008) Terrestrial laser scanning in geology: data acquisition, processing and accuracy considerations. J Geol Soc 165:625–638

8. Canuti P, Casagli N, Catani F, Fanti R (2000) Hydrogeological hazard and risk in archaeological sites: some case studies in Italy. J Cult Herit 1(2):117–125

9. Canuti P, Casagli N, Fanti R, Iotti A, Pecchioni E, Santo AP (2004) Rock weathering and failure of the "Tomba della Sirena" in the Etruscan necropolis of Sovana (Italy). J Cult Herit 5(3):323–330

10. Canuti P, Margottini C, Fanti R, Bromhead EN (2009) Cultural heritage and landslides: research for risk prevention and conservation. In: Sassa K, Canuti P (eds) Landslides—disaster risk reduction. Springer, Berlin. pp 401–433

11. Casagli N, Pini G (1993) Analisi cinematica della stabilità di versanti naturali e fronti di scavo in roccia. Geologia Applicata e Idrogeologia 28:223–232

12. Conticelli S, Francalanci L, Manetti P, Peccerillo A (1987) Evolution of Latera Volcano, Vulsinian district (Central Italy): stratigraphical and petrological data. Per Mineral 56:175–199

13. Crescenzi R, Piro M, Vallesi R (1995) Le cavità sotterranee a Roma. In: Funiciello R (ed) La geologia di Roma: il Centro Storico. Memorie descrittive della Carta Geologica d'Italia. pp 250–278

14. Evangelista A (1991) Cavità e dissesti nel sottosuolo dell'area napoletana. In: Atti del Convegno su Rischi naturali ed impatto antropico nell'area metropolitana napoletana, Facoltà di Ingegneria, Napoli, 7–8 giugno. pp 195–218

15. Fanti R (2005) Slope instability of San Miniato hill (Florence, Italy): possible deformation patterns. Landslides 3(4):323–330

16. Feng QH, Roshoff K (2004) In-situ mapping and documentation of rock faces using a full-coverage 3-D laser scanning technique. Int J Rock Mech Mining Sci 41:139–144

17. Giachi (XVIII century) Pianta della Terra di Pitigliano. Ms. available at National Central Library of Florence

18. Gigli G, Casagli N (2011) Semi-automatic extraction of rock mass structural data from high resolution LIDAR point clouds. Int J of Rock Mech Min Sci 48:187–198

19. Gigli G, Mugnai F, Leoni L, Casagli N (2009) Brief communication: analysis of deformations in historic urban areas using terrestrial laser scanning. Nat Hazards Earth Syst Sci 9:1759–1761

20. Gigli G, Fanti R, Canuti P, Casagli N (2011) Integration of advanced monitoring and numerical modeling techniques for the complete risk scenario analysis of rockslides: the case of Mt. Beni (Florence, Italy). Eng Geol 18:48–59

21. Gigli G, Frodella W, Mugnai F, Tapete D, Cigna F, Fanti R, Intrieri E, Lombardi L (2012) Instability mechanisms affecting cultural heritage sites in the Maltese Archipelago. Nat Hazards Earth Syst Sci (in press)

22. Goodman RE, Bray JW (1976) Toppling of rock slopes. In: Proc. speciality conference on rock engineering for foundations and slopes. ASCE, Boulder (Colorado), 2, pp 201–234

23. Gordon S, Lichti D, Stewart M (2001) Application of a high-resolution, ground based laser scanner for deformation measurements. In: Proceedings of 10th International FIG Symposium on Deformation Measurements, Orange, California, USA, 19–22 March 2001, pp 23–32

24. Grimstad E, Barton N, 1993. Updating the Q-system for NMT. In: Proceedings of Int. Symp. on Sprayed Concrete, Fagernes, Norwegian Concrete Association, Oslo, Norway 1993, pp 46–66

25. Hoek E, Bray JW (1981) Rock slope engineering, Revised Third Editionth edn. Institution of Mining and Metallurgy, London

26. Hudson JA, Harrison JP (1997) Engineering rock mechanics: an introduction to the principles and Applications. Pergamon ed., Oxford

27. International Society for Rock Mechanics (ISRM) (1978) International Society for Rock Mechanics (ISRM), Commission on standardization of laboratory and field tests: suggested methods for the quantitative description of discontinuities in rock masses. Int J of Rock Mech and Min Sci Geomech Abstr 15(6):319–368

28. Lato M, Diederichs MS, Hutchinson DJ, Harrap R (2009) Optimization of LiDAR scanning and processing for automated structural evaluation of discontinuities in rock masses. Int J Rock Mech Mining Sci 46:194–199

29. Lim M, Petley DN, Rosser NJ, Allison RJ, Long AJ, Pybus D (2005) Combined digital photogrammetry and time-of-flight laser scanning for monitoring cliff evolution. Photogramm Rec 20:109–129

30. Lollino G, Audisio C (2006) UNESCO World Heritage sites in Italy affected by geological problems, specifically landslide and flood hazard. Landslides 3(4):311–321

31. Lombardi L (2007) Nuove tecnologie di rilevamento e di analisi di dati geomeccanici per la valutazione della sicurezza. Ph.D. thesis, University of Florence, Italy

32. Matheson GD (1983) Rock stability assessment in preliminary site investigations - Graphical Methods. Transport and Road Research

Laboratory Report 1039. Transport and Road Research Laboratory, Crowthorne, Berkshire.

33. Oppikofer T, Jaboyedoff M, Blikra L, Derron M-H, Metzger R (2009) Characterization and monitoring of the Aknes rockslide using terrestrial laser scanning. Nat Hazards Earth Syst Sci 9:1003–1019

34. Palladino DM, Agosta E (1997) Pumice fall deposits of the western Vulsini Volcanoes (central Italy). J Volc Geoth Res 78(1–2):77–102

35. Palmström A (1995) RMi—a rock mass characterization system for rock engineering purposes. Ph.D. thesis, University of Oslo

36. Palmström A (2000) Recent developments in rock support estimates by the RMi. J of Rock Mech and Tunn Technol 6(1):1–19

37. Palmström A (2009) Combining the RMR, Q, and RMi classification systems. Tunn Undergr Space Technol 24(4):491–492

38. Palmström A, Broch E (2006) Use and misuse of rock mass classification systems with particular reference to the Q-system. Tunn and Undergr Space Technol 21(6):575–593

39. Priest SD (1993) Discontinuity analysis for rock engineering. Chapman & Hall, London

40. Ribacchi R, Sciotti M, Tommasi P (1988) Stability problems of some towns in central Italy: geotechnical situations and remedial measures. In: Marinos PG, Koukis GC (eds) The engineering geology of ancient works, monuments and historical sites. Preservation and protection. Proceedings of an international symposium. Athens, 19–23 September 1988. Vol. 1: Engineering geology and the protection of historical sites and monuments, pp 27–36.

41. RIEGL (2010a) Data sheet of long range & high accuracy 3D terrestrial laser scanner LMS-Z420i. Available fromhttp://www.riegl.com/uploads/tx_pxpriegldownloads/10_DataSheet_Z420i_03-05-2010.pdf. Accessed 18 October 2010

42. RIEGL (2010b) Operating & processing software RiSCAN PRO for RIEGL 3D laser scanners. Available fromhttp://www.riegl.com/uploads/tx_pxpriegldownloads/11_DataSheet_RiSCAN-PRO_22-09-2010.pdf. Accessed 18 October 2010.

43. Rosser NJ, Petley DN, Lim M, Dunning SA, Allison RJ (2005) Terrestrial laser scanning for monitoring the process of hard rock coastal cliff erosion. Q J Eng Geol 38(4):363–375

44. Slob S, van Knapen B, Hack R, Turner K, Kemeny J (2005) Method for automated discontinuity analysis of rock slopes with three-dimensional laser scanning. Transp Res Record 1913:187–194

45. Sturzenegger M, Stead D (2009a) Quantifying discontinuity orientation and persistence on high mountain rock slopes and large landslides using terrestrial remote sensing techniques. Nat Hazards Earth Syst Sci 9(2):267–287

46. Sturzenegger M, Stead D (2009b) Close-range terrestrial digital photogrammetry and terrestrial laser scanning for discontinuity characterization on rock cuts. Eng Geol 106(3–4):163–182

47. Tommasi P, Pellegrini P, Boldini D, Ribacchi R (2006) Influence of rainfall regime on hydraulic conditions and movement rates in the overconsolidated clayey slope of the Orvieto hill (central Italy). Can Geotech J 43(1):70–86

48. Varekamp JC (1980) The geology of the Vulsinian area, Lazio, Italy. Bull Volcan 43(3):489–503

49. Vezzoli L, Conticelli S, Innocenti F, Landi P, Manetti P, Palladino DM, Trigila R (1987) Stratigraphy of the Latera Volcanic Complex: proposals for a new nomenclature. Per Mineral 56:89–110

50. Yastikli N (2007) Documentation of cultural heritage using digital photogrammetry and laser scanning. J Cult Herit 8(4):423–427

Chapter 6

Development of methodology for quantitative landslide risk assessment—Example Göta river valley

Yvonne Andersson-Sköld[1,2,3*], Stefan Falemo[1,4], Marius Tremblay[1,5]

[1]Swedish Geotechnical Institute (SGI), Göteborg, Sweden

[2]Department of Earth Sciences, University of Gothenburg, Göteborg, Sweden

[3]COWI AB, Göteborg, Sweden

[4]ÅF Infrastructure, Göteborg, Sweden

[5]Tyréns, Geotechnical Engineering, Göteborg, Sweden

ABSTRACT

Effective landslide risk management requires knowledge of the landslide risks. This paper presents a risk assessment methodology for semiregional scale. The landslide probability is assessed taking into account expected climatechange in the case study area (the Göta river valley). Climate change is expected to result in increased erosion and water fluctuations. There are large areas with marine clays, often quick clay, in the area and the landslide process can be rapid with extensive damages and casualties. The consequence methodology includes a wide range of consequences assessed by monetary valuation. The consequences and the landslide probability are combined as pairs of values in a risk matrix and the risk is also presented on a map. The map has been used as discussion and decision bases in the municipalities in the Göta river valley, in the county administration and on governmental level to estimate the needs of risk mitigation and to make priorities.

KEYWORDS

Landslide; Risk Assessment; Methodology; Climate Change

1. INTRODUCTION

Climate change is expected to cause severe impacts on the natural and built environment. In addition to the temperature increase per se, climate change is expected to profoundly influence the hydrology, e.g. [1]. In Norway, Sweden and Northern Europe, the annual precipitation is expected to increase by up to 30% in the period 2071- 2100 compared to the reference period 1961-1990 [2-4]. The river runoff is in general expected to be higher in winter, followed by a less pronounced snowmelt peak and lower summer flows [4]. The summers will be dryer but, regardless of season, the predictions indicate an increase of intensive rainfall and extreme flows [3,4]. The changes will lead to increased erosion rate and also other impacts that will affect the slope stability such as water level changes and changes in vegetation, e.g. [1,5]. One of the conclusions in the Swedish climate and vulnerability inquiry [6] was that the increased risks related to erosion and landslides will increase in the coming decades to such an extent that adaptation measures are needed to be taken already today.

Land slide events, such as other natural hazard events, are not disasters per se but the consequences may besevere and even disastrous [7-13]. In technical risk analysis, the landslide risk is therefore referred as a function of the probability of a landslide event and the consequences thereof and the risk can be mitigated by either reducing the probability, the consequences of an event or both, e.g. [8,14-17].

For a well-functioning risk mitigation strategy, the risk mitigation requirement needs to be determined. The risk mitigation requirement is a function of the relation between the results of a risk assessment and the accepted level of risk, e.g. [8,17,18]. The risk assessment is done by assessing the probabilities of the unwanted events, e.g. landslides, and assessing the consequences of each event.

The causes of landslides are site specific and complex. The cause can be both primary (caused by factors that are long lasting and inherent in the constituent rock and the soil) and caused by triggers (factors that are varying or very short lived), e.g. [13]. Due to the wide range of landslide processes, the annual probability and the extension of a potential landslide are difficult to assess and the environmental and societal context is crucial for determining the specific effects on the vulnerability as pointed out by Andersson-Berry & King [19] and Roberts et al. [17]. The examples of quantitative assessments of landslide probability reported in the literature have been sited specific and based on statistical information combined with ground properties and parameter uncertainty analysis [11,20-25].

The consequence is a function of the amount of elements at risk (the numbers exposed) and the vulnerability of the affected element or system at risk [22]. Vulnerability is the degree to which a system is susceptible to, or unable to cope with the situation. It can be defined as a function of the character, magnitude and rate of the event to which a system is exposed, its sensitivity, and its adaptive capacity [5,26], or in other words the exposure to the event (perturbation and stresses) combined with the sensitivity and resilience and adaptive capacity of the system [13,14,16,26-28]. Even though the landslide probability is complex to determine the vulnerability, it is even more complex that there is no generic method for assessing the sensitivity, resilience and adaptive capacity for all types of elements at risk. For some impacts, quantitative assessment is possible while for others only qualitative assessments are preferable [8].

Landslide risk (R) is, as described above, referred to as a function of probability and the consequence of a landslide. Recently it often also includes the costs related to the event, i.e. $R = H*V*E$ where, H is the land hazard (landslide) probability, V is the physical vulnerability in case of an event and E is the cost of the particular elements at risk, e.g. [23]. To involve the cost in the risk, assessment function is relevant when assessing the costs and benefits of risk mitigation. As pointed out by van Westen et al. [29] and Crovelli and Coe [23], although the risk equation is simple, it is difficult to be applied in practise due to difficulties in determining H, V and E. Yet, the ability to forecast losses is critical for effective landslide risk management.

Despite the difficulties, cost benefit analyses (CBA) have recently been performed on numerous risk management options, i.e. physical measures [30], landslide susceptibility mapping [20], warning systems and physical measures, the efficiency of incorporating risk mitigation in local comprehensive plans [9] and risk reduction programs [31]. Crovelli and Coe [23] estimated the number of future landslides for a given time period and the economic losses from those landslides, based on historical landslide costs (losses) to assess the expected costs in the San Francisco bay region. The results can be, and were, applied for future prognosis but also to relate the costs to other regions. Similar future cost estimates have been done as presented in [10,20,22,32-36]. In such analyses, the consequence assessments are generally based on loss of life and material (buildings) including reconstruction and renovation costs. In some assessments, losses of income due to delays or closures are included [37]. In general, sensitivity and time-dependent exposure are not taken into account. In some countries, like Sweden, there are not enough data that can be applied for statistically based landslide prognoses [38]. Instead the probability needs to be assessed based on field measurements of the current situation combined with slope stability calculations. On the other hand, Sweden has a wealth of data when it comes to consequence analysis.

The aim of this paper is to present a quantitative landslide risk assessment methodology developed for semiregional scale such as a large run-off area and taking into account a wide range of consequences. The method has been applied for assessing the risk, in the Göta river valley, a large river system in one of Sweden's most landslide-prominent areas [39,40]. The work was initiated by the Swedish Government in response to the results of the Swedish climate and vulnerability inquiry [6]. The risk assessment is in Swedish referred to as Göta älvutredningen [41].

2. METHOD

2.1. The Study Area

The Göta river valley runs from Lake Vänern in thenorth to Göteborg in the south (Figure 1). The valley is one of the most frequent landslide valleys in Swede [39,40] with a number of landslides occurring every year. In general the landslides are fairly small, shallow and caused by erosion. Larger landslides have also occurred in the river valley resulting in human casualties and extensive property damage [39].

Geologically the prerequisites for landslides formed during and following the latest glaciation period, whendeep layers of clay formed in the river valley which wassubmerged in the sea during this period. Since the material was mainly deposited in a marine environment, quick clay is common in the area. Quick clay is a soil with high water content and weak bindings between the clay particles. Vibrations, or a small initial landslide, can cause a quick clay layer to collapse and liquefy, resulting in a large rapid landslide [42]. The river flow is regulated by the power dam in Lilla Edet and the so called Vänern regulation, i.e. the regulation of the power dam tapping from Lake Vänern, which has been applied since 1937. Already in 1963 Sundborg and Norrman mentioned that the regulation caused increased erosion and frequency of river beachfront landslides [41,43]. Climate change is expected to increase the frequency of extreme flows (high and low) from Lake Vänern further contributing to increased erosion [44,45].

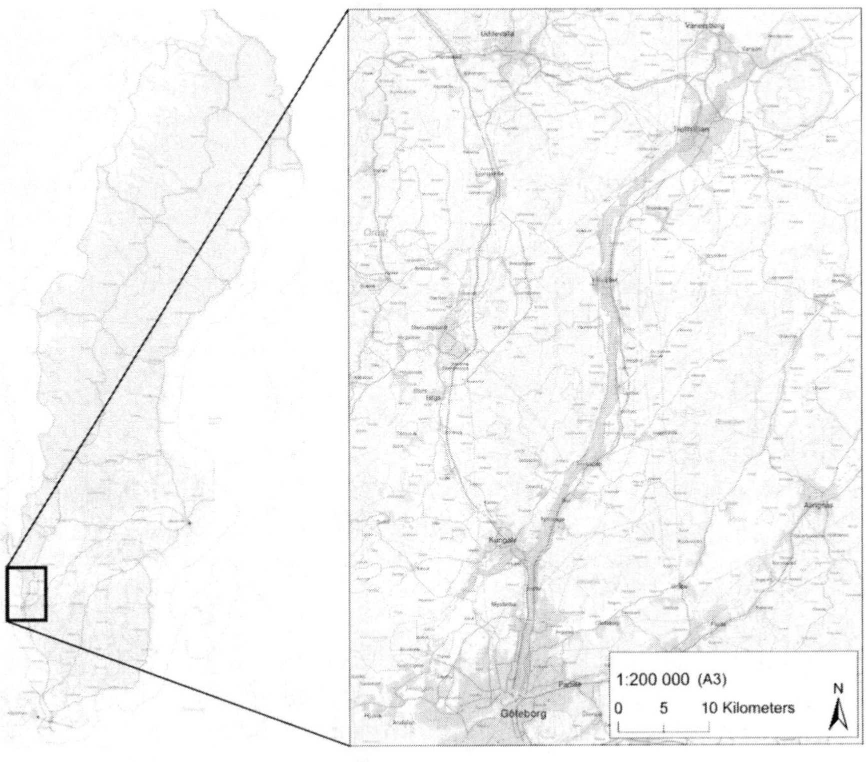

Figure 1. The Göta river valley (study area marked in grey). © SGI, Lantmäteriet, Geodatasamverkan.

The Göta river valley has a long history of anthropogenic activities such as settlements, shipping, harbours, industry, contaminated soil and infrastructure including large roads and railroads, e.g. [46]. Consequently the consequences of a large landslide can be severe.

2.2. Methodology for Quantitative Landslide Risk Assessment

2.2.1. Landslide Probability

The landslide probability was assessed based on slopestability factor estimates along the investigated area. The slope stability was calculated applying an overall security profile in accordance with the national code for practice instructions [47] by Slope/W 2007:7:13 (or later versions) method Morgenstern-Prices with "optimized" sliding surface, or the software GeoSuiteStabilitet 4 (or later versions) method Beast 2003. In all calculations hard crust and half way water filled cracks was assumed. Road and railroad loads were based on TK Geo [48] and building loads were assumed 10 kPa per estate floor. The calculations were done for both drained and undrained analyses, and the lowest resulting safety factor was used. Water levels were set as the lowest allowed level based on tapping and sinking limits of the river system and the average low sea water level as described in more detail in [49]. The slope stability calculations were based on the following information:

- Compilation on geology and stratigraphy in the valley based on recent investigations [50-53].

- GIS based terrain model in SWEREF99TM, RH 2000 [54] based on recent air and sea based laser scanning by Vattenfall ABin 2007, complemented by multibeam echosounding of the river bed within this investigation [55].

- Geotechnical field measurements for in total 240 sections perpendicular to the river. Along each section up to 6 bore holes have been investigated regarding: Total pressure soundings (664), CPT (804), soil-rock soundings (12), impact sounding (21); remoulded sample (333); vane shear testing (379); piston sampling (291); pore water pressure (557 in 70 stations). In total ca 2500 geotechnical analyses were done, and were complemented with information from previous geotechnical investigations in the area.

- Laboratory investigations of field samples from the investigated sections were done in accordance with SIS-CEN ISO/TS 17892-nr including earth measurements and type of soil estimates (3900), bulk density (3600), water content (3500), cone liquid limit (3100), cone tests undisturbed soils (3000), remoulded soils (2200), CRS (755), direct shear tests (106) and triaxial tests (7).

- Current ground water conditions based on field measurements (34 stations) combined with a compilation of previous investigations [56].

- Current erosion in the river was assessed based on investigations in the area [49].

Based on the information described above the landslide probability factor, P_f, was calculated. $P_f$was calculated for in total 80 sections in 39 different type areas (based on topography and geological conditions including geometry, undrained shear strength and soil density). The specific conditions for each type area were based on a previous classification by Millet [57]. The landslideprobability factor was calculated by First Order Reliability Method (FORM) [58]. The marginal to slope failure, the so called safety marginal, M, which is the difference between the load capacity and the load, is central in this method. If M > 0 the load capacity is larger than the load effect and for M < 0 the load capacity is less than the load effect. The landslide probability factor is the probability for slope failure, Equation (1):

$$P_f = p\ (M < 0) \tag{1}$$

The safety marginal, as applied here, was calculated as a function of the safety factor, F, Equation (2):

$$M = \ln F = \ln\left(\eta_{model}\, Nc/P_d\right) \tag{2}$$

F was calculated in accordance with Janbu [59] based on the model error, η_{model}, average shear strength along the sliding surface, c, the stabilitynumber, N, and the driving force pressure, P_d. The model error is applied to compensate for the deviation caused by the idealization of the modeled slope [59]. The stability numbers were taken from Janbu [59] and Alén [60]. The driving force pressure (P_d) was calculated by Equation (3):

$$P_d = \gamma \cdot H + q_e - \gamma_w \cdot H_w \,. \tag{3}$$

where γ is the average pressure caused by the soil layers above the slope foot, H is the vertical distance between the slope crest and foot, q_e is the equivalent surface load behind the slope top, γ_w is the pressure caused by the water in the water course and Hw is the distance between the water level in the water course and the slope foot. Including Equation (2) into Equation (1) results in Equation (4):

$$\begin{aligned} P_f &= p(\ln F < 0) \\ &= p\left[\left(\ln \eta \text{model} + \ln N + \ln c - \ln P_d\right) < 0\right]. \end{aligned} \tag{4}$$

Which, when taking into account the variation of data information, can be expressed as Equation (5):

$$P_f = \Phi(-\beta) \tag{5}$$

where Φ is the standardized normal distribution and β is the safety index which is calculated based on the average safety factor (μF) and the safety factor variation coefficient (V_F) in accordance with Equation (6):

$$\beta = \mu \ln F / \sigma \ln F \approx \ln(\mu F)/V_F \,, \tag{6}$$

where $V_F \approx$ the standard deviation for the safety marginal, σlnF.

The results from the calculations are the landslide probability factor P_f and the safety factor, F. These parameters were used to express the landslide probability in five classes, S1 - S5, in accordance with previous Swedish classification system and previous risk classification system for the Göta älv river system [61] synchronised with Eurocode, i.e. S3 corresponds to Eurocode's lowest accepted security level, as described in detail in Göta älvutredningen [41]. The resulting classification system is presented in Figure 2.

The probability factors and safety factors from the type sections were then applied on the additionally investigated sections, ca 200, to estimate the landslide probability class for each of those sections based on the estimated slope stability.

To include the potential extension of a landslide, thequick clay distribution in the investigated area was identified by total pressure and CPT as described by Möller and Bergdahl [62] and recently applied by Rankka et al. (2004). The Swedish definition of quick clay [42, 63], i.e. sensitivity (St) > 50 and remoulded shear strength (τR) < 0.4 KPa, was applied.

2.2.2. Consequences

The development of a method for consequence classification has previously been described [46]. The starting point of the consequence analysis was landslide consequence identification, followed by an inventory based on interviews, previous investigations and events. The inventory also included GIS based data on vulnerable elements including: Human life and personal injuries; Property; Potential income losses (industry and other commercial enterprises); Environmentally hazardous activities, i.e. elements under the emission register EMIR and Seveso classified elements; Contaminated land areas; Infrastructure of national importance: roads, railroads, seatraffic, heating, electricity, communication (tele, IT) and water supplies and sewage systems; Natural and cultural environments. The results of the inventories of each sec tor were collected in a GIS raster layer with resolution 100 × 100 m covering the area under investigation [46].

As the consequence in case of an event do not only depend on the number of elements that can be exposed but also the exposure at the time and the sensitivity and adaptive capacity of the system exposed, exposure factors and vulnerability factors were estimated. The exposure factor varies from 0 (no exposure) and 1 (full exposure). The landslide exposure factor for stationary objects such as buildings, roads and railroads, contaminated land and cultural artefacts was set to 1. The exposure factor for human life depend on how people divide their time between work, home, school, etc. at different physical addresses. Human exposure was estimated by applying the results from a time usage investigation for year 2000/01 [64] resulting in an annual average exposure factor of 0.69 for people in their homes. Exposure factors at work and in schools were estimated to an annual average of 0.24 and 0.14 respectively [46].

The vulnerability factor was applied to describe the combined effects of the sensitivity and adaptive capacity. The vulnerability factor for a property can be described as the degree of loss or of damages, and for people the mortality (probability of death) caused by the landslide. The vulnerability factor ranges from 0 (no loss) to 1 (total loss) for primary landslide impacts. The vulnerability factor for human life was determined from previous landslides in Sweden and Norway, i.e. the Swedish Natural Hazards Information System [65] and the Norwegian database Skrednett [66] and was expressed as a normal distribution with $\mu = 0.16$ and $\sigma = 0.09$ 0.16 ± 0.09 [46]. The vulnerability related to a landslide on contaminated land could, for example, be estimated based on the changed risks for human or environment related to the changed exposure of contaminants in the soil both in a short and long time perspective. There is, however, too scarce information for such an estimate. Therefore an alternative approach was applied, in which the impacts were valued using the extra costs related to activities to make the landslide area safe to work on after a landslide event. Similarly, flat values of potential extra costs wereapplied for environmentally hazardous activities. For roads, railroads and shipping the impacts on delays, traffic closure and changes of routes were included, as well as reconstruction costs. Costs related to traffic delays and impacts from redirecting road and railroad traffic were estimated through simulations applying different traffic models, i.e. Sampers[1] and Eva[2], and the results were implemented in the model applying flat rate values. The vulnerability factor, for a building or any other stationary object, was set to 1 [46].

The methods for monetary valuation are summarized in Table 1. Details can be found in Andersson-Sköld and Falemo [46]. Once the potential consequence had been estimated based on the inventory, exposure and vulnerability factors, i.e. Consequence = Inventory*Exposure factor*Vulnerability factor*Monetary value, the consequences were calculated in monetary terms foreach sector and 100 × 100 m grid square GIS of the investigated area. The raster layers were summed up in a GIS raster calculation resulting in total estimated monetaryvalue of the consequences in each grid square. The raster cells were then summed up in rows perpendicular to the river as the basis for creating consequence class-isolines as shown in Figure 3. Five consequence classes were used to map the consequences, K1 - K5 (Table 2).

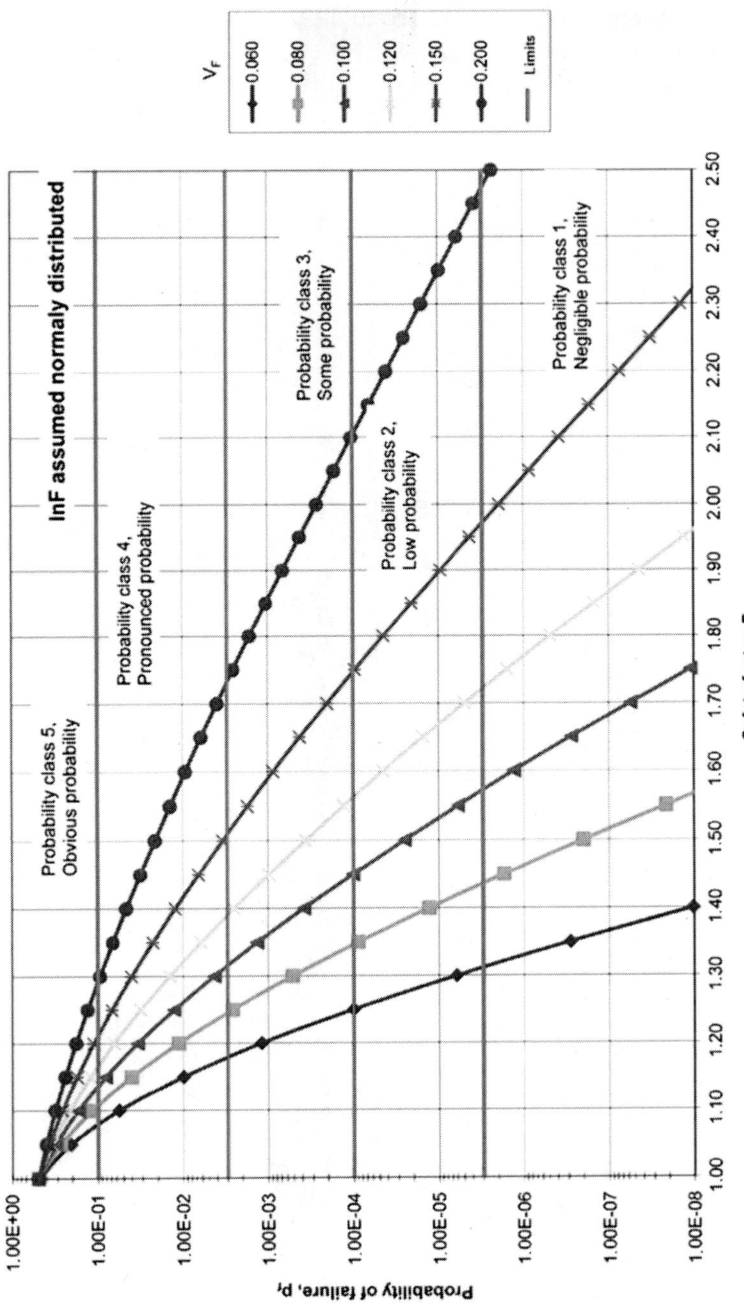

Figure 2. Correlation between probability failure, P_f, and average safety factor, F, for different safety factor variation coefficient, V_F values and the five probability classes, S1 - S5 (modified from Berggren et al. [58]).

Table 1. Methods used for socio-economic valuation of potential landslide consequences [46].

Social sector (element at risk)	Valuation method
Human life	Value of a statistical life
Property	Market value
Production losses (industry/commerce)	Average salary per capita Sweden
Environmentally hazardous activities	Extra security and clean-up costs (function of Seveso & EMIR classification)
Contaminated sites	Expected increased cost for remediation and securing the area (function of MIFO classification)
Roads, Railroads, Sea-traffic	Reconstruction cost + value of traffic delays (including increased accidents and maintenance)
Heating and electricity	Reconstruction costs
Water/sewage system	Reconstruction costs
Communication (telephone, IT)	Reconstruction costs
Natural and cultural environments	Only qualitatively

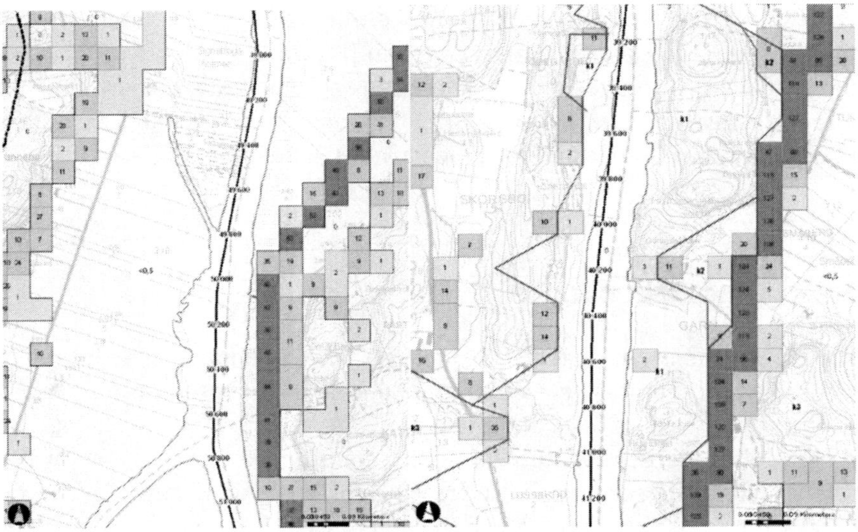

Figure 3. Examples of a resulting consequence map (left) and consequence class iso-lines (right, red lines). The consequences are shown as five classes and the total monetary value within each grid cell: grey (EUR65 million). © SGI, Lantmäteriet, Geodatasamverkan.

As the monetary valuation is site/context specific the values represented within each of the classes need to be decided among stakeholders/experts with knowledge about the specific investigated area. In this investigation the consequence class values were based on an existing qualitatively based, consequence classification system previously applied in the area [39,61,67]. The resulting landslide consequence map, displaying the resulting consequence classes, illustrates the estimated loss should that entire cell be directly affected by a landslide [46].

2.2.3. Risk Classes and Risk Mapping

The consequence map was then over-layered with the probability map and combined into risk classes. A risk matrix was used to classify consequences and probabilities applying the five classes of each, i.e. S1 - S5 and K1 - K5 as summarised in Table 2.

The risk is presented as pair of values, i.e. the failure probability class and the consequence class, in a matrix system (Figure 4). In order to facilitate the readability into a resulting risk map, the resulting risk was divided into three classes, i.e. low risk, medium and uncertain risk, and high risk, coloured yellow, orange and red respectively (Figure 4).

2.2.4. Time Dependence

The slope stability changes with time due to continuous change in the landslide prerequisites (e.g. erosion and changes in precipitation which alter the river surface level and the pore water pressure). These changes in prerequisites were estimated through scenario based calculations of the safety factor. The scenarios included changes in erosion, water levels and pore-water pressure. Future groundwater scenarios were achieved by combining the current groundwater situation with future climate parameters [68]. The surface water, i.e. the river water levels and river water flow for future conditions were based on scenario calculations by SMHI [45] including the expected tapping from Lake Vänern to Göta River [44]. Erosion changes were estimated based on available erosion measurements; water flow scenarios under conditions where current tapping regulation is applied; and for conditions with higher tapping volumes, by applying expert judgments by hydrologists, experts in sedimentology and geotechnical experts. The estimates of the erosion are, consequently, very uncertain and include rough assumptions and rough parameterization.

The results from the erosion calculations shall therefore only be used as indicators of potential magnitudes of changes. In reality the landslide prerequisites may change also due to changes in land use or other man made load changes, such changes have not been taken into account. When assessing the risk over a longer period, such as 100 years, in principle not only the landslide probability but also the time dependence of the consequences shall be included. There are several uncertainties in the changes related to the consequences, including both the social development in the area where the uncertainties of the demographic changes, industrial development and socio-economic development are large as are the variations within the investigated region, as are the uncertainties and variations of the current and forthcoming activities and ambition levels regarding physical planning to mitigate the risks etc. Spatial planning in the time perspective of 100 years is expressed in visions rather than in comprehensive plans. Further, the consequence costs estimated in this investigation are not complete due to lack of available knowledge and information, such as costs and vulnerability factors related to human injuries and general knowledge on landslide impacts on the natural environment [46]. Due to the large uncertainties, the time dependence, and consequently any discount rates of consequence costs have not been estimated in this investigation.

Table 2. Landslide probability and consequence class intervals.

Probability/consequence class	1	2	3	4	5
Failure probability	<3.0E−4	3.0E−6 - 1.0E−4	1.0E−4 - 3.0E−3	3.0E−3 - 1.0E−1	>1.0E−1
Socio-economic cost (MSEK)*	<6	6 - 35	35 - 150	150 - 650	>650

*The values within each of the classes need to be decided among the stakeholders/experts for each site specific area.

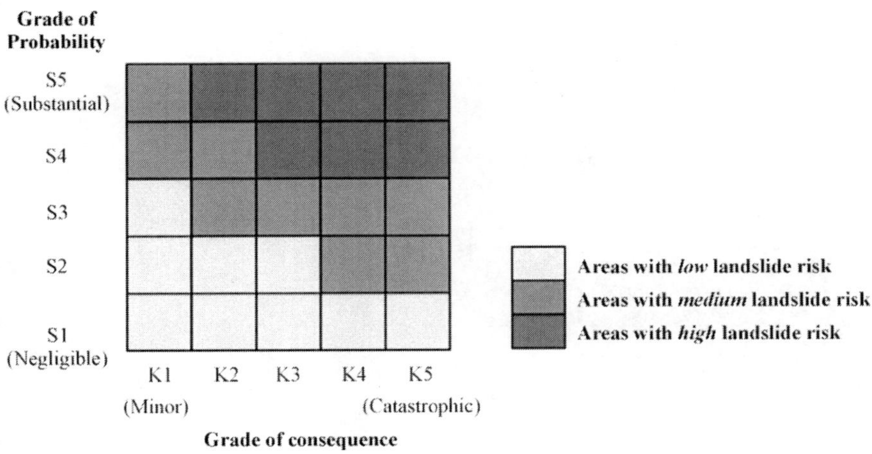

Figure 4. Risk matrix based on landslide probability and consequence classes.

3. RESULTS

The major result of the work described here is the development of a methodology that can be applied for landslide risk assessment and mapping anywhere where sufficient GIS data is available. The major steps involved in the methodology are described in Figure 5.

The methodology presented in Figure 5 is generic. For each of the individual steps, e.g. in the landslide probability assessment, several methods can be applied. The basis for the assessment can be based on available information or on new inventories and field measurements. In this paper the results from applying the above described methods in the Göta river valley are presented. The resulting risk map is shown in Figure 6. The map is meant to be used as a basis for risk management in the area on local to regional scale. The consequences have a resolution of one hectare and the slope stability analyses were done for sections spaced between 400 and 1500 m, and the map accordingly has to be used for such large scale risk management. For example the high risk class areas (red) need to be investigated in more detail prior to finalizing strategies of physical risk mitigation actions in specific areas. Medium risk class areas (orange) also need to be investigated further. The inclusion of the pair of values can be used to identify whether the more detailed investigations shall be focused on the elements at risk, the slope stability assessment, or both.

Landslide probability • Compile available information (geotechnical, soil type, hydrological, hydraulic, topography/geometry) ○ Complement with field investigations when needed • Calculate the slope stability • Calculate the landslide probability factor, P, the safety factor, F, for type sections • Apply the resulting probability and safety factors to express the landslide probability in five classes • Apply the probability factors and safety factors from the type sections on the remaining sections to estimate the landslide probability class for each of those based on the available information on slope stability • Compile the results in a GIS raster (the resolution depends on the needs for the actual investigation).	**Risk** • Define risk classes (low, medium, high) based on the consequence and risk classes • Combine in GIS the resulting landslide probability classes with the resulting consequence calluses into the resulting risk classes in the investigated area (See **Figure 4**).
Consequence analysis • Identify potential consequences in case of landslide event • Make an inventory of elements at risk (the resolution depends on the needs for the actual investigation) • Estimate the exposure factors (or apply already available ones such as the ones presented in [46]. • Decide on which method to apply for monetary evaluation of each of the sectors/type of elements at risk (methods are summarized in **Table 1** and described in more detail in [46]. • Apply GIS to calculate the monetary value for each of the sectors/type of elements at risk within the investigated area (the resolution depends on the needs for the actual investigation) • Apply GIS to calculate the total monetary value in each grid cell within the investigated area • Apply the results to define consequence classes to be applied in the GIS presentation	

Figure 5. Stepwise model for GIS based risk map on semi regional scale (river valley scale).

3.1. Climate Change

The map does not only include information on the current situation but also on expected impacts of climate change. The main impact is due to changes in erosion and thereby the geometric changes of the slopes in and above the river shore line. The erosion depends on the water flow and the water levels, which in turn depend on the maximum tapping and its duration. In the water flow and erosion simulations different future tapping scenarios were included. The maximum tapping is currently regulated to 1030 m^3/s, while in the future either the duration of such tapping will need to increase or the regulation needs to be changed to allow higher maximum tapping. The maximum tapping and potential duration for different potential tapping scenarios is presented in Table 3.

The erosion under current conditions and regulations is 0.05 m per year in most parts of the river, with a few exceptions with up to 0.15 m per year.

The erosion under current regulation but in a future climate is expected to be of the same annual mean as today, which is 0.4 to 0.5 m in the northern parts of the river and 1.0 to 1.5 m in the other parts over the next 90 years [49]. For the tapping scenarios 1 - 3, shown in Table 3, the erosion will increase, the higher the water flow the higher the erosion. As the most likely tapping will be a mix of the different scenarios the average increase in erosion for the tapping alternatives 1 - 3 is 0.8 to 2 m in the northern parts of the river and up to be- tween 2 - 3 meters in the southern parts of the river the nearest 90 years. The details regarding the erosion estimates are further described in Rydell et al. [49].

The impact of erosion on slope stability is estimated to be marginal for areas where the erosion is less than 1m in 100 years. In areas where the erosion is expected to be 1.5 m the slope is expected to move 4 m inland compared to current situation and the safety factor will be reduced by 4%. For the areas and tapping scenarios where the erosion is simulated to be 2 - 3 meters in 100 years, the safety factor will be reduced to 14% on average. In Figure 7 all sections where the slope stability would be affected by 1 m and 3 m erosion are presented along with the impact on the stability class.

As can be seen from Figure 7 the impact of erosion is more pronounced in areas where there currently is a low landslide probability compared to where it is high and in general the impact is higher for the higher (3 m) erosion. The potential impact of erosion in the investigated area is shown on the risk map by a raster marking when there may be potential medium or large impacts on the slope stability the nearest 90 years due to climate change and related needs of changed tapping regulations (Figure 6).

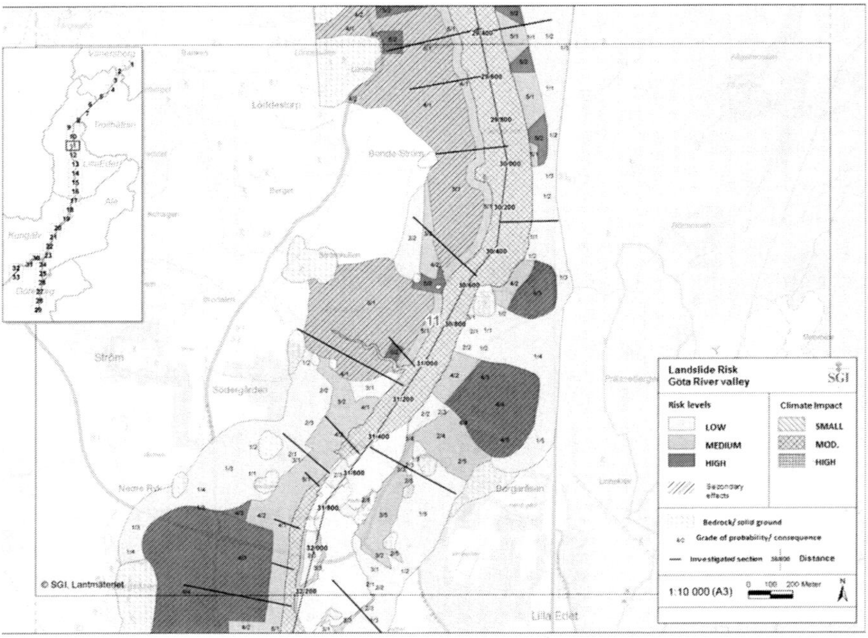

Figure 6. Example of risk map (from GÄU, 2013, part 3 [41]). © SGI, Lantmäteriet, Geodatasamverkan.

Table 3. Maximum tapping and duration under current conditions3 and simulated for different tapping scenarios [45].

Scenario	Flow rate (m^3/s) (max)	Duration (month)	Flow rate (m^3/s)	Duration (month)	Flow rate (m^3/s) (low)	Duration (month)
0	1030	1	550		170	2
1			780	12		
2	1030	6	780	6		
3	1250	4	780	6	550	2

3.2. Cost Assessment

The result from the risk map was used to estimate thevalue that would be affected in case of a landslide event in the entire area of the investigation (excluding rockyground). The total value was estimated to around EUR17 billion (150 BSEK). The corresponding value within high risk class areas (red in Figures 4 and 6) were calculated to ca EUR 900 million (8 BSEK) under current conditions. The high risk classed areas (red) will expand due to the shift to higher landslide probability classes in many parts of the investigated area as a consequence of climate change causing higher water flows and thereby increased erosion.

3.3. Mitigation Measures

In areas with high risk (red), risk mitigation measures were suggested by expertise within SGI and a rough cost assessment of the suggested measures was done based on current market prices and experience. The mitigation measures ranged from erosion protection, excavations and embankment to increased maintenance and monitoring and more detailed geotechnical investigations [41]. The estimated cost for suggested measures to decrease the landslide probability by 20% (compared to current landslide probability) in the high risk class areas (red) was EUR550 - 700 million (5 - 6 BSEK) with operational costs around EUR10 million (900 MSEK) per year.

The cost hence is less, but in the same range, as the protected value in the protected areas. The difference would be higher if all damages such as personal injuries, personal and municipal administration costs, affection values, cost of stress and sorrow etc. had been accounted for.

3.4. Communication with Stakeholders

The resulting risk map was communicated with the key stake holders in the region such as municipalities along the river, the regional road and railroad administration, the county administration board, the waterpower industry and the Swedish maritime administration and on national level including for example Swedish Civil Contingencies Agency, the Ministry of the Environment and the Ministry of Health and Social Affairs. Stakeholder communication was initiated by the start of the project and with occasional communication through the course of the project. By the end of the project, the final results were presented and discussed in stakeholder workshops and seminars. The final maps were made available to stakeholders and the public in March 2012. SGI experts had in advance booked meetings with all municipalities along the river to discuss the impacts of the results within each municipality. The maps have been applied as discussion and decision bases in the municipalities, on county and on governmental level to estimate the needs of risk mitigation and to make priorities.

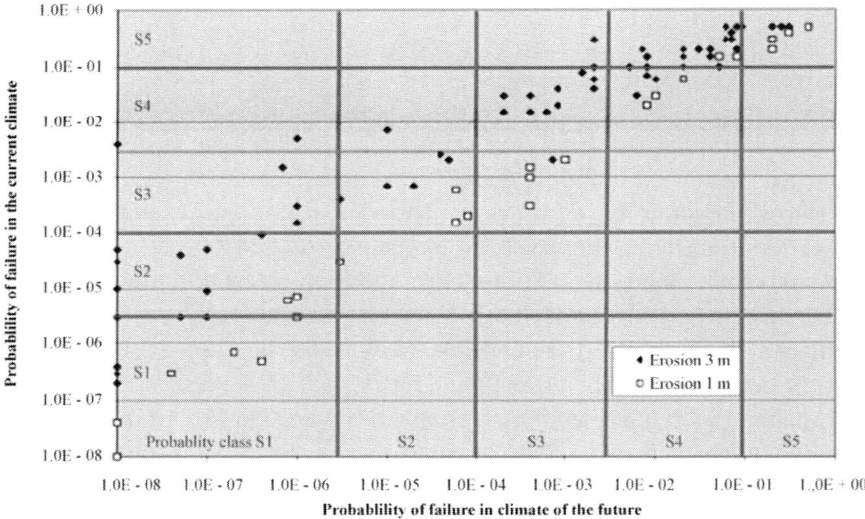

Figure 7. Impact on slope stability class due to increased erosion compared to current conditions.

4. DISCUSSION

The main result of the project presented here was a step wise method for landslide risk mapping including landslide probability assessment and monetary valuation of the potential consequences. In the landslide probability assessment applied in the Göta river valley, the upto-now most thorough field investigation of the area was combined with traditional slope stability analyses, model estimates of current and future erosion and water flows and levels, quick clay mapping and quick modelling to assess the current and future stability and the potential extension of a landslide. The consequence analysis covered more elements at risk than normally included in landslide risk assessments, taking into account both direct and indirect consequences. The reason for such a thorough analysis was the need of more detailed understanding of the risks in the area both under current conditions and taking into account climate change. For efficient mitigation strategies selected areas also need even more detailed investigations and assessments. In such cases the generic method shown in Figure 5 can, and will be used, but some steps already are described well enough for the purpose. We accordingly recommend the stepwise method to be used in an iterative process focusing on the needs at the moment and aiming to find which areas need further and deeper assessment.

Despite being regarded as a detailed and thorough assessment the results are based on many rough estimates and assumptions, and many uncertainties and unknowns still exist. Among the most important uncertainties are the estimates of current and future erosion and the quick clay distribution which are needed to assess the current and future stability and the potential extension of a landslide. For more accurate assessments several knowledge gaps needs to be filled regarding sediment transport and erosion processes for the different soils in the river. The mapping of quick clay is costly and methods and increased knowledge for faster and cheaper mapping is requested for more thorough estimates.

There are also large uncertainties in the values of the consequences and for some consequences, such as the nature environment, not even the magnitude and whether it is positive or negative impact can be estimated [46]. Of special concern are landslides on contaminated land or enterprises that may cause emissions of contaminants to air, soil and water as there is not enough information to describe the short and long terms impacts, and therefore they cannot even be evaluated. In the same way the impact on biodiversity and various biotopes of a landslide may be both positive and negative and both are in general impossible to quantify with current knowledge and thereby not available for monetary valuation. Human injuries are omitted and costs related to increased administration and soft impacts are also omitted.

The resulting map has been communicated and applied in the landslide risk management by stakeholders and municipalities in the investigated area and despite the uncertainties the resulting map is regarded as very useful and important among the stakeholders. The application of GIS, and that the resulting map is available on the web, is also highly appreciated among the stakeholders. This is in agreement with previous results on the perception of risk mapping and the communication with experts on how to interpret their information [38,69]. The map is useful as it indicates landslide prone areas with high population densities and important vulnerable infrastructure areas that shall be further investigated and regarded for further risk reduction activities. In areas identified as high risk areas, the consequence values are higher, but of the same magnitude as the costs estimated for preventive measures in the same areas. This indicates that preventive measures are beneficial, but also that more detailed site specific analyses are needed prior to identifying and implementing the most cost efficient risk reduction strategies. One of the major reasons for the relative low monetary value in high risk areas are previous landslide prevention measures taken. An example is EKA Nobel where in 1994 land reinforcements for EUR8 million (70 MSEK) were undertaken. Another example is active land use planning to reduce the vulnerability in the river valley [38]. This is in agreement with an iterative process, i.e. that the obvious needs can be treated without as thorough investigation as the one presented here, while in cases where the value of expected losses are close to the costs of risk mitigation measures more detailed analyses are needed. The generic step wise method and GIS based information and data is then easy to apply and combine with the information of the risk map for the larger region, thereby providing information that can be used to develop a more holistic risk management approach and strategy in the entire investigated area.

5. CONCLUSIONS

Risk maps are important tools in communication with stakeholders that may be impacted by potential landslides and/or are responsible for landslide risk management and prevention [38,69].

The level of detail of the risk map must depend on the scale investigated, and to achieve the most cost effective information level, a stepwise method is recommended to be used in an iterative process focusing on the needs at the moment and aiming to find which areas need further and deeper assessment. The method presented here (Figure 5) has been found applicable and useful in the preventive work in the investigated area.

There are many uncertainties related to understanding the risks both under current climate conditions and, not least, taking into account climate change.

There is a need for a deeper understanding of the process that may cause serious exposure, and other consequences, to health and environment due to landslides in order to make better quantitative assessments of the consequences.

In general, the understanding of the consequences could be increased if there were more systematically available data with regard to previous events including the number of people exposed and number of injuries in addition to the number of fatalities most commonly presented.

The knowledge on social and individual levels, and the social costs, related to landslides and other severe hazards needs to be better understood.

REFERENCES

1. IPCC (2007) Climate change 2007: Impacts, adaptation and vulnerability. Parry, M.L., Ed., Cambridge University Press, New York.

2. Andréasson, J., Bergström, S., Carlsson, B., Graham, P. and Lindström, G. (2004) Hydrological change: Climate change impact simulations for Sweden. A Journal of the Human Environment, 33, 228- 234.

3. Lawrence, D., Graham, L.P., den Besten, J., Andréasson, J., Bergström, S. and Engen-Skaugen, T. (2012) Climate change impacts and uncertainties in flood risk management: Examples from the North Sea Region. A Report of Working Group 1, SAWA Interreg IVB Project, 2012, Report No. 05.

4. Olsson, J., Yang, W., Graham, L.P., Rosberg, J. and Andréasson, J. (2011) Using an ensemble of climate projections for simulating recent and near-future hydrological change to Lake Vänern in Sweden. Tellus A, 63, 126-137.

5. IPCC (2001) Climate change 2001: The scientific basis. contribution of working group I to the third assessment report of the intergovernmental panel on climate change. Houghton, J.T., Ding, Y.,

Griggs, D.J., Noguer, M., van der Linden, P.J., Dai, X., Maskell, K. and Johnson, C.A., Eds., Cambridge University Press, Cambridge, United Kingdom and New York, 881.

6. SOU 2007:60. Sweden facing climate change: Threats and opportunities. Final Report, Fritze, Stockholm

7. Berke, P.R. (1998) Reducing natural hazard risks through state growth management. Journal of the American Planning Association, 64, 76-87.

8. Dai, F.C., Lee, C.F. and Ngai, Y.Y. (2002) Landslide risk assessment and management: An overview. Engineering Geology, 64, 65-87.

9. Srivastava, R. and Laurian, L. (2006) Natural hazard mitigation in local comprehensive plans: The case of flood, ildfire and drought planning in Arizona. Disaster Prevention and Management, 15, 461-483.

10. Sterlacchini, S., Frigerio, P. and Giacomelli, M. (2007) Brambilla, landslide risk analysis: A multi-disciplinary methodological approach. Natural Hazards and Earth System Sciences, 7, 657-675.

11. Friele, P., Jakob, M. and Clague, J. (2008) Hazard and risk from large landslides from Mount Meager volcano, British Columbia, Canada. Georisk, 2, 48-64.

12. Holcombe, E. and Anderson, M. (2010) Tackling landslide risk: Helping land use policy to reflect unplanned housing realities in the Eastern Caribbean. Land Use Policy, 27, 798-800.

13. Singh, A.K. (2010) Landslide management: Concept and philosophy. Disaster Prevention and Management, 19, 119-134.

14. Brooks, N. (2003) Vulnerability, risk and adaptation: A conceptual framework, tyndall centre for climate change research. University of East Anglia, Norwich.

15. Sarewitz, D., Pielke Jr., R. and Keykhah, M. (2003) Vulnerability and risk: Some thoughts from a political and policy perspective. Risk Analysis, 23, 805-810.

16. Füssel, H.M. and Klein, R.J. (2006) Climate change vulnerability assessments: An evolution of conceptual thinking. Climate Change, 75, 301-329.

17. Roberts, N.J., Nadim, F. and Kalsnes, B. (2009) Quantification of vulnerability to natural hazards. Georisk, 3, 164-173.

18. King, D. (2008) Reducing hazard vulnerability through local government engagement and action. Natural Hazards, 47, 497-508.

19. Anderson-Berry, L. and King, D. (2005) Mitigation of the impact of tropical cyclones in Northern Australia through ommunity capacity enhancement. Mitigation and Adaptation Strategies for Global Change, 10, 367-392.

20. Remondo, J., González, A., Díaz de Terán, J.R., Cendrero, A., Fabbri, A. and Chung, C.J.F. (2003) Validationof landslide susceptibility maps; examples and applications from a case study in northern Spain. Natural Hazards, 30, 437-449.

21. Uzielli, M. and Lacasse, S. (2007) Scenario-based probabilistic estimation of direct loss for geohazards. Georisk, 1, 142-154.

22. Cassidy, M.J., Uzielli, M. and Lacasse, S. (2008) Probability risk assessment of landslides: A case study at Finneidfjord. Canadian Geotechnical Journal, 45, 1250- 1267.

23. Crovelli, R.A. and Coe, J.A. (2009) Probabilistic estimation of numbers and costs of future landslides in the San Francisco Bay region. Georisk, 3, 206-223.

24. Ho, K.K.S. and Ko, F.W.Y. (2009) Application of quantified risk analysis in landslide risk management practice: Hong Kong experience. Georisk: Assessment and Management of Risk for Engineered Systems and Geohazards, 3, 134-146.

25. Kanungo, D.P., Arora, M.K., Sarkar, S. and Gupta, R.P. (2009) A fuzzy set based approach for integration of thematic maps for landslide susceptibility zonation. Georisk, 3, 30-43.

26. Rekacewicz, P. (2005) Vital climate change graphics, UNEP/GRID-arendal maps and graphics library.

27. Turner, B.L., Kasperson, R.E., Matsone, P.A., McCarthy, J.J., Corell, R.W., Christensen, L., Eckley, N., Kasperson, J.X., Luerse, A., Martello, M.L., Polsk, C., Pulsipher, A. and Schiller, A. (2003) A framework for vulnerability analysis in sustainability science. Proceedings of the

National Academy of Sciences of the United States od America, 100, 8074-8079.

28. Li, Z., Farrokh, N., Hongwei, H., Uzielli, M. and Lacasse, S. (2010) Quantitative vulnerability estimation for scenariobased landslide hazards. Landslides, 7, 125-134.

29. Van Westen, C.J., Castellanos, E. and Kuriakose, S.L. (2008). Spatial data for landslide susceptibility, hazard, and vulnerability assessment: An overview. Engineering Geology, 102, 112-131.

30. Schuster, R.L. and Highland, L.M. (2007) The third hans cloos lecture. Urban landslides: Socioeconomic impacts and overview of mitigative strategies. Bulletin of Engineering Geology and the Environment, 66, 1-27.

31. Anderson, M.G., Holcombe, E., Esquivel, M., Toro, J. and Ghesquiere, F. (2010) The efficacy of a programme of landslide risk reduction in areas of unplanned housing in the Eastern Caribbean. Environmental Management, 45, 807-821.

32. Plate, E.J. (2007) Flood risk management for setting priorities in decision making, in Extreme Hydrological Events: New Concpepts for Security. In: Vasiliev, O.F., van Gelder, P.H.A.J.M., Plate, E.J. and Bolgov, M.V., Eds., Extreme Hydrological Events: New Concpepts for Security, 78, 21-44.

33. Uzielli, M., Farrokh, N., Lacasse, S. and Kaynia, A.M. (2008) A conceptual framework for quantitative estimation of physical vulnerability to landslides. Engineering Geology, 102, 251-256.

34. Hinkel, J., Nicholls, R.J., Vafeidis, A.T., Tol, R.S.J. and Avagianou, T. (2010) Assessing risk of and adaptation to sea-level rise in the European Union: An application of DIVA. Mitigation and Adaptation Strategies for Global Change, 15, 703-719.

35. Zeng, J., Zhu, Z.Y., Zhang, J.L., Ouyang, T.P., Qiu, S.F., Zou, Y. and Zeng, T. (2012) Social vulnerability assessment of natural hazards on county-scale using high spatial resolution satellite imagery: A case study in the Luogang district of Guangzhou, South China. Environmental Earth Sciences, 65, 173-182.

36. Holcombe, E., Smith, S., Wright, E. and Anderson, M.G. (2012) An integrated approach for evaluating the effectiveness of landslide risk reduction in unplanned communities in the Caribbean. Natural Hazards, 61, 351-385.

37. Eder, S., Poscher, G. and Prager, C. (2006) Risk analyses and risk management—Slope instabilities in Alpine environments. ECI Geohazards Volume P07-Technical, Economical and Social Risk Evaluation.

38. Andersson-Sköld, Y., Bergman, R., Johansson, M., Persson, E. and Nyberg, L. (2013) Landslide risk management: A brief overview and example from Sweden of current situation and climate change. International Journal of Disaster Risk Reduction, 3, 44-61.

39. Hultén, C., Andersson-Sköld, Y., Ottosson, E., Edstam, T. and Johansson, Å. (2007) Case studies of landslide risk due to climate change in Sweden. Proceedings of International Geotechnical Conference on Climate Change and Landslides, Ventnor, 21-24 May 2007, 149-157.

40. Jonsson, A., Hjerpe, M., Andersson-Sköld, Y., Glaas, E., André, K. and Simonsson, L. (2012) Cities' capacity to manage climate vulnerability: Experiences from participatory vulnerability assessments in the lower Göta Älv Catchment, Sweden. Local Environment, 17, 735-750.

41. GÄU (2012) Skredrisker i Göta älvdalen i ett förändrat klimat. Swedish Geotechnical Institute, Slutrapport. Göta älvutredningen 2009-2011.http://www.swedgeo.se/templates/SGIStandardPage____2634.aspx?epslanguage=SV)

42. Andersson-Sköld, Y., Torrance, J.K., Lind, B., Odén, K., Stevens, R.L. and Rankka, K. (2005) Quick clay: A case study of chemical perspective in Southwest Sweden. Engineering Geology, 82, 107-118.

43. Sundborg, Å. and Norrman, J. (1963) Göta älv—Hydrlologi och morfologi med särskild hänsyn till erosionprocesserna. Svergiges Geologiska Undersökning, SGU. Rapport Ca 43.

44. Bergström, S., Andréasson, J., Asp, M., Caldarulo, L., German, J., Lindahl, S., Losjö, K. and Stensen, B. (2010) Fördjupad studie rörande översvämningsriskerna för Vänern—Slutrapport. SMHI Rapport nr 2010-85.

45. Bergström, S., Andréasson, J., Losjö, K., Stensen, B. and Wern, L. (2011) Hydrologiska och meteorologiska förhållanden i Göta älvdalen—Slutrapport. Statens geotekniska institut, SGI. Göta älvutredningen, GÄU. Delrapport 27. Linköping. Även publicerad som SMHI Rapport nr 2010-81.

46. Andersson-Sköld, Y. and Falemo, S. (2013) Ex ante natural hazard consequence and vulnerability analysis for local to regional scale—Method development focusing on landslide events. Manuscript submitted Natural Hazards.

47. Skredkommissionen (1995) Anvisningar för släntstabilitetsutredningar, ingenjörsvetenskapsakademien, IVA. Skredkommissionen. Rapport 3:95, Linköping.

48. Vägverket (2009) TK Geoär en teknisk beskrivning som anger Vägverkets och Banverkets krav och råd vid nybyggnad och förbättring av geokonstruktioner. Vägverket, VV. Publikation 2009:46/Banverket, BV. BVS 1585.001, Borlänge.

49. Rydell, B., Persson, H. and Blied, L. (2011) Klimateffekt på vattennivå, erosion och grundvattenförhållanden i Göta älv. Underlag till styrdokument 99ST003. Göta älvutredningen, GÄU. Statens geotekniska institut, SGI. Varia 624:1, Linköping.

50. Fredén, C. (1984) Beskrivning till jordartskartan Vänersborg SO. Sveriges Geologiska Undersökning, Serie Ae nr 48.

51. Fredén, C. (1986) Beskrivning till jordartskartan Göteborg NO. Sveriges Geologiska Undersökning, Serie Ae nr 40.

52. Klingberg, F. (2011) Bottenförhållanden i Göta älv. Statens geotekniska institut, SGI. Göta älvutredningen, GÄU. Delrapport 6. Även publicerad som SGU-rapport 2010:7, Linköping.

53. Klingberg, F., Påsse, T. and Levander, J. (2006) Bottenförhållanden och geologisk utveckling i Göta älv. Sveriges Geologiska Undersökning, SGU. Rapport K43, Uppsala.

54. Hedvall, T. (2010) Bearbetning av Terrängmodell över Göta och Nordre Älv. Vattenfall Power Consultant AB.

55. Marin, M. (2011) Ytgeologisk undersökning med backscatter-analys för Göta älv och Nordre älv. arin Miljö- analys AB, U304-0909. Statens geotekniska institut, SGI. Göta älvutredningen, GÄU. Delrapport 5 and Delrapport 34, Linköping.

56. Persson, H., Bengtsson, P.E., Lundström, K. and Karlsson, P. (2011) Bedömning av grundvattenförhållanden för slänter längs Göta älv. Allmän vägledning. Statens geotekniska institut, SGI. Göta älvutredningen, GÄU. Delrapport 7, Linköping.

57. Millet, D. (2011) River erosion, landslides and slope development in Göta river—A study based on athymetric data and general limit equilibrium slope stability analysis. Chalmers University of Technology. department of Civil and Environmental Engineering. GeoEngineering. Geology and Geotechnical Engineering. Master thesis 2011:131, Göteborg.

58. Berggren, B., Alén, C., Bengtsson, P.E. and Falemo, S. (2011) Metodbeskrivning sannolikhet för skred: Kvantitativ beräkningsmodell (Methodology for assessment of landslides probability: Quantitative model). Statens Geotekniska Institute, SGI. Report28, Linköping.

59. Janbu, N. (1954) Stability analysis of slopes with dimensionless parameters. Doctoral Thesis. Harvard University, Cambridge.

60. Alén, C. (1998) On probability in geotechnic. Random calculation models exemplified on slope stability analysis and ground-superstructure interaction. Doctoral Thesis, Department of Geotechnical Engineering, Chalmers University of Technology, Göteborg.

61. Alén, C., Bengtsson, P.E., Berggren, B., Johansson, L. and Johansson, Å. (2000) Skredriskanalys i Göta älvdalenMetodbeskrivning (Landslide risk analysis in Göta rivermethodology). Statens Geotekniska Institut, SGI. Report 58, Linköping.

62. Möller, B. and Bergdahl, U. (1982) Estimation of the sensitivity of soft clays from static and weight sounding tests. Proceedings of the European Symposium on Penetration Testing, 2, ESPOT2, Amsterdam, 1, 291-295.

63. Rankka, K., Andersson-Sköld, Y., Hulten, C., Larsson, R., Leroux, V. and Dahlin, T. (2004) Quick clay in Sweden. Swedish Geotechnical Institute, Report 65, Linköping.

64. SCB (2003) Time for daily activities and living conditions (Tid för vardagsliv, Levnadsförhållanden, in Swedish). Statistiska Centralbyrån SCB 99.

65. MSB (2011) National database of natural hazards, Swedish Civil Contingencies Agency. http:/ndb.msb.se

66. NGU och NVE (2010) "Skrednett". Norges vassdragsog energidirektorat. www.skrednett.no

67. Berggren, B., Fallsvik, J., Hintze, S. and Stille, H. (1991) Lerslänters stabilitetsförhållanden. Riskvärdering och eslutsteori. Förslag till metod för riskvärdering. Statens Geotekniska Institut, SGI. Varia 333, Linköping.

68. Blied, L. and Persson, H. (2011) Bedömd förändring av maximala grundvattennivåer i Göta älvdalen till följd av förändrat klimat. Statens Geotekniska Institut. SGI. Göta Älvutredningen, GÄU. Delrapport 9, Linköping.

69. Kjellgren, S. (2013) Exploring local risk managers use of flood hazard maps for risk communication purposes in Baden-Württemberg. Natural Hazards & Earth System Sciences, 13, 1857-1872.

Chapter 7

Landslide susceptibility mapping using multi-criteria evaluation techniques in Chittagong Metropolitan Area, Bangladesh

Bayes Ahmed

Institute for Risk and Disaster Reduction (IRDR), Department of Earth Sciences, University College London (UCL), Gower Street, London, WC1E 6BT, United Kingdom (UK)

ABSTRACT

Landslides are a common hazard in the highly urbanized hilly areas in Chittagong Metropolitan Area (CMA), Bangladesh. The main cause of the landslides is torrential rain in short period of time. This area experiences several landslides each year, resulting in casualties, property damage, and economic loss. Therefore, the primary objective of this research is to produce the Landslide Susceptibility Maps for CMA so that appropriate landslide disaster risk reduction strategies can be developed. In this research, three different Geographic Information System-based Multi-Criteria Decision Analysis methods—the Artificial Hierarchy Process (AHP), Weighted Linear Combination (WLC), and Ordered Weighted Average (OWA)—were applied to scientifically assess the landslide susceptible areas in CMA. Nine different thematic layers or landslide causative factors were considered. Then, seven different landslide susceptible scenarios were generated based on the three weighted overlay techniques. Later, the performances of the methods were validated using the area under the relative operating characteristic curves. The accuracies of the landslide susceptibility maps produced by the AHP, WLC_1, WLC_2, WLC_3, OWA_1, OWA_2, and OWA_3 methods were found as 89.80, 83.90, 91.10, 88.50, 90.40, 95.10, and 87.10 %, respectively. The verification results showed satisfactory agreement between the susceptibility maps produced and the existing data on the 20 historical landslide locations.

KEYWORDS

Landslide Susceptibility Mapping (LSM); Artificial Hierarchy Process (AHP); Weighted Linear Combination (WLC); Ordered Weighted Average (OWA); Relative Operating Characteristic (ROC); Geographic Information System (GIS)

1. INTRODUCTION

Chittagong Metropolitan Area (CMA) is highly vulnerable to landslide hazard, with an increasing trend of frequency and damage. Devastating landslides have hit CMA (Figure1) repeatedly in recent years (Table 1). The major recent landslide events were related to extreme rainfall intensities having short period of time. All the major landslide events occurred as a much higher rainfall amount compared to the monthly average. Moreover, rapid urbanization, increased population density, improper landuse, alterations in the hilly regions by illegally cutting the hills, indiscriminate deforestation, and agricultural practices are aggravating the landslide vulnerability in CMA (Khan et al. 2012).

In addition, there is no strict hill management policy within CMA. This has encouraged many informal settlements along the landslide-prone hill-slopes in Chittagong. These settlements are being considered as illegal by the formal authorities, while the settlers claim themselves as legal occupants. This is how there is acute land tenure conflict among the formal authorities, the settlers, and the local communities over the past few decades. This kind of conflict has also weakened the institutional arrangement for reducing the landslide vulnerability in Chittagong City (Ahammad 2009).

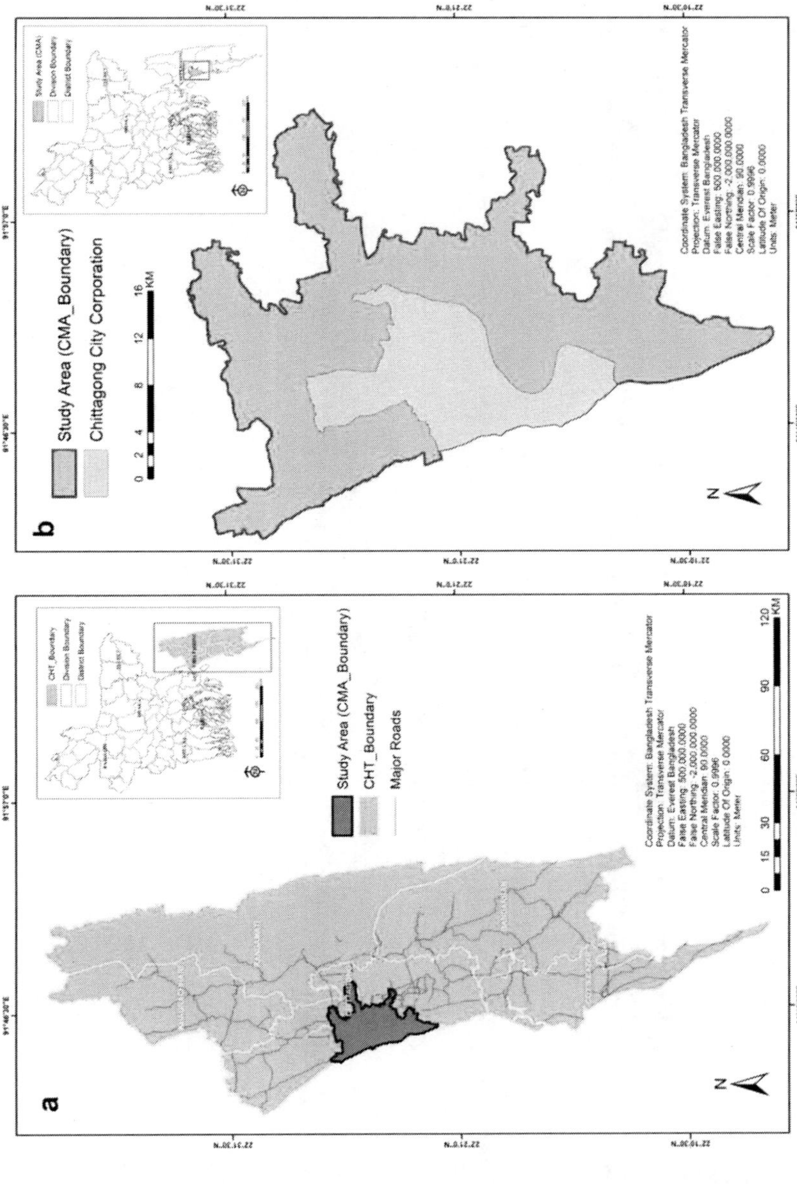

Figure 1. a Location of the study area in Chittagong hill tracts and **b** location of CMA

Table 1. Major landslide events in CMA in recent years*

Date	Location	Rainfall sequence (cumulated rainfall)	Consequences
13 August 1999	Gopaipur, Kotwali Thana, Chittagong	435 mm – 12 days 2 – 13 Aug 1999	10 people killed
24 June 2000	Chittagong University Campus	108 mm – 8 days 17 – 24 June 2000	13 people killed and 20 injured
29 June 2003	Patiya, Chittagong	658 mm – 10 days 20 – 29 June 2003	4 people killed
3 August 2005	Nizam Road Housing Society, Panchlaish area	25 mm – 2 days 2–3 August 2005	2 people killed and 12 injured
11 June 2007	Matijharna Colony,Lalkhan Bazar	610 mm – 8 days 4 – 11 June 2007	128 people killed and 100 injured
10 September 2007	Nabi Nagar, Chittagong	452 mm – 7 days 4 – 10 Sept 2007	2 people killed
18 August 2008	Matijharna, Chittagong	454 mm – 11 days 8 – 18 August 2008	11 people killed and 25 injured
26 June 2012	Lebubagan area and Foys lake surroundings	889 mm – 8 days 19 – 26 June 2012	90 people killed and 150 injured

*Source: Comprehensive Disaster Management Programme-II (CDMP-II), Ministry of Food and Disaster Management, Bangladesh; and Field Survey, (September 2013).

At this drawback, it is therefore essential to determine the landslide prone areas in CMA (Figure 1b) so that appropriate landslide disaster risk reduction strategies can be developed. Producing up-to-date and accurate landslide susceptibility maps can ensure safety to people and property at risk and avoid extensive economic loss (Kavzoglu et al. 2013).

2. LITERARY WORKS ON LANDSLIDE SUSCEPTIBILITY MODELLING

Landslides are one of the most significant natural damaging disasters in hilly environments (Ayala et al. 2006). Social and economic losses due to landslides can be reduced by the means of effective planning and management (Rajakumar et al.2007). Landslide hazard assessment is generally based on the concept that 'the present and the past are keys to the future'. This is why, most landslide hazard analyses take into account an up-to-date landslide inventory that represents the fundamental tool for identifying the hill-slope instability factors in triggering landslides (Lee and Sambath 2006).

Various geo-structural as well as causative-factor based approaches are already available for landslide susceptibility zoning. But Geographic Information System (GIS) modelling of landslide phenomena has taken precedence in recent time. Geospatial technologies like the use of GIS, Global Positioning System (GPS), and Remote Sensing (RS) are useful in the hazard assessment, risk identification, and disaster management for landslides. GPS is a space-based global navigation satellite system which provides the information of position and time anywhere in the world in all weather conditions (Akbar and Ha 2011). Previous studies showed the application of GPS for mapping and identifying landslide zones. GIS is used for data collection, storage, and analysis of processes where geographic information is involved. The use of GIS for landslide mapping is common in various studies. Remote sensing is the science in which information is acquired about the surface of earth without physically being in contact with it. RS is also used for monitoring and mapping of landslides (Akbar and Ha 2011).

Mapping the areas that are susceptible to landslides is essential for proper land use planning and disaster management for a particular locality or region. Throughout the years, different techniques and methods have been developed and applied in the literature for landslide susceptibility mapping.

Landslide susceptibility maps can be produced using both the quantitative or qualitative approach (Park et al. 2013).

Qualitative maps weight each factors affecting the landslides based on the practical experience and expertise of the researcher (Park et al. 2013). Qualitative methods simply portray the hazard zoning in descriptive terms (Guzzetti et al. 1999). But because of the developments in computer programming and geospatial technologies, quantitative techniques have become popular in recent decades. Moreover, it incorporates the causes of landslides (instability factors) and probabilistic methods (Bai et al. 2010).

There are mainly four methods available to map landslide susceptibility, namely landslide inventory based probabilistic, deterministic, heuristic, and statistical techniques (Guzzetti et al. 1999). Landslide inventory-based probabilistic techniques involve the inventory of landslides, construction of databases, geomorphological analysis, and producing the susceptibility maps based on the collected data (Duman et al. 2005). Deterministic techniques (quantitative methods) involve the estimation of quantitative values of stability variables and require the creation of a map that displays the spatial distribution of input data (Godt et al. 2008; Isik Yilmaz 2009).

Heuristic analysis (qualitative method) is based on the intrinsic properties of the geo-morphologists to analyse aerial photographs or to conduct field surveying (Yilmaz and Yildirim 2006). In this kind of analysis, researchers establish the susceptibility and the analyst uses expert knowledge to assign weights to a series of parameters for preparing the qualitative map (Isik Yilmaz 2009). Statistical analysis is used to analyse factors affecting landslides in areas with environmental conditions similar to those where past landslides have been reported (Park et al. 2013). These methods use sample data based on the relationship between the dependent variable (the presence or absence of landslides), and the independent variables (landslides triggering/causative factors). Through these statistical techniques, quantitative predictions are possible to make for areas where there is no landslides and with similar conditions (Isik Yilmaz 2009).

Within these techniques, the probabilistic and statistical methods have been commonly used in recent years. These methods have become more popular, assisted by GIS and RS techniques (Lee and Sambath 2006). The probabilistic (non-deterministic) models like frequency ratio, bivariate analysis, multivariate analysis, and Poisson probability model (Bui et al. 2013) are more frequently used to determine the landslide susceptibility zones (Zêzere et al. 2008). Among the widely used statistical method is the logistic regression (Isik Yilmaz 2009). Many researchers have also used different techniques such as heuristic approach (Karsli et al. 2009) and deterministic models (Avtar et al. 2011).

Moreover, GIS-based Multi Criteria Decision Analysis (GIS-MCDA) provides powerful techniques for the analysis and prediction of landslide hazards. GIS-MCDA belongs to heuristic analysis. These include the Analytic Hierarchy Process (AHP), the Weighted Linear Combination (WLC), the Ordered Weighted Average (OWA), etc. (Feizizadeh and Blaschke 2013). Most recently, new non-parametric techniques like cellular automata, fuzzy-logic, artificial neural networks (Poudyal et al. 2010), support vector machines, and neuro-fuzzy models have also been used for landslide susceptibility modelling (Park et al. 2013).

The primary objective of this paper is to apply GIS-MCDA techniques for the Landslide Susceptibility Mapping (LSM) in CMA, Bangladesh. The reason behind choosing GIS-MCDA is that it is being widely used in LSM in recent years (Feizizadeh and Blaschke2013; Park et al. 2013; Ahammad 2009; Kayastha et al. 2012).

3. STUDY AREA PROFILE

Chittagong is the second-largest and main seaport of Bangladesh. The city is comprised of small hills and narrow valleys, bounded by the *Karnaphuli* River to the south-east, the Bay of Bengal to the west, and *Halda* River to the north-east (Figure 1b). The city has a population of about 5 million and is constantly growing (Community Report, Chittagong District 2012). The study area, CMA, is situated within 22° 14′ and 22° 24′ 30″ north latitude and between 91° 46′ and 91° 53′ east longitude (Figure 1b). The total area of CMA is approximately 775 km^2 (using Bangladesh Transverse Mercator projection). CMA is also known as 'Chittagong Metropolitan Master Plan (CMMP) Surveyed Area'.

CMA is a part of the Chittagong Hill Tracts (CHT) region (Figure 1a). Therefore, it is important to know about the characteristics of CHT to understand the causes and geological reasons of landslides in CMA. CHT is originated as a result of the collision between India and Asia. Later, India broke apart from Australia and started to drift north-easterly. That is the time when the history began for CHT (Chowdhury 2012). The weather of CHT (Figure 1a) is characterised by tropical monsoon climate with mean annual rainfall nearly 2,540 mm in the north-east and 2,540 to 3,810 mm in the south-west. The monsoon season is from June to October, which is warm, cloudy, and wet (Chowdhury 2012). Moreover, due to climate change, CMA is experiencing higher intensity of rainfall in recent years which is making the landslide situation worse (Mangiza 2011). A gradual upward shift in precipitation is noted in the last five decades (1960–2010), with an abrupt fluctuation in the mean annual precipitation levels (Figure 2).

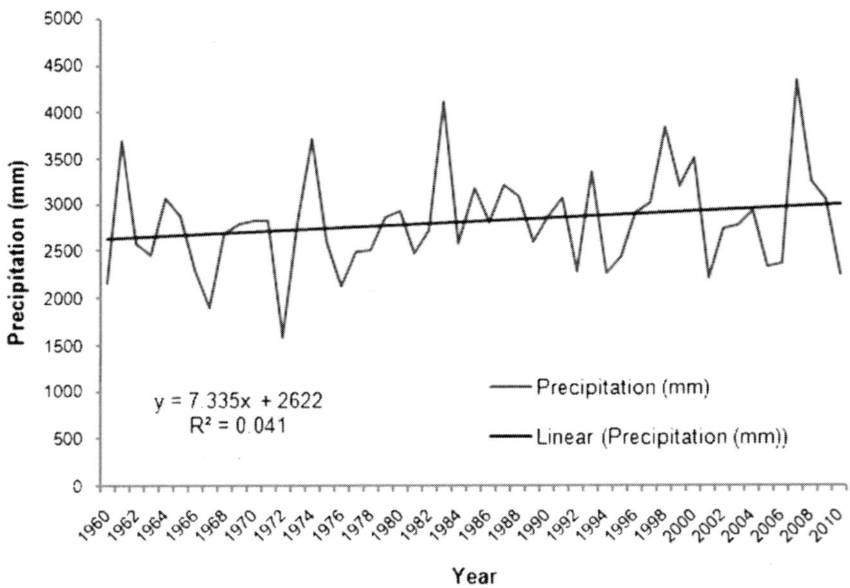

Figure 2. Annual rainfall pattern of Chittagong City from 1960 to 2010. (Data source: Bangladesh Meteorological Department, 2013)

The hilly soils in CHT are mainly yellowish brown to reddish brown loams which grade into broken shale or sandstone as well as mottled sand at a variable depth. According to the physiography of Bangladesh, CHT falls under the Northern and Eastern Hill unit and the High Hill or Mountain Ranges sub-unit. At present, all the mountain ranges of CHT are almost hogback ridges. They rise steeply, thus, looking far more impressive than their height would imply. Most of the ranges have scarps in the west, with cliffs and waterfalls. The region is characterised by a huge network of trellis and dendritic drainage consisting of some major rivers draining into the Bay of Bengal (Chowdhury2012).

In general, the geological structures and soils are weak in CMA. Moreover, the hills have steep slopes that are vulnerable to landslides (Khan et al. 2012). The general soil type in CMA is termed as 'brown hill soil'. The soils consist of hard red clay with a mixture of fine sand of the same colour and nodules containing a large percentage of sesquioxides. The soils are moderately to strongly acidic. Major limitations include very steep slopes, heavy monsoon rainfall, erodibility of most soils, difficulty of making terrace, generally low soil fertility, and rapid permeability. The soils are highly leached and have a low natural fertility. Hills are mainly under natural and plantation forests. Shifting cultivation is practiced in some places (Islam 2012). The landslides in CMA were classified as 'earth slides' since those consist of 80 % sand and finer particles. These landslides were shallow in nature and occurred just during/after the rainfall. It has been stated that the rainfall intensity and duration play very important roles in producing these shallow landslides in CMA (Ahammad 2009). Figure 3 depicts how people of *Matijharna*, an informal residential area within CMA, are living at the risks of landslide hazards. On 11 June 2007, about 128 people died and 100 others were injured in this area due to landslides triggered by heavy rainfall for continuous 8 days (CDMP-II, Ministry of Food and Disaster Management, Bangladesh 2012).

4. DATA COLLECTION

To produce landslide susceptible maps, it is important to know the causative factors and to prepare the necessary thematic layers. For this research purpose, nine different GIS layers have been produced for LSM. All the raster images (30 m × 30 m) were projected to 'Bangladesh Transverse Mercator (BTM)' using 'Everest Bangladesh' datum. Moreover, where necessary, the maps were classified using Natural Breaks (Jenks) method with 5 classes. Natural breaks classes are based on natural groupings inherent in the data. It identifies break points by picking the class breaks that best group similar values and maximize the differences between classes. The features are then divided into classes whose boundaries are set where there are relatively big jumps in the data values. Natural breaks are data-specific classifications; this is why this method is chosen (ArcGIS® 10 Help 2012). The details of the data collection procedure and ways of preparing the thematic layers are described as follows:

Figure 3. Landslide vulnerable areas in *Matijharna*, CMA. (Source: Field visit, September 2013)

4.1 Landslide inventory map

A total of 20 landslide locations were identified in CMA through field visits. The latitude and longitude values were collected using a GPS device. The Digital Elevation Model (DEM) image (acquired on 29 November 2013) was collected from the Advanced Space-borne Thermal Emission and Reflection Radiometer-Global Digital Elevation Model web-portal (Tachikawa et al. 2011). The 20 observed landslide locations in CMA are represented in Figure 4.

4.2 Land cover map

Landsat Thematic Mapper (TM) satellite images were used for the land cover mapping (2010) of CMA. The images were acquired from the Global Visualization Viewer of the United States Geological Survey (USGS). The land cover classification methodology was based on the Object Based Image Analysis (Uddin 2013; Ahmed and Rubel2013). Finally, five broad land cover types (urban area, semi-urban area, water body, vegetation, and bare soil) were identified (Figure 5).

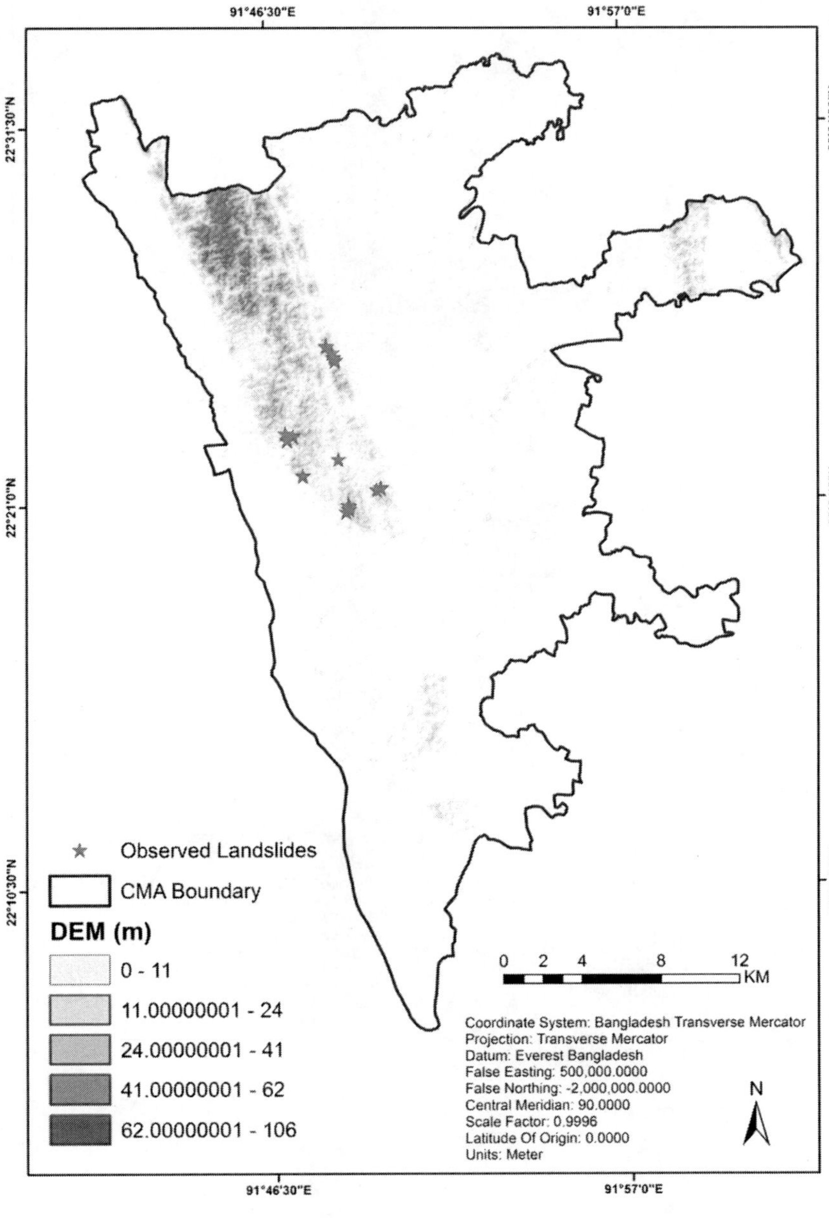

Figure 4. Landslide inventory map

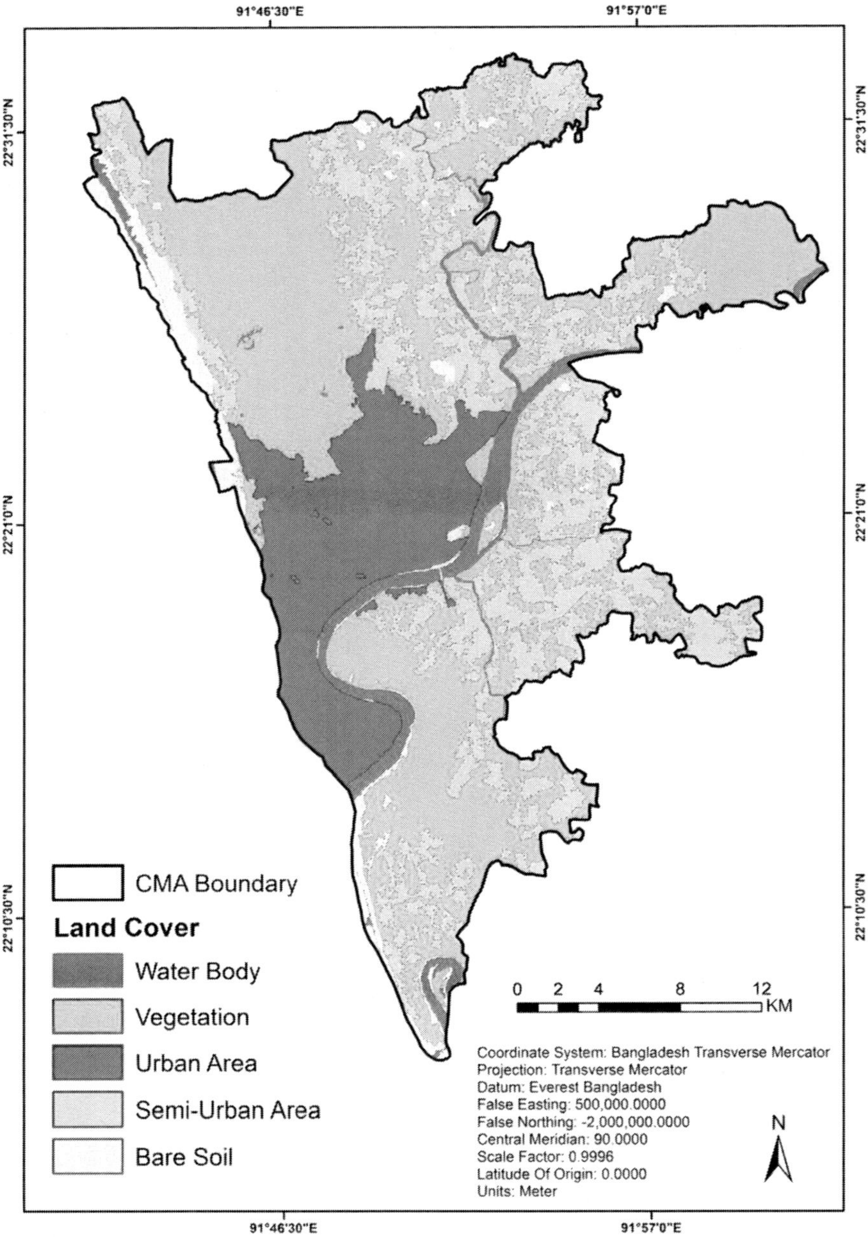

Figure 5. Land cover map

Now, it is important to perform accuracy assessment. As it was not possible to ground truth the classified image, therefore, a total of 300 reference pixels were generated using stratified random sampling method. Then, the reference pixels were compared with the base map (2010) collected from the Chittagong Development Authority (CDA). The producer's, users, and overall accuracy were found as 84.88, 87.67, and 86.80 %, respectively (Ahmed and Ahmed 2012).

4.3 Precipitation map

The daily observed precipitation data (1960–2010) were collected from the Bangladesh Meteorological Department (BMD). Moreover, the average annual precipitation map (Figure 6) was extracted from the maps collected from the Geological Survey of Bangladesh (GSB) and the USGS website (Persits et al. 2001).

Normalized difference vegetation index map

The Normalized Difference Vegetation Index (NDVI) is a standardized index that allows generating an image displaying greenness (relative biomass). The NDVI equation is as follows (ArcGIS® 10 Help 2012):

$$\mathrm{NDVI} = ((IR - R)/(IR + R)) \tag{1}$$

where IR = pixel values from the infrared band (band 4) and R = pixel values from the red band (band 3). This index output values between −1.0 and 1.0, mostly representing greenness, where any negative values are mainly generated from clouds, water, and snow, and values near zero are mainly generated from rock and bare soil. Very low values of NDVI (0.1 and below) correspond to barren areas of rock or sand. Moderate values (0.2 to 0.3) represent shrub and grassland, while high values (0.6 to 0.8) indicate temperate and tropical rainforests (Ahmed et al. 2013a; ArcGIS® 10 Help 2012). In this research, the Landsat 4–5 TM images from the same season (dry and summer) were acquired from the official website of the USGS. Finally, the NDVI map (Figure 7) of CMA was prepared by analysing band 3 and band 4.

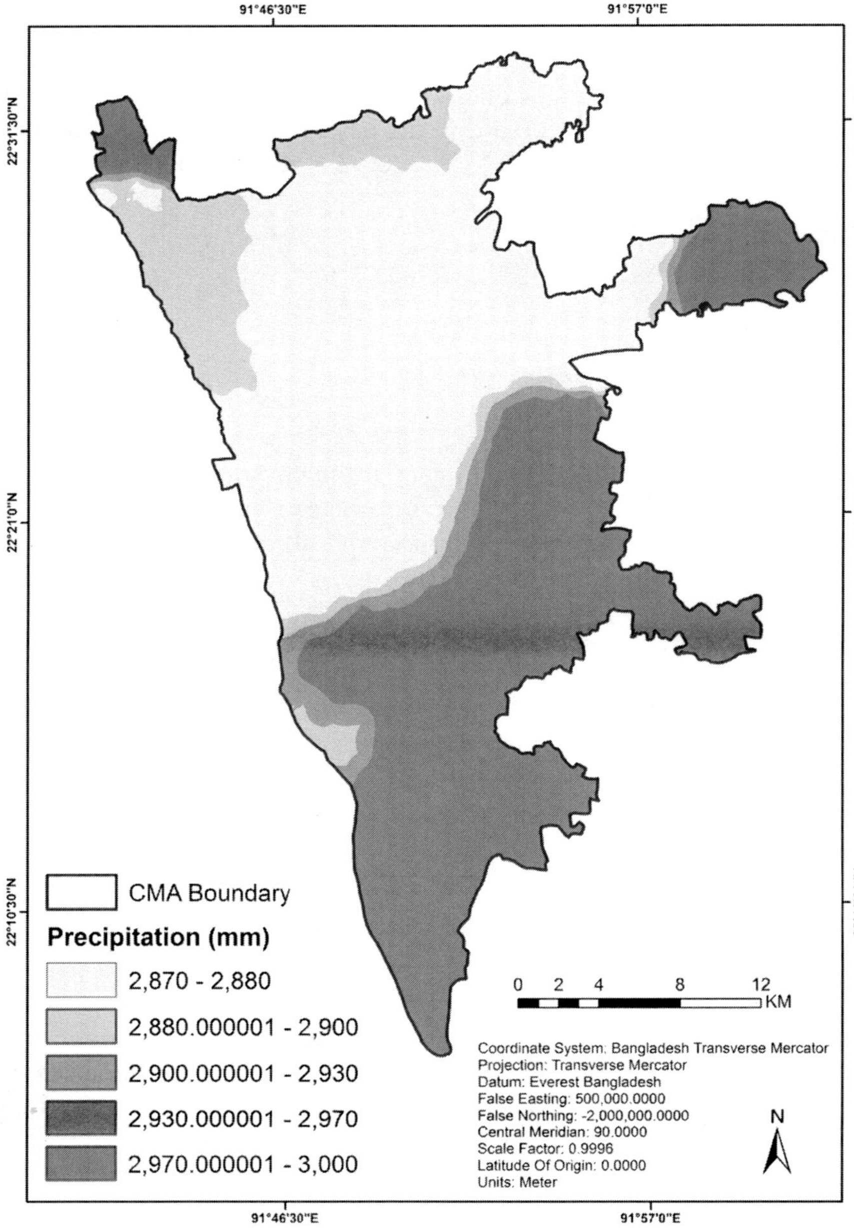

Figure 6. Precipitation map

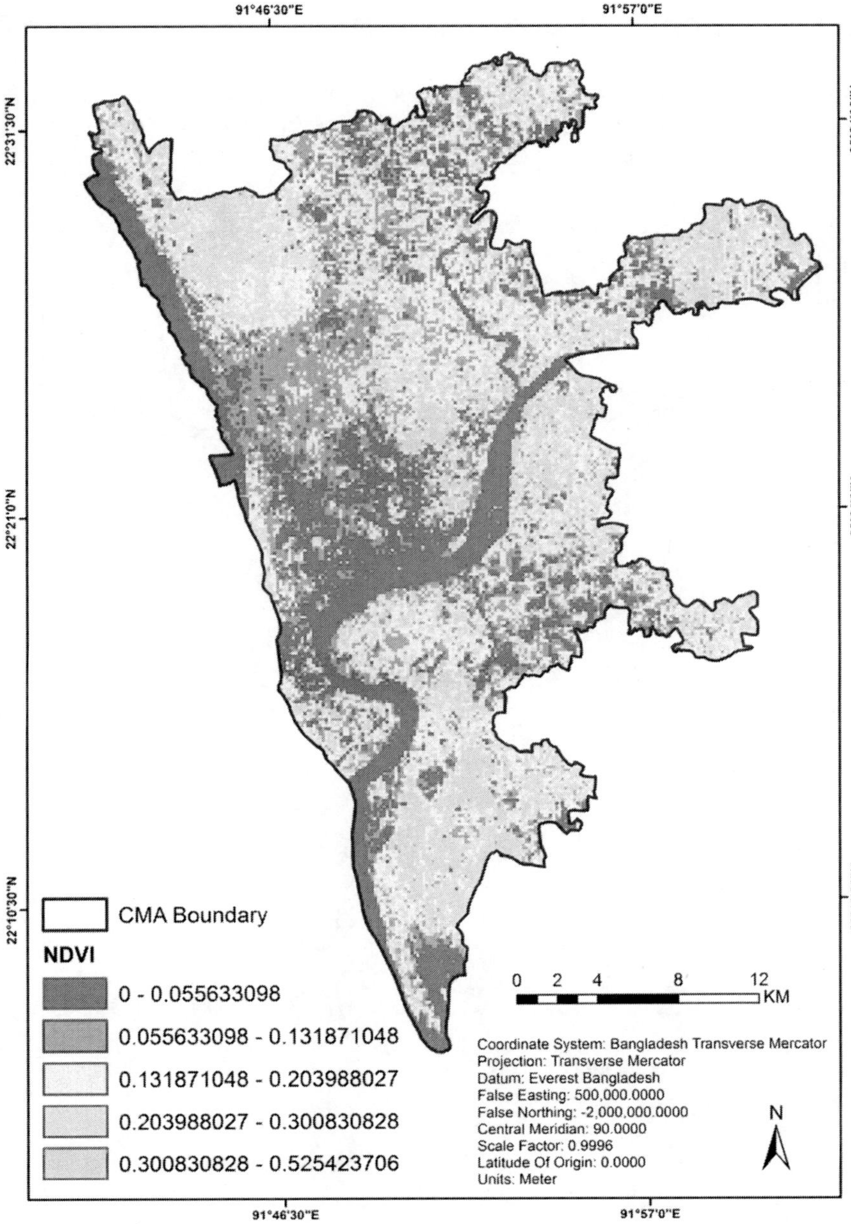

Figure 7. NDVI map

4.4 Elevation and slope map

Elevation and slope maps were produced from DEM layer. The maps were then classified using Natural Breaks (Jenks) method (ArcGIS® 10 Help 2012) with 5 classes (Figs. 8 and 9).

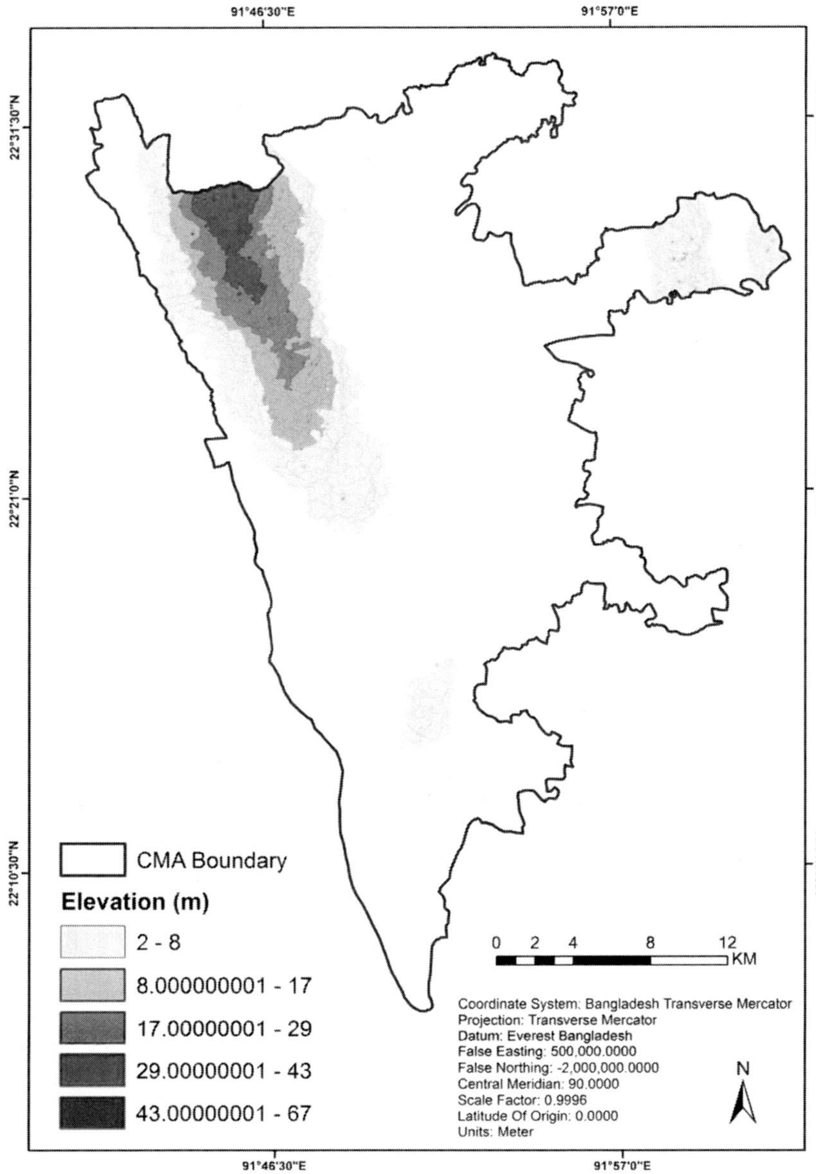

Figure 8. Elevation map

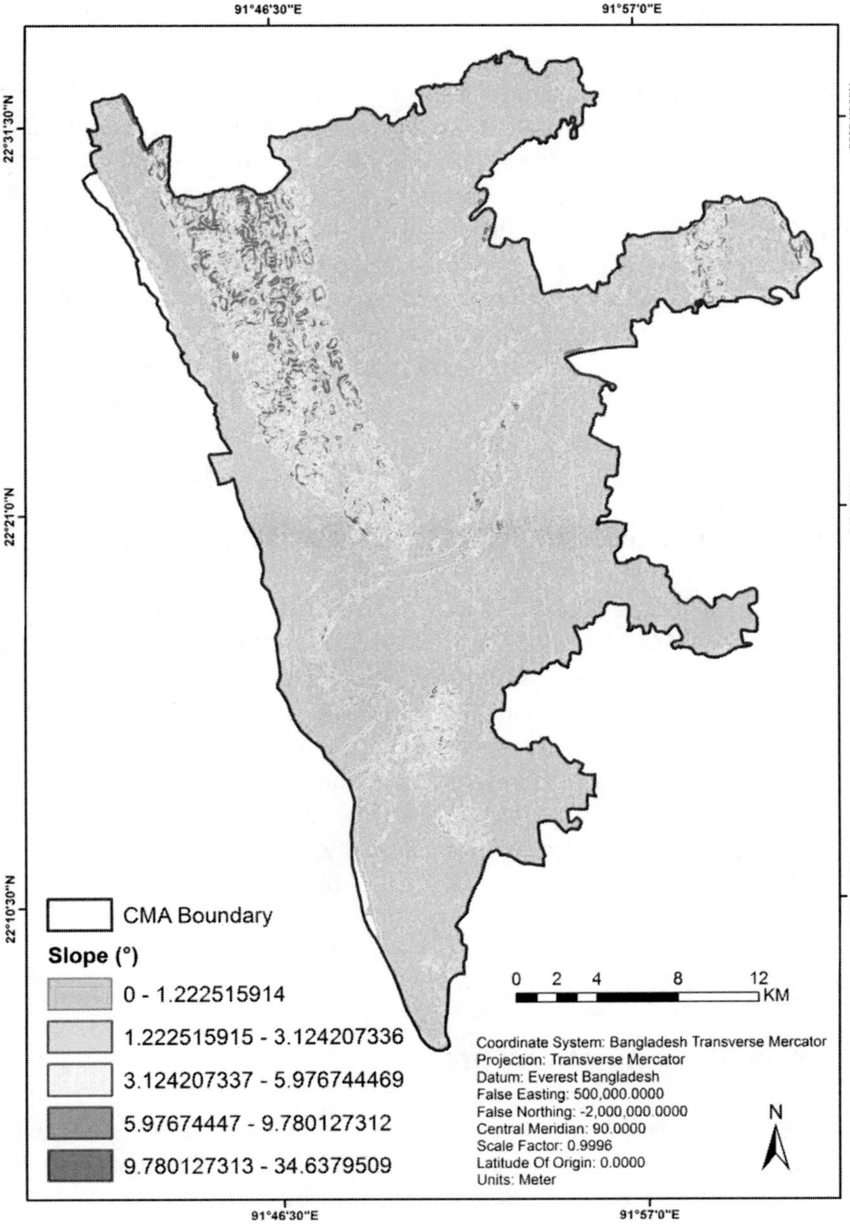

Figure 9. Slope map

4.5 Other layers

The road network, drainage network, and water body layers were collected from CDA. The distance images from all these layers were prepared using 'Euclidean Distance' technique (ArcGIS® 10 Help 2012), which gives the distance from each cell in the raster to the closest source (Figs. 10, 11, and 12). The Euclidean distance tools give information according to Euclidean or straight-line distance. Euclidean distance is calculated from the centre of the source cell to the centre of each of the surrounding cells. True Euclidean distance is calculated in each of the distance tools (ArcGIS® 10 Help 2012). The soil permeability map (Figure 13) was collected from GSB and USGS (Persits et al. 2001).

5. LANDSLIDE SUSCEPTIBILITY-MAPPING METHODS

In this research, three MCDA methods were used for LSM. These were AHP, WLC, and OWA. MCDA is a GIS-based overlaying method used to combine a set of criteria to achieve a single composite basis for a decision according to a specific objective (Eastman 2012). MCDA technique is useful for overlaying large data/maps and easy to understand/implement. The expert's or decision maker's preferences get importance in these methods. Therefore, the main drawback of MCDA is only failing to choose appropriate assumptions/criteria for suitability analysis (Malczewski 2004). In this study, IDRISI® Selva and ArcGIS® 10.2 software were used for LSM in CMA.

5.1 Analytical hierarchy process

The AHP method is used to derive the weights associated with suitability/attribute map layers (Saaty 1977). Later the weights can be combined with the attribute map layers (Malczewski 2004). AHP can deal with complex decision making and also useful for checking the consistency of the evaluation measures as suggested by the decision makers. The input of this method can be price, weight etc. AHP builds a hierarchy of decision criteria through pairwise comparison of each possible criterion pair (Poudyal et al. 2010; Feizizadeh and Blaschke 2013).

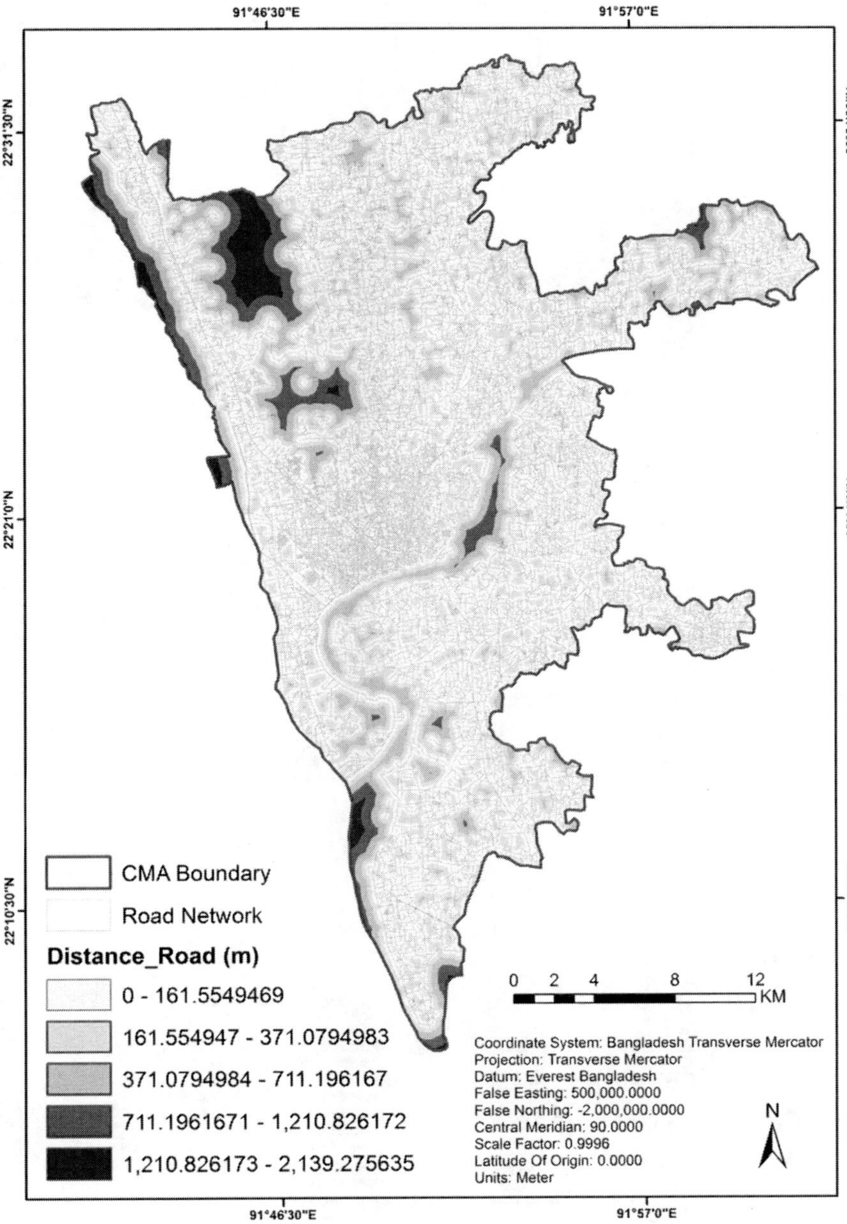

Figure 10. Distance to road map

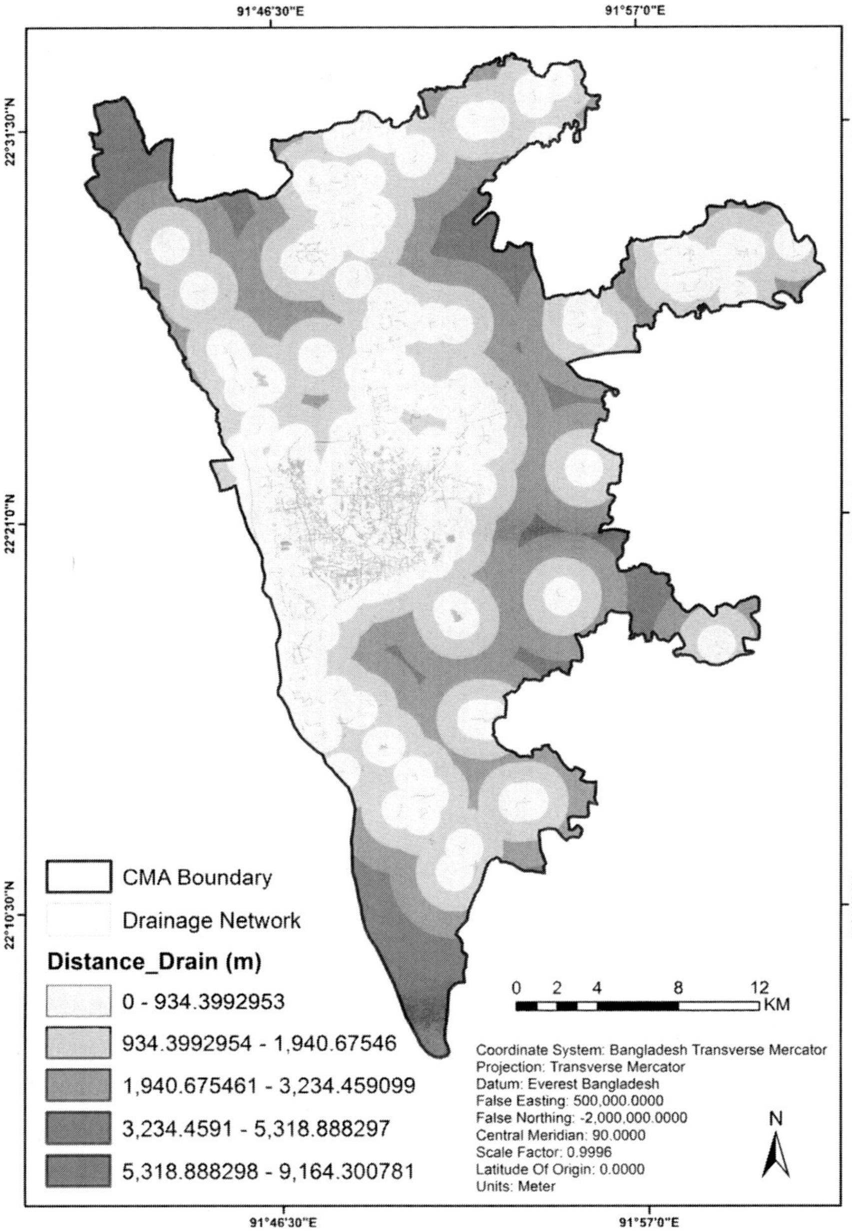

Figure 11. Distance to drain map

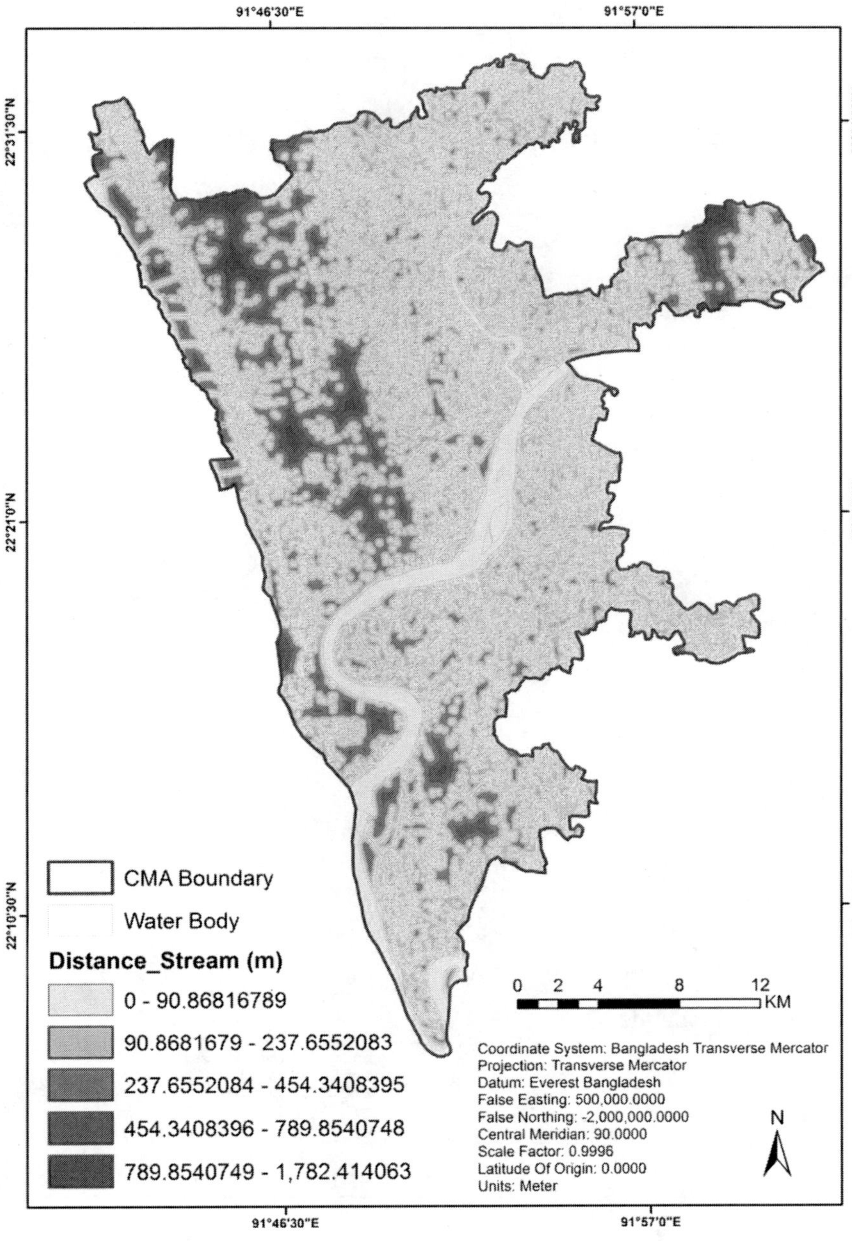

Figure 12. Distance to stream map

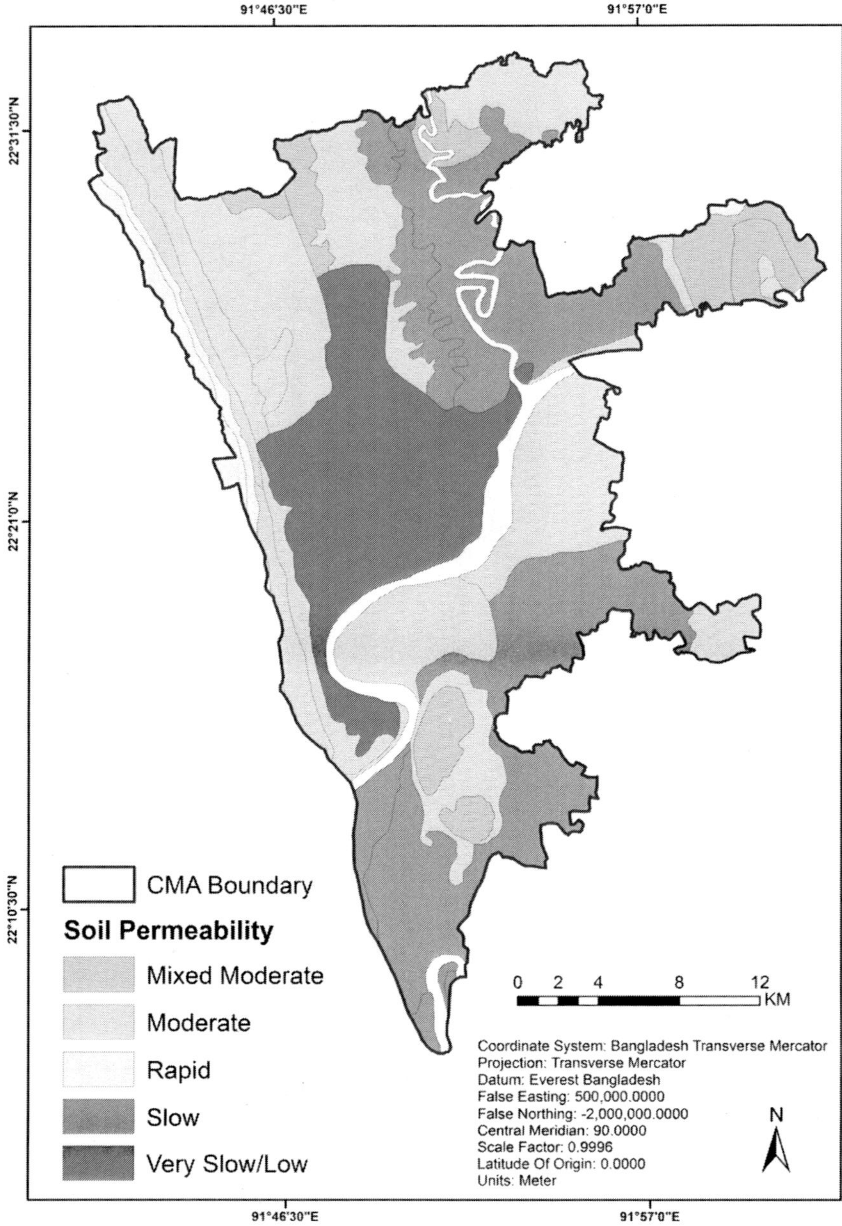

Figure 13. Soil permeability map

The weights can be derived by taking the principal eigenvector of a square reciprocal matrix of pairwise comparisons between the criteria. It is also necessary that the weights sum to one. The comparisons concern the relative importance of the two criteria involved in determining suitability for the stated objective. Ratings are provided on a 9-point continuous scale: (1/9, 1/8, 1/7,1/6, 1/5, 1/4, 1/3, 1/2, 1, 2, 3, 4, 5, 6, 7, 8, 9). The values range from 1/9 representing the least important, to 1 for equal importance and to 9 for the most important, covering all the values in the set (Eastman2012). It is also possible to determine the degree of consistency that has been used in developing the ratings (Eastman 2012). It is a procedure by which an index of consistency, known as a consistency ratio (CR), can be produced. The CR indicates the probability that the matrix ratings were randomly generated. It is stated that matrices with CR ratings greater than 0.10 should be re-evaluated (Saaty 1977).

5.2 Weighted linear combination

In WLC or simple additive weighting method, the decision maker directly assigns the weights of relative importance to each attribute map layer. A total score is then obtained for each alternative by multiplying the importance weight assigned for each attribute by the scaled value given to the alternative on that attribute and summing the products over all attributes. When the overall scores are calculated for all of the alternatives, the alternative with the highest overall score is chosen (Malczewski2004). Due to the criterion weights being summed to one, the final scores of the combined solution are expressed on the same scale (Feizizadeh and Blaschke 2013). In this case, the higher the factor weight the more influence that factor has on the final suitability map (Saaty 1977).

5.3 Ordered weighted averaging

The OWA method was introduced by Yager (1988). It involves both criterion importance weights and order weights. An importance weight is assigned to a given criterion/factor for all locations in a study area to indicate its relative importance according to the decision maker's preferences (Malczewski 2004). Order weights are a set of weights assigned not to the factors themselves but to the rank order position of factor values for a given location/pixel (Eastman 2012). The number of order weights is equal to the number of criteria and must sum to one (Jiang and Eastman2000).

6. RESULTS

Calculating factor weights has a crucial role in the production of landslide susceptibility maps by applying GIS-MCDA methods (Kavzoglu et al. 2013). The calculations of relative weights of the factors and order were based on the expert opinion surveying, analysing the landslide inventory map, and local knowledge obtained from field surveying.

6.1 LSM using AHP method

To apply the AHP method, first it is necessary to construct a pairwise matrix. Then, both the weight values of sub-criteria of the criterions and the datasets/factors were calculated (Tables 2 and 3). In the next step, the CR was calculated in order to determine whether the pairwise comparisons were consistent or not (Saaty 1977). In this research, the resulting CR for all the cases was found less than 0.10 (Tables 2and 3). It means the relative weights were appropriate and the comparisons were consistent (Saaty 1977; Feizizadeh and Blaschke 2013).

It was found that the highest weight was assigned to soil permeability map. Slope, elevation, land cover, and NDVI factors were also found effective. The other layers (i.e., precipitation, distance to drain, road, and stream) were identified as less important (Table 3).

After applying the AHP generated weights in the data layers, the resulting map was reclassified into three meaningful levels as: low, medium, and high susceptibility zones (Figure 14). This is helpful for presentation and evaluation purposes. An expert knowledge-based classification was used to define the class intervals. This technique of landslide susceptibility zoning was implemented for the rest two methods (WLC and OWA).

6.2 LSM using WLC method

Based on expert opinion surveying and using local knowledge, three different combinations of factor weights were generated (Table 4). At first, the factor maps multiplied the weights from the pairwise comparison matrix and all the weighted factor maps were then aggregated. Finally, the maps were reclassified to produce the WLC generated landslide susceptibility maps (Figure 15).

Table 2. Pairwise comparison matrix, consistency ratio and weights of the sub-criteria of the data layers

Factors	(1)	(2)	(3)	(4)	(5)	Eigen values
Distance to drain (m)						
(1) 0 – 934.3992953	1					0.0448
(2) 934.3992954 – 1,940.67546	2	1				0.0699
(3) 1,940.675461 – 3,234.459099	3	2	1			0.1098
(4) 3,234.4591 – 5,318.888297	6	4	3	1		0.2408
(5) 5,318.888298 – 9,164.300781	7	6	5	4	1	0.5346
Consistency ratio: 0.04						
Elevation (m)						
(1) 2 – 8	1					0.0501
(2) 8.000000001 - 17	3	1				0.0964
(3) 17.00000001 - 29	4	2	1			0.1521
(4) 29.00000001 – 43	6	5	4	1		0.4548
(5) 43.00000001 – 67	4	3	2	1/2	1	0.2465
Consistency ratio: 0.03						
Land cover						
(1)Water body	1					0.0434
(2) Vegetation	3	1				0.1196
(3) Urban area	7	6	1			0.5019
(4) Semi-urban area	5	4	1/3	1		0.2537
(5) Bare soil	3	1/3	1/5	1/3	1	0.0814
Consistency ratio: 0.08						
NDVI						
(1) 0 – 0.055633098	1					0.4380
(2) 0.05563098 – 0.131871048	1/2	1				0.2913
(3) 0.131871048 – 0.203988027	1/4	1/3	1			0.1544
(4)0.203988027 – 0.300830828	1/5	1/4	1/3	1		0.0881
(5) 0.300830828 – 0.525423706	1/9	1/8	1/7	1/6	1	0.0282
Consistency ratio: 0.07						
Precipitation (mm)						
(1) 2,870 – 2,880	1					0.0618
(2) 2,880.000001 – 2,900	2	1				0.0973
(3) 2,900.000001 – 2,930	3	2	1			0.1599
(4) 2,930.000001 – 2,970	4	3	2	1		0.2625
(5) 2,970.000001 – 3,000	5	4	3	2	1	0.4185
Consistency ratio: 0.02						
Distance to road (m)						
(1)0 – 161.5549469	1					0.4185
(2) 161.554947 – 371.0794983	1/2	1				0.2625
(3) 371.0794984 - 711.196167	1/3	1/2	1			0.1599
(4) 711.1961671 – 1,210.826172	1/4	1/3	1/2	1		0.0973
(5) 1,210.826173 – 2,139.275635	1/5	1/4	1/3	1/2	1	0.0618

Consistency ratio:0.02						
Slope (°)						
(1)0 – 1.222515914	1					0.0515
(2) 1.222515915 – 3.124207336	2	1				0.0718
(3) 3.124207337 – 5.976744469	3	4	1			0.1565
(4) 5.97674447 – 9.780127312	7	6	5	1		0.4869
(5) 9.780127313 – 34.6379509	4	3	2	1/2	1	0.2333
Consistency ratio:0.04						
Soil permeability						
(1)Mixed moderate	1					0.0385
(2) Moderate	2	1				0.0522
(3) Rapid	3	4	1			0.1088
(4) Slow	5	4	3	1		0.1900
(5) Very slow/ low	9	8	7	6	1	0.6105
Consistency ratio:0.08						
Distance to stream (m)						
(1)0 – 90.86816789	1					0.3999
(2) 90.8681679 – 237.6552083	1/2	1				0.2427
(3) 237.6552084 – 454.3408395	1/3	1/2	1			0.1592
(4) 454.3408396 – 789.8540748	1/3	1/2	1/2	1		0.1200
(5) 789.8540749 – 1,782.414063	1/4	1/3	1/2	1/2	1	0.0783
Consistency ratio:0.02						

Table 3. Pairwise comparison matrix, factor weights and consistency ration of the data layers

Factors	(1)	(2)	(3)	(4)	(5)	(6)	(7)	(8)	(9)	Eigen values
(1) Distance to drain	1									0.0469
(2) Elevation	5	1								0.1989
(3) Land cover	3	1/3	1							0.0975
(4) NDVI	2	1/4	1/2	1						0.0706
(5) Precipitation	1/3	1/7	1/4	1/3	1					0.0366
(6) Distance to road	1/3	1/8	1/6	1/5	1/2	1				0.0243
(7) Slope	5	1	5	4	3	6	1			0.1989
(8) Soil permeability	7	2	5	6	8	9	2	1		0.3074
(9) Distance to stream	1/2	1/7	1/6	1/5	1/4	1/3	1/7	1/8	1	0.0190
Consistency ratio: 0.07										

6.3 LSM using OWA method

OWA method uses order weights in addition to criterion/factor weights. Order weights control the manner in which the weighted factors are aggregated (Eastman and Jiang1996). In traditional WLC method, criteria weights determine how factors trade-off relative to each other (Jiang and Eastman 2000). Trade-off is the degree to which one factor can compensate for another; how they compensate is governed by a set of ‘factor weights'/'trade-off weights'. A factor with a high factor/trade-off weight may compensate for low suitability in other factors that have lower factor/trade-off weights (Eastman 2012). However, the level of trade-off is not adjustable in WLC. But, in the case of OWA method, the criteria weights can be adjusted according to the level of trade-off by using the order weights (Jiang and Eastman 2000).

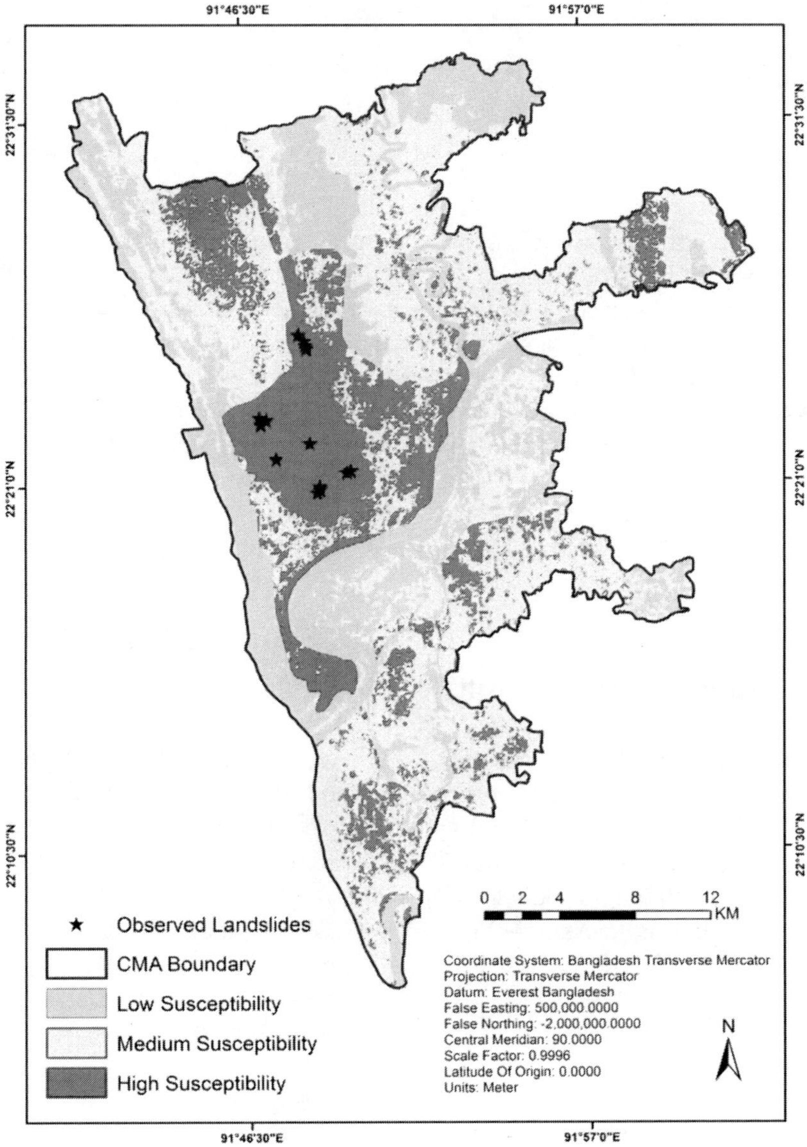

Figure 14. Landslide susceptibility map derived from AHP method

Table 4. Factor weights for WLC analysis

Factors	Weight_1	Weight_2	Weight_3
Distance to drain	0.05	0.10	0.10
Elevation	0.10	0.10	0.15
Land cover	0.10	0.15	0.15
NDVI	0.10	0.10	0.10
Precipitation	0.05	0.05	0.05
Distance to road	0.05	0.05	0.05
Slope	0.10	0.15	0.15
Soil permeability	0.40	0.25	0.20
Distance to stream	0.05	0.05	0.05
Total	1.00	1.00	1.00

The factor with the lowest suitability score, after factor weights are applied, is given the first order weight. The factor with the next lowest suitability score is given the second order weight, and so on. Applying order weights has the effect of weighting factors based on their rank from minimum to maximum value for each location. When ordered weights are the same, OWA creates a result that is identical to WLC. This indicates that WLC is a special case of OWA (Eastman 2012). In this research, the factor and order weights (Tables 5 and 6) were obtained from the AHP and WLC methods, respectively, as described above. This is how, three different OWA generated landslide susceptibility maps were produced (Figure 16).

6.4 Analysis of the results

At first, the landslide susceptibility maps were evaluated qualitatively. It helps to select the most appropriate method of LSM for a particular area (Feizizadeh and Blaschke2013). In the case of the AHP method, high susceptibility zones cover about 23 % of the total area, while about 54 % area was classified as medium susceptible, and the remaining 24 % area was classified as low susceptible zone (Table 7). About 42, 20, and 2 % areas fall within the high susceptible zone for the WLC_1, WLC_2, and WLC_3, respectively. Similarly, about 20, 3, and 1 % areas were classified as high susceptible zones for the OWA_1, OWA_2, and OWA_3 methods, respectively (Table 7).

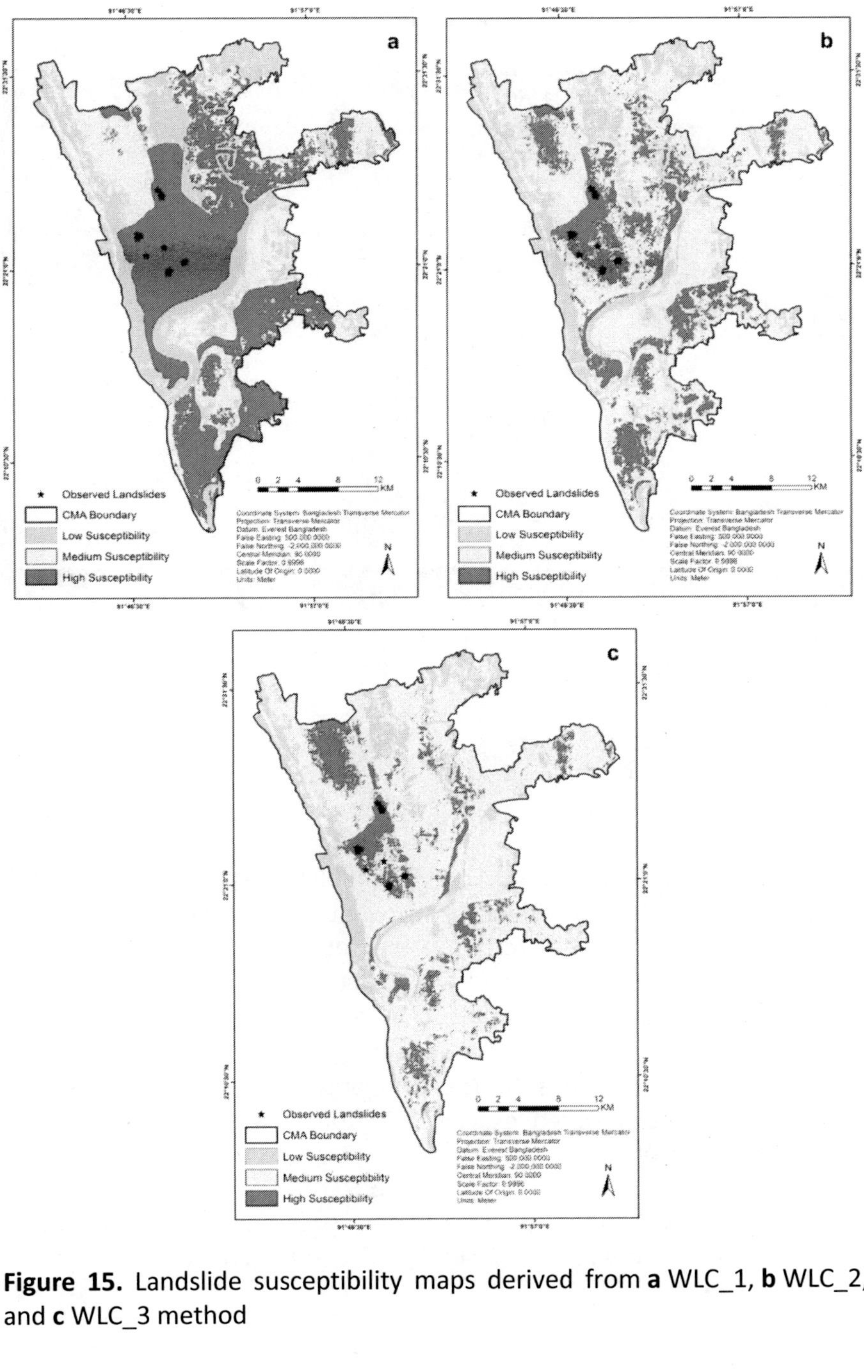

Figure 15. Landslide susceptibility maps derived from **a** WLC_1, **b** WLC_2, and **c** WLC_3 method

Table 5. Factor weights for OWA analysis

Factors	Factor weight
(1) Distance to stream	0.0190
(2) Distance to road	0.0243
(3) Precipitation	0.0366
(4) Distance to drain	0.0469
(5) NDVI	0.0706
(6) Land cover	0.0975
(7) Elevation	0.1989
(8) Slope	0.1989
(9) Soil permeability	0.3074

Table 6. Order weights for OWA analysis

Order weight	Weight 1	Weight 2	Weight 3	Weight 4	Weight 5	Weight 6	Weight 7	Weight 8	Weight 9
OWA_1	0.05	0.05	0.05	0.05	0.10	0.10	0.10	0.10	0.40
OWA_2	0.05	0.05	0.05	0.10	0.10	0.15	0.10	0.15	0.25
OWA_3	0.05	0.05	0.05	0.10	0.10	0.15	0.15	0.15	0.20

Then the accuracies of the landslide susceptibility maps were determined quantitatively. To do this, the landslide inventory map with the 20 known landslide events was compared with the respective susceptibility maps derived from the AHP, WLC, and OWA methods (Table 7). For the AHP method, the comparison shows that 100 % of the known landslides fall into the high susceptibility zone. No known landslide event was observed in the remaining categories (Table 7). The comparisons showed that the high susceptibility zones covered exactly 100, 100, and 90 % of the known landslides for the WLC_1, WLC_2, and WLC_3, respectively. Lastly, the high susceptible zones covered 100, 90, and 45 % of the known landslides for the OWA_1, OWA_2, and OWA_3 methods, respectively. In all the cases, no landslide was observed in the low susceptibility zones (Table 7).

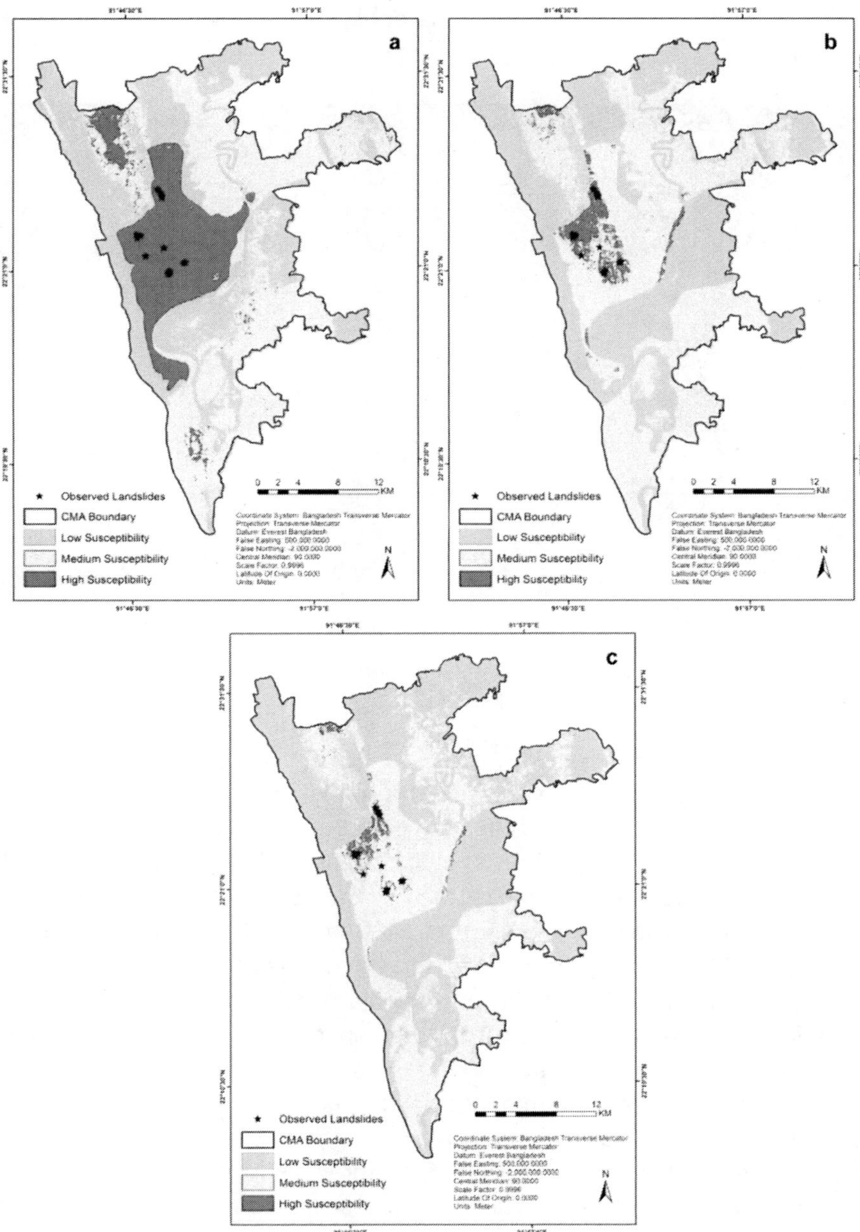

Figure 16. Landslide susceptibility maps derived from **a** OWA_1, **b** OWA_2, and **c** OWA_3 method

Table 7. Comparison of landslide susceptibility maps with the observed landslides

Methods	Susceptible zones	Area under category (%)	Number of landslides	Comparison with landslide area (%)
AHP	High susceptibility	22.713	20	100
	Medium susceptibility	53.609	0	0
	Low susceptibility	23.677	0	0
WLC_1	High susceptibility	41.890	20	100
	Medium susceptibility	36.791	0	0
	Low susceptibility	21.319	0	0
WLC_2	High susceptibility	20.365	20	100
	Medium susceptibility	66.079	0	0
	Low susceptibility	13.556	0	0
WLC_3	High susceptibility	1.5520	18	90
	Medium susceptibility	96.911	2	10
	Low susceptibility	1.5370	0	0
OWA_1	High susceptibility	19.565	20	100
	Medium susceptibility	46.149	0	0
	Low susceptibility	34.286	0	0
OWA_2	High susceptibility	3.4750	18	90
	Medium susceptibility	52.381	2	10
	Low susceptibility	44.144	0	0
OWA_3	High susceptibility	1.3410	9	45
	Medium susceptibility	46.115	11	55
	Low susceptibility	52.544	0	0

High susceptibility zone covering 100 % of known landslides does not always mean that the results are accurate. In some cases, high susceptibility zone occurred in the flat areas with moderate or mixed moderate soil permeability indicates that the results obtained using the MCDA methods also have some errors. MCDA methods are generally based on weighting the factor maps and finally overlaying those layers. As a result, any incorrect perception on the role of the different slope-failure criteria can be easily conveyed from the expert's opinion into the weight assignment (Kritikos and Davies 2011). This can cause errors in the final outputs. In this research, a total of 9 factor maps each with 5 classes were considered. It is difficult to assign criteria weights for all these sub-factors and develop a proper combination. Therefore, it is important to keep the factor map layers and their classes into reasonable numbers for getting better results. Errors can also occur due to incorrect pairwise comparisons between the criteria, classifying the factor maps and defining the susceptibility zones qualitatively. Moreover, there might be errors in the GIS/remotely sensed datasets, problems while conducting questionnaire surveying for defining the weights, and taking the GPS values, etc. The main challenge of this kind of GIS-MCDA analyses is to keep the errors as less as possible. This can be achieved by defining the best combination of criteria weights.

AHP method uses pairwise comparison of each criterion, while WLC directly assigns the weights of relative importance to each attribute map layer and OWA involves two-step weighting (criterion and order weights). Each method follows its own way of assigning weights to factors or orders. Therefore, it is not possible to declare that one method is superior to other. As it is a trial and error/iterative process, the final output maps may give some errors as well. Still the MCDA methods give better accuracies, which are acceptable for producing real-world LSM in terms of landslide disaster risk reduction. Moreover, each MCDA method can produce different types of landslide susceptibility maps based on assigning different weights for the instability factors. The weights can be obtained through expert opinion surveying, or even it can be achieved from the participatory-based community surveying, or it can be a combination of both. Finally, it is the researcher's or policy makers' decision to choose the appropriate weighting combination and the output maps, as per the local context and research/project objectives.

6.5 Validation of the methods

In order to determine the statistical reliability of the results, it is important to perform validation of spatial results in a structured manner (Ahmed et al. 2013b). To do this validation, Relative Operating Characteristic (ROC) method was used in this research. The ROC analysis is useful for cases in which the scientist wants to see how well the suitability map portrays the location of a particular category, but does not have an estimate of the quantity of the category. For example, the ROC could be used to compare an image of modelled probability for landslides against an image of actual observed landslides (Eastman 2012). The area under ROC curves (AUC) constitutes one of the most common used accuracy statistics for the prediction models in natural hazard assessments (Ahmed and Rubel 2013). The minimum value of AUC is 0.5, which means no improvement over random assignment. The maximum value of AUC is 1 that denotes perfect discrimination (Nefeslioglu et al. 2008).

The comparison results are shown in Figure 17 as a line graph (threshold type is equal interval and number of thresholds is 25 %). The AUC values are indicating the accuracies of the methods used for LSM. The AUC values of the AHP, WLC_1, WLC_2, WLC_3, OWA_1, OWA_2, and OWA_3 methods were calculated as 0.898, 0.839, 0.911, 0.885, 0.904, 0.951, and 0.871, respectively (Figure 17). In general, the verification results showed satisfactory agreement between the susceptibility map produced and the observed landslide location map (AUC values ranged from 0.871 to 0.951). Finally, it can be stated that higher accuracy was found for all the MCDA methods applied. But the landslide susceptibility map produced by the OWA_2 method appeared to be slightly more accurate than those generated by applying the other methods (Figure 17).

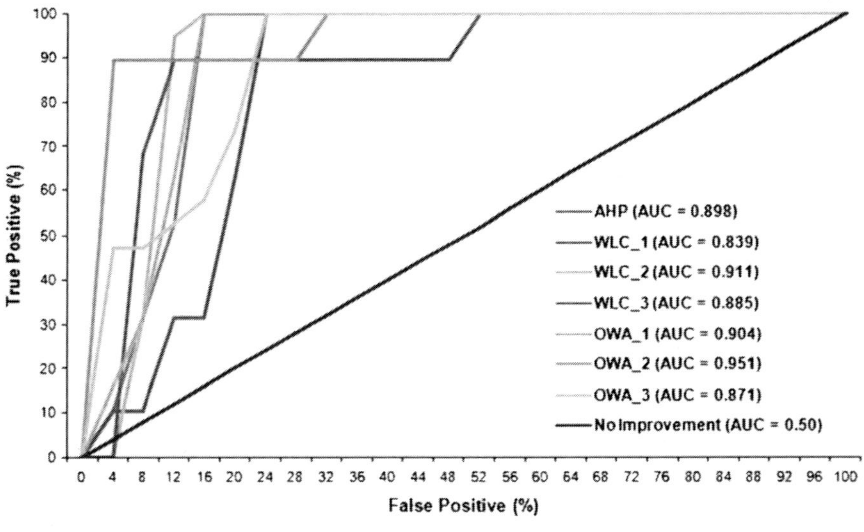

Figure 17. Assessment of model performance based on the ROC curves

7. DISCUSSIONS AND FUTURE RESEARCH

GIS-MCDA methods are based on the weighting of the factors and orders of the thematic layers. Therefore, the preferences of the decision makers play important role in producing the LSM. In this case, the factor/order weights should be assigned carefully after proper consultation with the relevant experts, local people, and communities at risk, and extensive field surveying is also required. From the analysis, it was evident that urbanized areas with very low soil permeability and steep slope were highly vulnerable to landslide hazards.

Till now, in all the cases, the researchers used the previous or existing land cover images to produce the landslide susceptibility maps. But land-use/land-cover maps can be treated as a dynamic variable because within the next few decades, the land use of a particular area can be abruptly changed (e.g. CMA). This change in land use pattern (e.g. deforestation, hill-cutting) can put triggering impacts on occurring landslide hazards (Mugagga et al. 2012). Therefore, while modelling landslide hazard maps, it is very important to take into account a new thematic layer—'the future projected land cover map'.

Same thing can happen for the trend in precipitation pattern. It is already indicated that due to climate change, a significant increase in annual and pre-monsoon rainfall in Bangladesh is observed (Shahid 2011). Moreover, the recent landslide events were related to the extreme rainfall intensities within short period of time (Khan et al. 2012). Therefore, it is also important to consider future rainfall pattern while producing landslide susceptibility maps. These are the current research gaps in landslide susceptibility modelling.

In this drawback, future research should take into consideration the dynamic variables like the simulated land cover map and future rainfall pattern to predict the future landslide scenario in CMA. This will be a new kind of analysis and interesting outcomes are expected. It will help answering whether using the simulated dynamic variables will give better accuracy/results or not. It is also expected that preparing landslide hazard maps, using this kind of new modelling concept will add new knowledge to landslide susceptibility mapping in modern science. The flowchart of this new concept is depicted in Figure 18.

On the other hand, the term 'Natural Disaster' refers to extreme natural events like tsunami, earthquakes, landslides, floods, cyclones, etc. But it has been argued that these events are not disasters until a vulnerable group of people is exposed (Wisner et al. 2003). Therefore, it is important to understand the underneath reasons for this progression of urban landslide vulnerability. It is also evident that even being aware of the landslide risks people are living in those risky areas. This situation needs to be taken into consideration. Just producing some automated hazard and susceptibility maps, using different scientific modelling techniques, cannot reduce the landslide risks. The vulnerability of the people at risk and human adaptation to landslides needs to be analysed to address this problem properly (Alexander 2005; Alexander 2000).

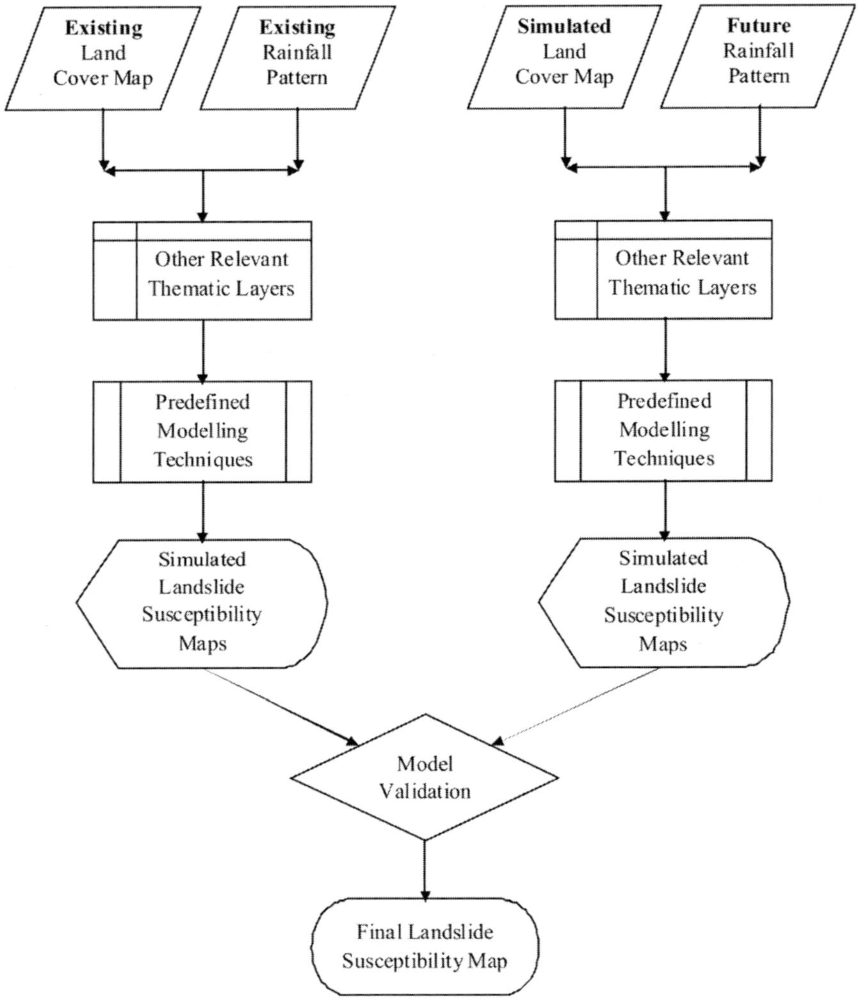

Figure 18. Flowchart of the new concept for landslide susceptibility mapping

8. CONCLUSIONS

Landslides are a common hazard in highly urbanized hilly areas in CMA, especially during the rainy season (Khan et al. 2012; Ahmed and Rubel 2013). Therefore, the aim of this research is to produce acceptable landslide susceptibility maps for CMA, so that appropriate landslide disaster mitigation strategies can be developed. In this research, three different GIS-

MCDA methods—the AHP, WLC, and OWA—were applied to scientifically assess the landslide susceptible areas in CMA. Nine different thematic layers or landslide causative factors were considered. Then seven different landslide susceptibility scenarios were generated based on the three weighted overlay techniques.

Later, the performances of the MCDA methods were validated using the area under the ROC curves. The accuracy of the landslide susceptibility maps produced by the AHP, WLC_1, WLC_2, WLC_3, OWA_1, OWA_2, and OWA_3 methods were found as 89.80, 83.90, 91.10, 88.50, 90.40, 95.10, and 87.10 %, respectively. The verification results showed satisfactory agreement (Nefeslioglu et al. 2008; Feizizadeh and Blaschke 2013) between the susceptibility maps produced and the existing data on the 20 historical landslide locations. It is important to mention that due to the dynamic nature of the land use and precipitation patterns, the landslide susceptibility maps are subject to vary. Hence, these maps should be updated and modified, time-to-time.

The preparation of landslide susceptibility map is the first step towards the reduction of landslide hazard. But it is also important to create awareness among the local people based on the predictive landslide susceptibility maps. Moreover, developing early warning system, increasing cooperation among different public/autonomous/non-governmental organizations, launching public awareness campaign, arranging relevant seminars and workshops, and generating facilities for proper evacuation system in crisis moments are highly recommended.

The outcome of this research shall help the endangered local inhabitants/communities, urban planners, and engineers to reduce losses caused by existing and future landslides by means of prevention, mitigation, and avoidance. The results will also be useful for explaining the driving factors of the known historical landslides, for supporting emergency decisions, and for upholding the efforts on the mitigation of future landslide hazards in CMA, Bangladesh.

ACKNOWLEDGEMENTS

Bayes Ahmed is a Commonwealth Scholar funded by the UK government. The author is grateful to the IRDR, UCL, International Centre for Integrated Mountain Development (ICIMOD), the Association of American

Geographers (AAG), Chittagong University of Engineering and Technology (CUET), Professor Dr. David E. Alexander, Yiaser Arafat Rubel, Renato Forte, Kabir Uddin, Debasish Roy Raja, Rashed Jalal Himal, the relevant CDA officials, and the people of the landslide vulnerable areas who helped in numerous occasions. This article is a part of the PhD thesis that is being conducted by the corresponding author at IRDR, UCL. Finally, the author would like to thank the anonymous reviewers and the editorial board of the 'Landslides' journal for their constructive comments that improved the quality of the manuscript.

REFERENCES

1. Ahammad R (2009) Understanding institutional changes for reducing social vulnerability to landslides in Chittagong City, Bangladesh. Dissertation, Stockholm University, Sweden

2. Ahmed B, Ahmed R (2012) Modeling urban land cover growth dynamics using multi-temporal satellite images: a case study of Dhaka, Bangladesh. ISPRS Int J Geo-Inf 1:3–31

3. Ahmed B, Rubel YA (2013) Understanding the issues involved in urban landslide vulnerability in Chittagong metropolitan area, Bangladesh. The Association of American Geographers (AAG). https://sites.google.com/a/aag.org/mycoe-servirglobal/final-arafat. Assessed 17 January 2014

4. Ahmed B, Kamruzzaman M, Zhu X, Rahman MS, Choi K (2013a) Simulating land cover changes and their impacts on land surface temperature in Dhaka, Bangladesh. Remote Sens 5:5969–5998

5. Ahmed B, Ahmed R, Zhu X (2013b) Evaluation of model validation techniques in land cover dynamics. ISPRS Int J Geo-Inf 2:577–597

6. Akbar TA, Ha SR (2011) Landslide hazard zoning along Himalayan Kaghan Valley of Pakistan—by integration of GPS, GIS, and remote sensing technology. Landslides 8(4):527–540

7. Alexander DE (2000) Confronting catastrophe: new perspectives on natural disasters. Oxford University Press, UK

8. Alexander D (2005) Vulnerability to landslides. In: Glade T, Anderson MG, Crozier MJ (eds) Landslide hazard and risk, front matter. Wiley, Chichester, pp 175–198.

9. ArcGIS® 10 Help (2012) Environmental Systems Research Institute (ESRI). Redlands, CA, USA. http://help.arcgis.com/en/arcgisdesktop/10.0/help/. Accessed 7 November 2013

10. Avtar R, Singh CK, Singh G, Verma RL, Mukherje S, Sawada H (2011) Landslide susceptibility zonation study using remote sensing and GIS technology in the Ken-Betwa River Link area, India. Bull Eng Geol Environ 70:595–606

11. Ayala IA, Chavez OE, Parrot JF (2006) Landsliding related to land-cover change: a diachronic analysis of hillslope instability distribution in the Sierra Norte, Puebla, Mexico. Catena 65:152–165

12. Bai SB, Wang J, Lü GN, Zhou PG, Hou SS, Xu SN (2010) GIS-based logistic regression for landslide susceptibility mapping of the Zhongxian segment in the Three Gorges area, China. Geomorphology 115:23–31

13. Bui DT, Pradhan B, Lofman O, Revhaug I, Dick ØB (2013) Regional prediction of landslide hazard using probability analysis of intense rainfall in the Hoa Binh province, Vietnam. Nat Hazards 66:707–730

14. Chowdhury SQ (2012) Chittagong City. In: Islam N (ed) Banglapedia, National Encyclopaedia of Bangladesh, 2nd edn. Banglapedia Trust, Asiatic Society of Bangladesh, Dhaka, Bangladesh. http://www.bpedia.org/C_0215.php. Assessed 14 September 2013

15. Community Report, Chittagong Zila (2012) Population and Housing Census 2011, Bangladesh Bureau of Statistics, Statistics and Informatics Division, Ministry of Planning, Dhaka, Bangladesh

16. Comprehensive Disaster Management Programme-II (2012) Landslide Inventory & Land-use Mapping, DEM Preparation, Precipitation Threshold Value & Establishment of Early Warning Devices. Ministry of Food and Disaster Management, Disaster Management and Relief Division, Government of the People's Republic of Bangladesh

17. Duman TY, Can T, Emre O, Kecer M, Dogan A, Ates S, Durmaz S (2005) Landslide inventory of north-western Anatolia, Turkey. Eng Geol 77:99–114

18. Eastman R (2012) The IDRISI Selva Help. Clark Labs, Clark University 950 Main Street, Worcester MA 01610–1477 USA

19. Eastman JR, Jiang H (1996) Fuzzy measures in multi-criteria evaluation. In: Proceedings at the Second International Symposium on Spatial Accuracy Assessment in Natural Resources and Environmental Studies, May 21–23, Fort Collins, Colorado, 527–534.

20. Feizizadeh B, Blaschke T (2013) GIS-multicriteria decision analysis for landslide susceptibility mapping: comparing three methods for the Urmialake basin, Iran. Nat Hazards 65:2105–2128

21. Godt JW, Baum RL, Savage WZ, Salciarini D, Schulz WH, Harp EL (2008) Transient deterministic shallow landslide modeling: Requirements for susceptibility and hazard assessment in a GIS framework. Eng Geol 102:214–226

22. Guzzetti F, Carrara A, Cardinali M, Reichenbach P (1999) Landslide hazard evaluation: a review of current techniques and their application in a multi-scale study, Central Italy. Geomorphology 31:181–216

23. Isik Yilmaz I (2009) Landslide susceptibility mapping using frequency ratio, logistic regression, artificial neural networks and their comparison: a case study from Kat landslides (Tokat—Turkey). Comput Geosci 35:1125–1138

24. Islam MS (2012) Seven Soil Tracts. In: Islam N (ed) Banglapedia, National Encyclopaedia of Bangladesh, 2nd edn. Banglapedia Trust, Asiatic Society of Bangladesh, Dhaka, Bangladesh. http://www.banglapedia.org/HT/S_0251.htm. Assessed 18 June 2014

25. Jiang H, Eastman JR (2000) Application of fuzzy measures in multi-criteria evaluation in GIS. Int J Geogr Inform Sci 14:173–184

26. Karsli F, Atasoy M, Yalcin A, Reis S, Demir O, Gokceoglu C (2009) Effects of land-use changes on landslides in a landslide-prone area (Ardesen, Rize, NE Turkey). Environ Monit Assess 156:241–255

27. Kavzoglu T, Sahin EK, Colkesen I (2013) Landslide susceptibility mapping using GIS-based multi-criteria decision analysis, support vector machines, and logistic regression. Landslides.

28. Kayastha P, Dhital MR, Smedt FD (2012) Landslide susceptibility mapping using the weight of evidence method in the Tinau watershed, Nepal. Nat Hazards 63:479–498

29. Khan YA, Lateh H, Baten MA, Kamil AA (2012) Critical antecedent rainfall conditions for shallow landslides in Chittagong City of Bangladesh. Environ Earth Sci 67(1):97–106

30. Kritikos T, Davies TRH (2011) GIS-based Multi-Criteria Decision Analysis for landslide susceptibility mapping at northern Evia, Greece. Z dt Ges Geowiss 162:421–434

31. Lee S, Sambath T (2006) Landslide susceptibility mapping in the Damrei Romel area, Cambodia using frequency ratio and logistic regression models. Environ Geol 50:847–855

32. Malczewski J (2004) GIS-based land-use suitability analysis: a critical overview. Prog Plan 62:3–65

33. Mangiza NDM, Arimah BC, Jensen I, Yemeru EA, Kinyanjui MK (2011) Cities and climate change: global report on human settlements, 2011, United Nations Human Settlements Programme (UN-Habitat), Nairobi, Kenya

34. Mugagga F, Kakembo V, Buyinza M (2012) Land use changes on the slopes of Mount Elgon and the implications for the occurrence of landslides. Catena 90:39–46

35. Nefeslioglu HA, Gokceoglu C, Sonmez H (2008) An assessment on the use of logistic regression and artificial neural networks with different sampling strategies for the preparation of landslide susceptibility maps. Eng Geol 97:171–191

36. Park S, Choi C, Kim B, Kim J (2013) Landslide susceptibility mapping using frequency ratio, analytic hierarchy process, logistic regression, and artificial neural network methods at the Inje area, Korea. Environ Earth Sci 68:1443–1464

37. Persits FM, Wandrey CJ, Milici RC, Manwar A (2001) Digital Geologic and Geophysical Data of Bangladesh. The U.S. Geological Survey Open File Report 97-470H (CD-ROM). http://pubs.usgs.gov/of/1997/ofr-97-470/OF97-470H/. Assessed 18 November 2013

38. Poudyal CP, Chang C, Oh HJ, Lee S (2010) Landslide susceptibility maps comparing frequency ratio and artificial neural networks: a case study from the Nepal Himalaya. Environ Earth Sci 61:1049–1064

39. Rajakumar R, Sanjeevi S, Jayaseelan S, Isakkipandian G, Edwin M, Balaji P, Ehanthalingam G (2007) Landslide susceptibility mapping in a hilly terrain using remote sensing and GIS. J Indian Soc Remote Sensing 35(1):31–42

40. Saaty TL (1977) A scaling method for priorities in hierarchical structures. J Math Psychol 15:231–281

41. Shahid S (2011) Trends in extreme rainfall events of Bangladesh. Theor Appl Climatol 104:489–499

42. Tachikawa T, Hato M, Kaku M, Iwasaki A (2011) The characteristics of ASTER GDEM version 2, IGARSS. http://www.jspacesystems.or.jp/ersdac/GDEM/E/1.html. Assessed 23 October 2013

43. Uddin K (2013) Image classification: hands on exercise using eCognition. eCognition Community, Trimble Geospatial Imaging, Arnulfstrasse 126, 80636 Munich, Germany

44. Wisner B, Blaikie P, Cannon T, Davis I (2003) At Risk: Natural Hazards, People's Vulnerability and Disasters. Routledge, New York, USA

45. Yager RR (1988) On ordered weighted averaging aggregation operators in multicriteria decision making. IEEE Trans Syst Man Cybern 8:183–190

46. Yilmaz I, Yildirim M (2006) Structural and geomorphological aspects of the Kat landslides (Tokat-Turkey), and susceptibility mapping by means of GIS. Environ Geol 50(4):461–472

47. Zêzere JK, Garcia RAC, Oliveira SC, Reis E (2008) Probabilistic landslide risk analysis considering direct costs in the area north of Lisbon (Portugal). Geomorphology 94:467–495

Chapter 8

How to assess landslide activity and intensity with Persistent Scatterer Interferometry (PSI): The PSI-based matrix approach

Francesca Cigna[1,2*], Silvia Bianchini[1], Nicola Casagli[1]

[1]Department of Earth Sciences, University of Firenze, Via La Pira 4, 50121 Firenze, Italy

[2]British Geological Survey, Nicker Hill, Keyworth, NG12 5GG Nottingham, United Kingdom

ABSTRACT

We provide a step-by-step analysis and discussion of the 'PSI-based matrix approach', a methodology employing ground deformation velocities derived through Persistent Scatterer Interferometry (PSI) for the assessment of the state of activity and intensity of extremely to very slow landslides. Two matrices based on PSI data are designed respectively for landslides already mapped in preexisting inventories and for newly identified phenomena. Conversely, a unique intensity scale is proposed indiscriminately for both. Major influencing factors of the approach are brought to light by the application in the 14 km^2 area of Verbicaro, in

Northern Calabria (Italy). These include lack of PSI data within the landslide boundaries, temporal coverage of the available estimates, and need of field checks as well as the operative procedures to set the activity and intensity thresholds. For the area of Verbicaro, we exploit 1992–2011 PSI data from ERS1/2 and RADARSAT1/2 satellites, projecting them along the maximum slope directions. An activity threshold of ±5 mm/year is determined by applying the average projection factor of local slopes to the PSI data precision. The intensity threshold between extremely and very slow phenomena (16 mm/year) is reduced by ~20 % to account for temporal and spatial averages being applied to attribute representative velocities to each landslide. The methodology allows assessing the state of activity and the intensity for 13 of the 24 landslides premapped in the 2007 inventory and for two newly identified phenomena. Current limitations due to characteristics and spatial coverage of PSI data are critically tackled within the discussion, jointly with respective implications.

KEYWORDS

Landslide; Persistent Scatterer; Interferometry; State of activity; Intensity; SAR Interferometry; PSI-based matrix

1. INTRODUCTION

In many countries of the world, landslide hazard is one of the major causes of life losses, injuries, and damages to private and public structures and infrastructure (WP/WLI 1993). Identifying and mapping these phenomena is essential to reduce their social and economic impacts and for the consequent improvement of future urban expansion and planning strategies. Inventory maps address the identification of landslide processes and include location, geometrical parameters (e.g., volume and total length or area), type, causal factors, activity, and intensity of each landslide within an investigated territory. The information reported within these maps represent crucial and priority factors for every landslide hazard or risk study, since they facilitate the assessment of the spatial and temporal frequencies of landslides and a better understanding of susceptibility and vulnerability.

In hilly and mountainous areas, on-site methods for the analysis of slope instability are not always economically and practically suitable for carrying out a systematic investigation of landslide phenomena at the regional scale because of both huge extension of observed areas and frequently, difficult accessibility. Integrating and complementing conventional investigations with remote sensing techniques represent a valuable support for landslide mapping due to wide area coverage, noninvasiveness, and cost-effectiveness of remotely sensed data. Some innovative methodologies for the assessment of landslide activity through aerial photointerpretation have been carried out in the last years (Canuti and Focardi 1986; Gonzalez-Diez et al. 1999; Guzzetti et al. 1999; Parise and Wasowski 1999; Wieczorek 1984); however, they suffer from significant uncertainties due to the subjectivity in the activity evaluation and difficulties in the recognizability of landslide-related geomorphic features and in the quantification of the observed land changes, especially for urban and suburban environments. As well as for the state of activity, many difficulties are usually encountered during the evaluation of landslide intensity (Hungr 1997; i.e., destructiveness of a landslide, in terms of kinetic energy, volume, area, depth, movement velocity or total displacement, etc.), due to the lack of a single standardized procedure for the quantification of this parameter (e.g., Lateltin et al. 2005; Uzielli et al.2008). Moreover, when investigating very large areas affected by a variety of different phenomena, retrieving detailed information about volume, kinetic energy or velocity it is—more often than not—difficult and time-consuming.

A new methodology for the updating of landslide inventories has been recently experimented by the scientific community (e.g., Cascini et al. 2010; Farina et al. 2006; Herrera et al. 2010; Meisina et al. 2006; Notti et al. 2010; Righini et al. 2011). This methodology couples conventional thematic data (e.g., topographic, geological, land use maps, and optical images) and on-site investigations with ground deformation estimates obtained through Persistent Scatterer Interferometry (PSI) analyses (e.g., Crosetto et al. 2010). It generally operates at medium or large scales (i.e., 1:100,000–1:25,000 and 1:25,000–1:5,000, respectively; Fell et al. 2008), and its contributions consist in (Figure 1): (i) detection of geomorphologic phenomena not emerging from conventional analyses, field surveys, or bibliographic studies; (ii) verification or modification of landslide boundaries; (iii) assessment of landslide velocity and intensity; and (iv) evaluation of state of activity.

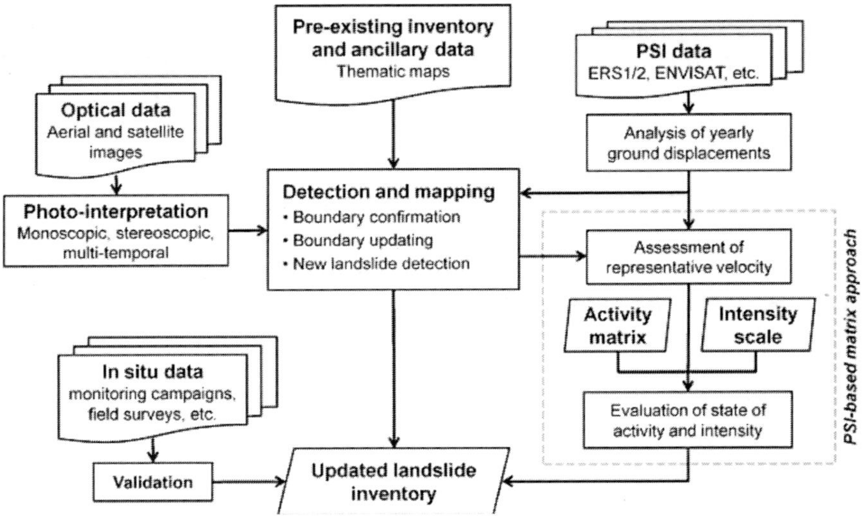

Figure 1. Methodology for the updating of inventory maps and the assessment of state of activity and intensity of slow-moving landslides exploiting the PSI-based matrix approach

Bianchini et al. (2012) give an overview of the whole methodology and assign it the name 'Landslide HotSpot Mapping' (LHSM) to refer to its capability of scanning wide areas and identifying narrow unstable zones characterized by higher landslide hazard. Treasuring of the above-mentioned applications and their findings, this study focuses on the improvement and standardization of the procedure for the assessment of the landslide state of activity and intensity, which represents a crucial phase of the LHSM methodology. To define this procedure, we coin the name 'PSI-based matrix approach' and provide a detailed and step-by-step analysis and discussion of its operational chain. The potentials of the methodology are also shown through the example of Verbicaro (Italy), for which the preexisting landslide inventory is updated using PSI-derived 1992–2011 data from ERS1/2 and RADARSAT1/2 satellites.

2. PSI-BASED MATRIX APPROACH

Since the beginning of the 21st century, Synthetic Aperture Radar Interferometry (InSAR) and PSI techniques are demonstrating their potential for detecting land motions through the analysis of long temporal series of satellite SAR images. Results of PSI processing are sparse grids of radar targets (the so-called Persistent Scatterers, PS), for which the following information are generally provided: Line Of Sight (LOS) estimates of yearly motion rates (i.e., millimeters per year), time series of deformation-acquisition by acquisition (i.e., millimeters), and quality parameters such as PS coherence as well as velocity and height standard deviations (e.g., Crosetto et al. 2010; Ferretti et al. 2001; Werner et al. 2003). Both historical archives of SAR images (e.g., ERS1/2 and JERS data) and currently operational missions (e.g., Envisat, Radarsat1/2, TerraSAR-X, and COSMO-SkyMed) can be exploited to analyze past and recent ground displacements and reconstruct the deformation history of the observed scene with millimeter precision.

PSI-based investigations of natural hazards and, in particular, landslide processes, are successfully being carried out in the last years, demonstrating the flexibility of these techniques to different landslide typologies and dimensions, thanks to their high spatial and temporal resolutions, multitemporal scalability, integrability, and interoperability with conventional on-site investigations (e.g., Bovenga et al. 2006; Cascini et al. 2010; Colesanti and Wasowski 2006; Farina et al. 2006; Herrera et al.2010; Hilley et al. 2004).

2.1 Definition and applicability

The 'PSI-based matrix approach' is a methodology for the assessment (or updating) of the landslide state of activity and intensity, which exploits PSI-derived multitemporal estimates of ground deformation as indicators of activity and magnitude of landslide processes. This approach uses a simplified version of the official classification of the states of activity defined in the multilingual landslide glossary (WP/WLI 1993), distinguishing the following four classes of activity:

- 'Stabilized' (S), not affected by their original causes anymore;
- 'Dormant' (D), potentially being reactivated;
- 'Active' (A), currently moving;
- 'Reactivated' (R), moving after being inactive.

The first two classes are included in the inactive one, comprising all phenomena which did not move in the last 12 months. The stabilized class includes both the artificially stabilized phenomena (artificially protected from their causes through remedial measures) and the naturally stabilized ones, i.e., abandoned (no longer affected by their original causes) and relict (developed under different climatic and geomorphologic conditions from the current ones) landslides. Suspended landslides (moved in the last 12 months but not moving at present) are incorporated in the active class.

With regard to the intensity (Hungr 1997), the PSI-based matrix approach considers the landslide potential to cause damage in terms of movement velocity, exploiting the velocity classification of Cruden and Varnes (1996). Due to the main intrinsic characteristic of PSI-derived estimates—some of which are analyzed critically within the discussion—the methodology is currently applicable just to the classes of Extremely Slow (ES) and Very Slow (VS) movements ($V < 16$ mm/year and 16 mm/year $\leq V < 1.6$ m/year, respectively). This limitation influences also the applicability of the PSI-based methodology in terms of landslide typology, restricting its range to the analysis of phenomena with very slow dynamics, such as deep-seated gravitational slope deformations, creep, and, just partially, rototranslational slides, flows, and complex landslides, as long as their velocities do not overcome the above-mentioned rates.

As observed by Colesanti and Wasowski (2006), landslide-induced displacements detectable through PSI techniques mainly concern postfailure residual deformation and landslide reactivations involving naturally degraded slopes and preexisting shear surfaces. These movements are very suitable to analysis through PSI techniques, since they usually evolve with very low velocities (few tens of centimeters per year) with respect to first-time activations and generally persist over long time periods (several years).

2.2 Landslide representative velocity

The evaluation of spatial distribution and boundaries of landslide phenomena within the investigated area represents the preparatory phase of the PSI-based matrix approach (Figure 1). This step, which is not tackled inside the present paper, can be carried out through conventional geomorphologic approaches, photointerpretation, and even the support from PSI data (e.g., Bianchini et al. 2012; Meisina et al. 2008; Righini et al.2011). Once this preparatory phase is completed, the PSI-based matrix approach starts with the estimation of landslide representative velocities.

The ideal situation for the implementation of this approach is the availability of two or more sets of PSI data acquired over different monitoring intervals; this requirement enables the analysis of the evolution of the deformational behavior of the observed phenomena by comparing velocities and trends between different time spans. The availability of SAR archives responds to this need by providing long temporal stacks of satellite radar imagery for many areas of the Earth's surface and thereby facilitating the understanding of landslide behaviors, distributions of activity (e.g., advancing, retrogressive, enlarging, and diminishing), and states of activity. For the identification of the latter, the possibility of comparing the status and evolution of ground motions in time is a fundamental requirement.

Assuming the availability of both past and recent PSI-derived data, two (or more) different time spans can be considered for the analysis: the historical and present intervals, whose respective temporal coverages can be defined as [D_{H1}–D_{H2}] and [D_{P1}–D_{P2}]. The historical and present representative velocities of each landslide, V_H and V_P, will be evaluated through the analysis of the available PSI data in the [D_{H1}–D_{H2}] and [D_{P1}–D_{P2}] periods respectively, e.g., using ERS1/2 [1992–2001] and ENVISAT [2002–2010] PSI data. Separately for the evaluation of the V_H and V_P, the spatial distribution, number, and density δ (i.e., Persistent Scatterers per square kilometer) of radar targets identified inside the boundary of each landslide are analyzed (Figure 1). A reliable and robust interpretation of landslide motion rates is usually focused only on those phenomena with a minimum density of targets of about 20–30 PS/km^2; this is done to increase the reliability of the representative velocity evaluations (Notti et al.2010). Anyway, this density threshold depends on the PSI data source and, in particular, the spatial resolution of the processed SAR images. In any case, regardless of the landslide spatial extension, at least three to five radar targets should be considered for such an analysis because data retrieved on single targets may not be indicative of a real landslide process but more likely due to single instable structures, e.g., building settlement (Meisina et al. 2008; Notti et al. 2010).

Usually, a simple average of the LOS deformation rates of the different PS belonging to a single phenomenon is used as its representative velocity; this is done by analyzing separately historical and recent PSI data and calculating $V_{H,LOS}$ and $V_{P,LOS}$ for each mapped landslide. Nevertheless, depending of the specific landslide at hand, the use of the maximum value of observed PS velocities (peak velocity)—rather than their average—could be more suitable to describe the analyzed area; this approach follows the criteria suggested by Cruden and Varnes (1996), which promote the use of peak velocities of substantial portions of the displaced masses for the evaluation of landslide intensity.

Using simplified geomorphologic schemes, such as those proposed by Cascini et al. (2010), the conversion of LOS deformation values (V_{LOS}) to the most probable direction of displacement can be performed. For instance, assuming a pure translational mechanism, LOS velocities can be projected along the direction of the maximum slope, i.e., the steepest slope direction, obtaining $V_{H,SLOPE}$ and $V_{P,SLOPE}$ for the historical and present velocities, respectively (e.g., Greif and Vlcko 2011; Herrera et al. 2010). This step was applied for the test area of Verbicaro, as described in the following sections. Alternatively, if PSI data are available for both ascending and descending acquisition modes, the combination of these independent estimations along the two opposite geometries can be performed to better constrain the landslide movements; under the assumption of the absence of N–S horizontal motion, this combination may be used to reconstruct both the vertical and the E–W horizontal deformation components. Subsequently, the highest between these components can be used as the representative velocity for the analyzed phenomenon. Although the theoretical potentials of the combination of ascending and descending data are higher than those associated with the use of a single acquisition geometry, real practice shows significant difficulties in mountainous and hilly areas where the chances of retrieving radar targets in both geometries (ascending and descending) are significantly low due to geometrical visibility and distortions (see discussions).

When the distribution of PS deformation rates inside the landslide boundary is inhomogeneous (i.e., significantly variable PS velocities in the same landslide area), the estimation of a single representative rate for the whole phenomenon may be inappropriate; in those cases, a fragmentation or partition of the landslide area into two or more sectors with different deformation behaviors (and respective representative velocities) may also be performed. This can be done by tracing the boundaries between moving and stable parts of the same landslide mass and/or identifying portions characterized by significantly different deformation velocities (e.g., Greif and Vlcko 2011; Notti et al. 2010; Righini et al. 2011).

2.3 Activity matrices and intensity scales

Once the representative velocities of each landslide have been evaluated, PSI-based activity matrices and intensity scales are exploited to identify its state of activity and intensity (Figure 1). Activity matrices and intensity scales are, respectively, 2D and 1D grids defined in terms of PS average annual velocity. They combine past and recent estimates of land deformation and may also include the information coming from the preexisting inventory map of the investigated area (Figs. 2 and 3). Such tools were used for the first time for medium scale applications by Righini et al. (2011) and Bianchini et al. (2012) and they are here standardized for a generic case study as well as improved through the consideration of the whole range of input data availabilities and combinations.

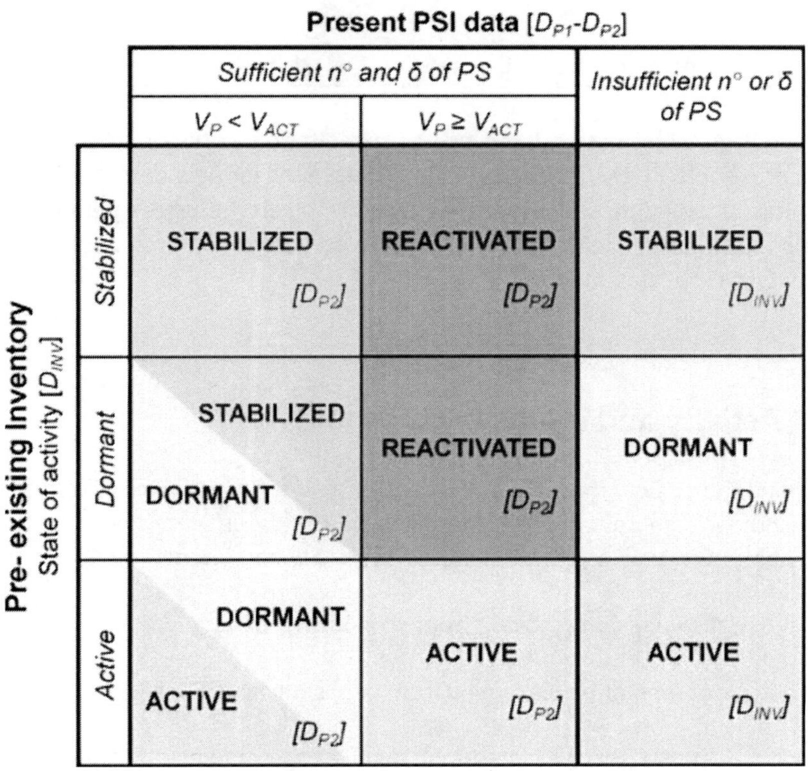

Figure 2. Activity matrix based on the exploitation of present [D_{P1}–D_{P2}] PSI data and time-referenced information about the state of activity (e.g., preexisting inventory [D_{INV}]), usually employed for premapped landslides. V_{ACT}deformation threshold discriminating states of activity

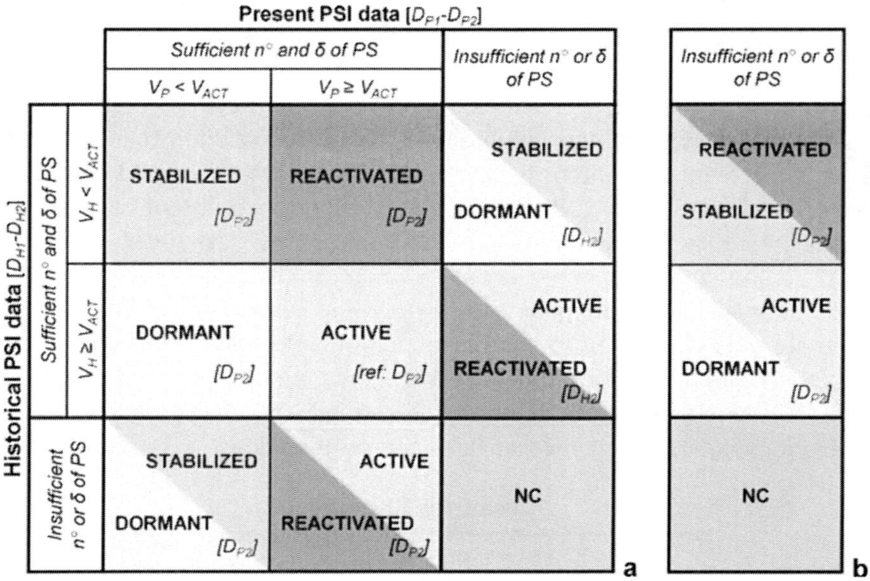

Figure 3. Activity matrix based on the exploitation of historical [D_{H1}–D_{H2}] and present [D_{P1}–D_{P2}] PSI data, usually employed for new detections. Third column of **a** may be substituted with **b** to refer all the assessments to the date [D_{P2}]. V_{ACT} deformation threshold discriminating states of activity, *NC* not classified

2.4 Activity and intensity thresholds

To build the activity matrices and intensity scales, two or more deformation thresholds have to be preliminarily fixed to discriminate moving/stable phenomena as well as different states of activity and intensities:

- V_{ACT}, distinguishing moving from nonmoving areas;
- V_{INT}, discriminating extremely from very slow landslides.

These values are strictly dependent on the case studies at hand and are empirically determined taking into account both the typology of deformation processes affecting the investigated area and the specific characteristics of the available PSI data, e.g., landslide geometry, LOS direction, measurement precision, and relativity of PSI estimates with respect to the reference point. Examples of V_{ACT} velocity thresholds exploited for recent landslide studies at the regional scale range between

±1.5 and ±4.0 mm/year, referred to LOS data (Bianchini et al. 2012; Cascini et al. 2010; Farina et al. 2006; Meisina et al. 2006, 2008; Righini et al. 2011), whereas an example of LOS velocity threshold distinguishing extremely slow from very slow phenomena, V_{INT}, is ±10 mm/year (Bianchini et al. 2012; Righini et al. 2011).

Criteria for choosing the different thresholds are tackled within the discussion and a simple operational methodology suggested for the selection of the thresholds is also experimented for the case study of Verbicaro which is analyzed as test area in the following section.

2.5 Activity matrix for premapped landslides

Past time-referenced information about the state of activity are generally already available for those landslides identified through past investigations, such as phenomena included in preexisting inventory maps; date of reference of their activity estimation is here called D_{INV}. In these cases, a matrix combining this past evaluation with more recent PSI-derived estimates of ground displacements can, therefore, be employed to update or confirm the activity of these phenomena (Figure 2). Input parameters of such a matrix thereby consist of:

- Time-referenced state of activity from a preexisting inventory, defined at date D_{INV};
- Representative velocity V_P, defined in the interval [D_{P1}–D_{P2}], subsequent to D_{INV}.

The nine different combinations that constitute this first typology of activity matrix, and their outcomes in terms of updated activities and respective temporal references, can be summarized as follows:

- Stabilized [D_{INV}] landslides, whose V_P is:

  a. Lower than V_{ACT} → Stabilized [D_{P2}]

  b. Higher than (or equal to) V_{ACT} → Reactivated [D_{P2}]

  c. Not assessable → Stabilized [D_{INV}]

- Dormant [D_{INV}] landslides, whose V_{P} is:
    a. Lower than V_{ACT} → Stabilized/Dormant [D_{P2}]
    b. Higher than (or equal to) V_{ACT} → Reactivated [D_{P2}]
    c. Not assessable → Dormant [D_{INV}]
- Active [D_{INV}] landslides, whose V_{P} is:
    a. Lower than V_{ACT} → Dormant/Active [D_{P2}]
    b. Higher than (or equal to) V_{ACT} → Active [D_{P2}]
    c. Not assessable → Active [D_{INV}].

As pointed out by Bianchini et al. (2012), this typology of activity matrices are based on a 'conservative' or 'cautionary' approach; in other words, even if PSI data register low movement rates (i.e., V_{P} is lower than the chosen threshold, V_{ACT}), the state of activity recorded in the preexisting inventory cannot be lowered (i.e., from dormant to stabilized, or from active to dormant), unless field evidences and in situ monitoring data confirm an actual lowering of activity (e.g., in Figure 2, premapped dormant and active landslides showing recent PSI-derived velocity lower than V_{ACT}). This explains the ambiguous (double) classification of the combinations 2a and 3a within the above-mentioned list, which reflects the uncertainty in the assessment of the state of activity, resolvable only through field validation.

As observable from the above described matrix, the temporal coverage of PSI data influences the final time reference of the updated inventory; this will correspond to the last acquisition of the PSI monitoring interval. Thus, for those phenomena whose updating is performed by using PSI estimates, the reference will be [D_{P2}]; for instance, the employment of Radarsat1 [2003–2009] data produces an updated inventory referenced to 2009. On the other hand, landslides with total absence or insufficiency of PSI-based deformation velocities (i.e., not assessable V_{P}) will retain the time reference of the inventory from which they belong to, i.e., [D_{INV}]. Causes and possible solutions to partial and total lack of PS within the landslide boundaries are tackled within the discussion.

2.6 Activity matrix for newly detected landslides

For newly detected landslides or those premapped phenomena lacking in past evaluation of activity, an activity matrix combining historical and recent deformation PSI data can be exploited. In this case, the input parameters of the matrix are:

- Historical representative velocity V_H, defined in the interval $[D_{H1}–D_{H2}]$;
- Present representative velocity V_P, defined in the interval $[D_{P1}–D_{P2}]$, subsequent to $[D_{H1}–D_{H2}]$.

The nine different combinations constituting this second type of activity matrix, and their respective outcomes, can be summarized as follows (Figure 3a):

- Landslides whose V_H is lower than V_{ACT} and whose V_P is:
  a. Lower than V_{ACT} → Stabilized $[D_{P2}]$
  b. Higher than (or equal to) V_{ACT} → Reactivated $[D_{P2}]$
  c. Not assessable → Stabilized/Dormant $[D_{H2}]$
- Landslides whose V_H is higher than (or equal to) V_{ACT} and whose V_P is:
  a. Lower than V_{ACT} → Dormant $[D_{P2}]$
  b. Higher than (or equal to) V_{ACT} → Active $[D_{P2}]$
  c. Not assessable → Active/Reactivated $[D_{H2}]$
- Landslides with not assessable V_H and whose V_P is:
  a. Lower than V_{ACT} → Stabilized/Dormant $[D_{P2}]$
  b. Higher than (or equal to) V_{ACT} → Active/Reactivated $[D_{P2}]$
  c. Not assessable → not classified (NC).

Ambiguous classification of the combinations 1c, 2c, 3a, and 3b (Figure 3a) are caused by the absence or insufficiency of PSI data for either the historical or the present monitoring interval; it reflects the uncertainty in the activity assessment due to the impossible comparison between movement conditions before and after a certain date (i.e., indiscernible history of deformation). As for the previous matrix, field checks and validation may clear up this uncertainty.

As well as for the first typology of matrices, the temporal coverage of PSI data also influences the time reference of the evaluations for newly detected phenomena; for those phenomena whose updating is performed by using historical estimates, the reference will be [D_{H2}], while for present data it will be [D_{P2}]. For instance, the employment of Envisat [2002–2010] data produces an updated inventory referenced to 2010, while the use of ERS1/2 [1992–2001] data references the updated phenomenon to 2001. In case of new phenomena lacking in both past and recent estimates, they will be attributed to the NC category (Figure 3a).

An alternative for the matrix in Figure 3a, is the substitution of its third column with the modified version which is shown in Figure 3b and corresponds to the typology of matrix used by Righini et al. (2011). In this case, when the present PSI data are insufficient within the landslide boundary, but the information on the historical landslide velocity is available, conditions 1c and 2c are modified as follows:

1c) V_H lower than V_{ACT} → Stabilized/Reactivated [D_{P2}]

2c) V_H higher than (or equal to) V_{ACT} → Dormant/Active [D_{P2}].

The difference with the original matrix in Figure 3a consists in the reference time of the assessment of the state of activity. While in the original matrix the determination is considered ambiguous between the states of stabilized and reactivated at time [D_{H2}], in the modified version shown in Figure 3b, the assessment of the activity is done at the reference time [D_{P2}] and the uncertainty is between the states of dormant and active continuous.

2.7 Intensity scale

The evaluation of landslide intensity is performed by comparing the PSI-derived landslide velocity not only with the activity threshold, V_{ACT}, but also with the intensity threshold, V_{INT}, which discriminates extremely from very slow velocities (Figure 4). The only input parameter for the employment of this scale is the representative velocity of each landslide, in the present (or historical) interval, i.e., V_P (or V_H). The four different combinations constituting the intensity scale, and their outcomes in terms of intensities and respective temporal references, can, therefore, be summarized as follows:

1 V_P (or V_H) is lower than V_{ACT} → Negligible [D_{P2}] (or [D_{H2}])

2 V_P (or V_H) is higher than (or equal to) V_{ACT} but lower than V_{INT} → Extremely Slow [D_{P2}] (or [D_{H2}])

3 V_P (or V_H) is higher than (or equal to) V_{INT} → Very Slow [D_{P2}] (or [D_{H2}])

4 Neither V_P nor V_H are assessable → Not Classified (NC).

Present (or **historical**) **PSI data** [D_{P1}-D_{P2}] (or [D_{H1}-D_{H2}])

Sufficient n° and δ of PS			*Insufficient n° or δ of PS*
V_P (or V_H) < V_{ACT}	$V_{ACT} \leq V_P$ (or V_H) < V_{INT}	V_P (or V_H) ≥ V_{INT}	
NEGLIGIBLE *[D_{P2}] or [D_{H2}]*	**EXTREMELY SLOW** *[D_{P2}] or [D_{H2}]*	**VERY SLOW** *[D_{P2}] or [D_{H2}]*	**NC**

Figure 4. Example of intensity scale based on the exploitation of present [D_{P1}–D_{P2}] or historical [D_{H1}–D_{H2}] PSI data. V_{ACT} deformation threshold discriminating states of activity, V_{INT} deformation thresholds discriminating intensities, *NC* not classified

As for the activity matrices, the coverage of PSI data influences the time reference of the intensity evaluation, which will correspond to the last acquisition of the PSI monitoring interval, i.e., [D_{P2}] when using recent data or [D_{H2}] when using historical data. Total absence or insufficiency of PSI data induce the analyzed phenomenon to be attributed to the NC category (Figure 4).

3. TEST AREA OF VERBICARO, CALABRIA (ITALY)

A sheet pile of basement thrust units piled up and overthrust toward the Apulian foreland during the Neogene characterizes the geological background of the Calabrian peninsula in Southern Italy (Cucci and Cinti 1998). Hercinian crystalline and metamorphic basement units crop out; these are overlaid by Mesozoic and Tertiary sedimentary rocks, covered by Upper Neogene and Quaternary sediments (Amodio Morelli et al. 1976). The whole peninsula shows mountainous and hilly morphologies and is highly susceptible to landslides due to high slope gradients, intense rainfalls, weathered and faulted crystalline-metamorphic rocks, and weakly consolidated sedimentary sequences. Associated with extensive human activities, such as urban sprawl and deforestation, these factors contribute to production of steep and unstable slopes (Carrara and Merenda 1976).

A small area in the province of Cosenza, in the northern sector of Calabria Region, was chosen as an example for the application of the PSI-based matrix approach: a sector extending ~14 km^2 and located within the municipality of Verbicaro. This choice was driven by the geomorphologic characteristics of this area and its assortment in terms of landslide activity and intensity as well as availability of PSI data for the different phenomena affecting its urban and suburban infrastructures.

The geological setting of Verbicaro is characterized by tectonostratigraphic units piled up and locally dislocated by normal faults. The Alpine basement assemblage, previously deformed, structurally overlies the Appennine carbonate succession and its Paleogene-Lower Miocene cover (Van Dijk et al. 2000; Ferranti et al. 2009). The outcropping lower unit is made of a thick Triassic dolomite interval and a Jurassic cherty limestone succession of variable thickness (Iannace et al. 2005). Dolomites and well-stratified limestones are in stratigraphic continuity and derive from the Mesozoic carbonate platform; this unit represents a deformed complessively 4,000-m thick sequence, which now forms the backbone of the Pollino Range (Cucci and Cinti1998) and outcrops eastward within the area of interest (Figure 5). Mesozoic limestones sometimes include basic volcanic elements. The carbonate succession is covered, in disconformity, by a Paleocene-Lower Miocene open shelf turbiditic calcarenites and coarse calcareous breccias. The upper element is composed of alpine metasedimentary rocks that mostly comprise schistes, fillads, and calceshistes (Scisti del Fiume Lao Formation); these deep basinal metapelitic rocks highly tectonized, outcrop throughout the investigated area and have been included in the ophiolite-bearing Frido Unit of the Calabrian borderland, overlying platform carbonates (Ferranti et al. 2009; Iannace et al. 2005; Van Dijk et al. 2000). Upper Pliocene-Pleistocene undisturbed terrigenous deposits fill the area, covered by recent Quaternary continental deposits (Cucci and Cinti 1998). Outcropping terrains are sliced by Quaternary high-angle strike-slip and extensional faults (Ferranti et al. 2009) that cut the ridges of the test area westward into fault-bounded blocks.

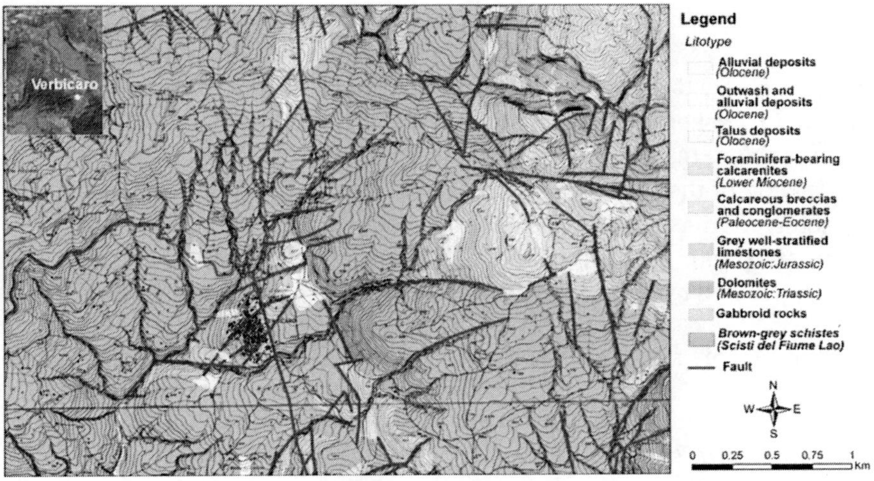

Figure 5. Geological setting (modified from geological map 1:25,000, Regional Cartographical Centre, 1969) and geographic location (*inset*) of Verbicaro, Northern Calabria (Italy), overlapped on a 1:10,000 topographic map

The preexisting inventories available for the test area are the Hydrogeological Setting Plan (Piano Stralcio per l'Assetto Idrogeologico, PAI) produced by the River Basin Authority of Calabria Region in 2001 with following modifications until 2006 and the Italian Landslide Inventory (Inventario Fenomeni Franosi in Italia, IFFI) produced by the Italian Institute for Environmental Protection and Research in 2007. Both PAI and IFFI inventories were created on a reference scale of 1:10,000 and employed aerial photointerpretation, field campaigns, and local databases. IFFI was compiled updating the PAI inventory; hence, it includes all the information of the PAI and was thereby used as the primary map to be analyzed for Verbicaro. Inside this inventory, three different states of activity are distinguished, i.e., active (including active, reactivated, and suspended), dormant, and stabilized (including artificially and naturally stabilized).

The municipality of Verbicaro has been classified by the PAI as a municipality affected by very high hydrogeological risk, determined on the basis of the effects on settlement, manufactures, and population, as defined by the Italian Decree Law no. 180/98 concerning urgent measures for risk prevention; the PAI zonation also estimates the extent of the high and very high risk areas of the municipality—whose total extent is ~32.6 km^2—as exceeding 0.15 km^2, thus collocating the municipality in the highest risk class. The whole municipality is characterized by a mountainous morphology with dense network of narrow rivers and waterways and is affected by several large landslides, especially in the urban and suburban areas of the main village, which were selected as test sites for the application of the PSI-based matrix approach. The last important activation of some of these phenomena dates back to February 2010 when the provincial road, which guarantees the main access to the village, was partially buried by debris moved downhill from the overlying slope.

Within the area selected as test site, a total of 24 landslides are mapped in the preexisting inventory of IFFI. For the sake of simplicity and to identify them easily, we have assigned these unstable areas a short code, going from L1 to L24 (Figure 6 and Table 1). Most of the identified phenomena are classified as rotational/translational slides (i.e., L6 to L13, and L16 to L21) and complex movements (i.e., L1 to L5, and L22 to L24), and involve mainly low-grade metamorphic rocks and occasionally well-stratified limestones. A dormant complex landslide, L22, constitutes the larger phenomenon affecting the area, with an extension of ~2.1 km^2. It involves the whole northwestern sector of the village of Verbicaro and is partially overlaid by smaller-scale active slides and complex landslides (e.g., L5 to L7) as well as a flowlike phenomenon characterized by N to S motion direction (L15) and, to the S, a complex landslide (L1) whose foot reaches the village downstream. The larger portion of Verbicaro village is affected by phenomena moving towards W and SW, mainly complex (L2 to L4) and rotational/translational slides (e.g., L10, L11, L17, and L18). The latter overlay a dormant complex landslide (L24) involving the whole left bank of the incision where L1 flows. Despite limited extension of these phenomena, generally less than 0.07 km^2, they represent a major cause of risk for local population and urban structures, since their main bodies involve the whole portion of the village located on the W and SW facing slopes. Only small portions of the village are excluded from the unstable zones identified by the PAI and IFFI inventories, such as the elongated zone between phenomena L4 and L18 and that between L2 and L14; however, the geomorphologic indicators and the proximity to active landslides suggest potential enlargement of the landslide boundaries to include also some sectors of the urban area, as described in the following sections.

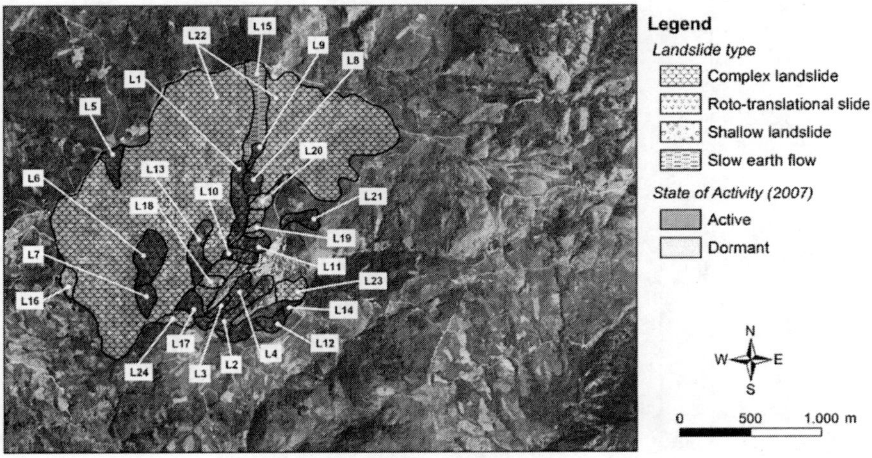

Figure 6. Preexisting landslide mapping in the area of Verbicaro, overlapped on a color orthophoto acquired in 2000

5. APPLICATION OF THE PSI-BASED MATRIX APPROACH

5.1 Input data

For the implementation of the PSI-based matrix approach to the area of Verbicaro, the IFFI inventory was used as the primary map to be updated, setting the D_{INV} as 2007, corresponding to the reference year of this inventory. As for the PSI data, the following three sets were employed (Figure 6):

Table 1. Summary and statistics of the PSI-based assessment of landslide states of activity and intensity in the area of Verbicaro, Italy, with respective reference years. *α* azimuth, *β* slope, *A* active, *R* reactivated, *D* dormant, *S*stabilized, *N* negligible, *ES* extremely slow, *VS* very slow, *NC* not classified, – absence of PS

ID	Preexisting information [2007]					ERS1/2 [1992–2000]		RSAT1 [2003–2010] and RSAT2 [2009–2011]			Updated inventory	
	Area (km^2)	Type	Activity	α (°)	β (°)	$V_{H,LOS}$ (mm/year)	$V_{H,SLOPE}$ (mm/year)	$V_{RSAT1,LOS}$ (mm/year)	$V_{RSAT2,LOS}$ (mm/year)	$V_{P,SLOPE}$ (mm/year)	Activity [year]	Intensity [year]
L1	0.056	Complex landslide	A	174	13.9	0.1	1.3	−0.2	−1.0	18.0	A [2011]	VS [2011]
L2	0.062	Complex landslide	A	227	12.8	–	–	−19.2	−20.0	38.8	A [2011]	VS [2011]
L3	0.009	Complex landslide	A	242	14.0	−16.6[a]	31.9[a]	–	−19.0[a]	30.2[a]	A [2007]	NC
L4	0.050	Complex landslide	A	216	14.1	−18.4	48.2	−19.9	−20.9	47.4	A [2011]	VS [2011]
L5	0.029	Complex landslide	A	202	14.6	–	–	–	–	–	A [2007]	NC
L6	0.064	Rototranslational slide	A	202	13.7	–	–	−10.5[a]	−11.1[a]	36.1[a]	A [2007]	NC
L7	0.025	Rototranslational slide	A	210	15.9	–	–	–	–	–	A [2007]	NC
L8	0.019	Rototranslational slide	A	170	16.1	–	–	–	–	–	A [2007]	NC
L9	0.010	Rototranslational slide	A	194	14.9	–	–	–	−0.2[a]	0.8[a]	A [2007]	NC
L10	0.03	Rototranslational slide	A	255	15.0	−16.1	27.7	−19.9	−21.9	31.2	A [2011]	VS [2011]
L11	0.069	Rototranslational slide	A	251	21.2	−9.7	14.8	−16.0	−14.5	21.4	A [2010]	VS [2010]
L12	0.020	Rototranslational slide	A	254	10.8	–	–	–	–	–	A [2007]	NC
L13	0.049	Rototranslational slide	A	182	15.0	−9.6	55.5	−9.0	−9.8	77.2	A [2010]	VS [2010]
L14	0.057	Shallow landslide	A	239	11.8	−7.6	16.1	−14.1	−12.0	24.7	A [2010]	VS [2010]
L15	0.079	Slow Earth flow	D	212	16.6	–	–	−1.0[a]	−0.4[a]	2.4[a]	D [2007]	NC
L16	0.019	Rototranslational slide	D	178	18.4	–	–	–	–	–	D [2007]	NC
L17	0.025	Rototranslational slide	A	257	18.0	−10.2[a]	16.2[a]	–	−17.0[a]	22.7[a]	A [2007]	NC
L18	0.014	Rototranslational slide	D	277	20.0	−13.7[a]	20.0[a]	−16.3	−17.3[a]	20.7	R [2010]	VS [2010]
L19	0.007	Rototranslational slide	D	264	16.9	−22.0[a]	34.9[a]	–	−30.5[a]	40.4[a]	D [2007]	NC
L20	0.015	Rototranslational slide	D	275	18.9	−14.9[a]	22.3[a]	−3.2	−7.0	8.7	R [2011]	ES [2011]
L21	0.034	Rototranslational slide	A	243	23.2	−4.6	7.0	−1.5	−13.0	17.6	A [2011]	VS [2011]
L22	2.107	Complex landslide	D	181	15.5	−5.4	30.7	−10.3	−6.8	89.2	R [2010]	VS [2010]
L23	0.031	Complex landslide	D	216	16.3	−1.0	2.5	−1.5	−1.3	3.3	D/S [2010]	N [2010]
L24	0.107	Complex landslide	D	267	16.9	−16.5	25.9	−18.6	−19.8	26.0	R [2011]	VS [2011]
L25	New detection			233	18.2	−6.8	12.8	−8.1	−9.9	16.0	A [2011]	VS [2011]
L26	New detection			256	13.1	−11.6	20.8	−16.7	−12.1	24.5	A [2010]	VS [2010]

[a]Less than 4 PS within the landslide boundary (unreliable estimation)

- ERS1/2 stack consisting of 78 scenes in descending mode, spanning the interval 17/05/1992–20/12/2000, VV polarization, look angle 23.4 °, and directional cosines: 0.38972 (E), −0.07976 (N), 0.91747 (Z).

- Radarsat1 Standard Beam S3 stack consisting of 80 scenes in descending mode, spanning the interval 19/03/2003–17/01/2010, HH polarization, look angle 34.5 °, and directional cosines: 0.55423851 (E), −0.11857239 (N), 0.82386907 (Z).

- Radarsat2 Standard Beam S3 stack consisting of 33 scenes in descending mode, spanning the interval 10/01/2009–06/04/2011, HH polarization, look angle 33.9 °, and directional cosines: 0.54939554 (E), −0.09516750 (N) and 0.83012510 (Z).

D_{H1} and D_{H2} were thereby set to 1992 and 2000, while D_{P1} and D_{P2} corresponded to 2003 and 2010 for the Radarsat1 stack, and 2009 and 2011 for the Radarsat2 stack.

Although the use of both ascending and descending data stacks is generally suggested to implement the approach, for Verbicaro, only data stacks with descending acquisition geometry were used. This is due to the local morphology of the area which is predominantly oriented in the SW, W, and NW directions and consequently prone to be analyzed with PSI descending data.

Historical ERS1/2 data were processed with the PSInSAR (Permanent Scatterers Interferometric SAR) technique by TRE S.r.l. (Ferretti et al. 2001) in the framework of the Piano Straordinario di Telerilevamento Ambientale (PST-A) project, while Radarsat1 and Radarsat2 stacks were processed—into two separate stacks—by e-GEOS with the Persistent Scatterers Pairs (PSP) technique (Costantini et al. 2000) in the framework of the European project Services and Applications for Emergency Response (SAFER). The resulting stacks of radar targets are characterized by similar spatial distribution and density, even given the different approaches employed for the processing of the available stacks, i.e., the Persistent Scatterers concept, used by the PSInSAR technique, and interferogram stacking concept, used by the PSP. Higher density of targets is retrieved for build-up sectors of Verbicaro and exposed rocks characterizing the eastern portion of the test area, while the densely vegetated zones prevented stable scatterers to be identified (Figure 7). Reference points of each PS stack (zero deformation points) were selected, taking into account not only the phase stability of the PS candidates analyzed during the processing but also the geological setting of these coherent points and verifying the absence of local-scale motions affecting the areas selected as reference. In particular, three different locations over postorogene Plio-Pleistocene deposits, between 30 and 60 km E of the area of Verbicaro, were selected; the distance of the three reference points from Verbicaro and the choice of three different locations are due to the PSI data belonging to stacks of imagery with slightly dissimilar spatial coverage (extension of the processing area) and different coherence variations in time and space.

PS velocities showed higher variability in space for the Radarsat2 stack in the whole area of Verbicaro (Figure 7c). This is justified by the lower number of scenes used to perform the PSP processing, the short temporal interval (only 2 years) and significant orbital and atmospheric errors, and the consequent lower precision in the estimated velocities. Standard deviations of Radarsat2 targets ranged between 1.1 and 4.1 mm/year, unlike the lower values recorded for the other two stacks, i.e., 0.6–1.0 mm/year for Radarsat1 and 0.4–0.8 mm/year for ERS1/2.

5.2 Preparatory phase

The ancillary data used to support photo and radar interpretation included topographic maps at scales 1:10,000 (produced in the 1950s by the Regional Cartographic Center) and 1:25,000 (1990s; Italian Military Geographic Institute), a geological map at 1:25,000 (1960s; Regional Cartographic Center), a Digital Terrain Model (DTM) with 20-m resolution and its derived maps (1990s; Italian Military Geographic Institute), and colored aerial orthophotos with 1-m resolution (1998–1999). The preexisting inventory of the IFFI project was first updated and improved combining a monoscopic photo interpretation of the aerial orthophotos with the radar interpretation of PSI-derived ground motion data in 1992–2000 (ERS1/2 descending) and 2003–2011 (Radarsat1 and Radarsat2 descending). The methodology employed to update the boundaries of premapped phenomena and to identify new landslides followed the approach of LHSM described by Bianchini et al. (2012). Orthophotos, DTM, and derived layers (aspect, slope, and hillshade) facilitated the identification of landslides in the vegetated areas outside the Verbicaro village, while PSI data were mostly used to update the landslide boundaries within the built-up areas. For eight premapped landslides, we applied modification to their boundaries (Figure 8); an example is the enlargement of the boundary of L11 in the E direction to include the clearly recognizable head, minor and major scarps (Figure 9), or the extension of L2 to the NNE direction to include a nonpreviously mapped portion of the urban area.

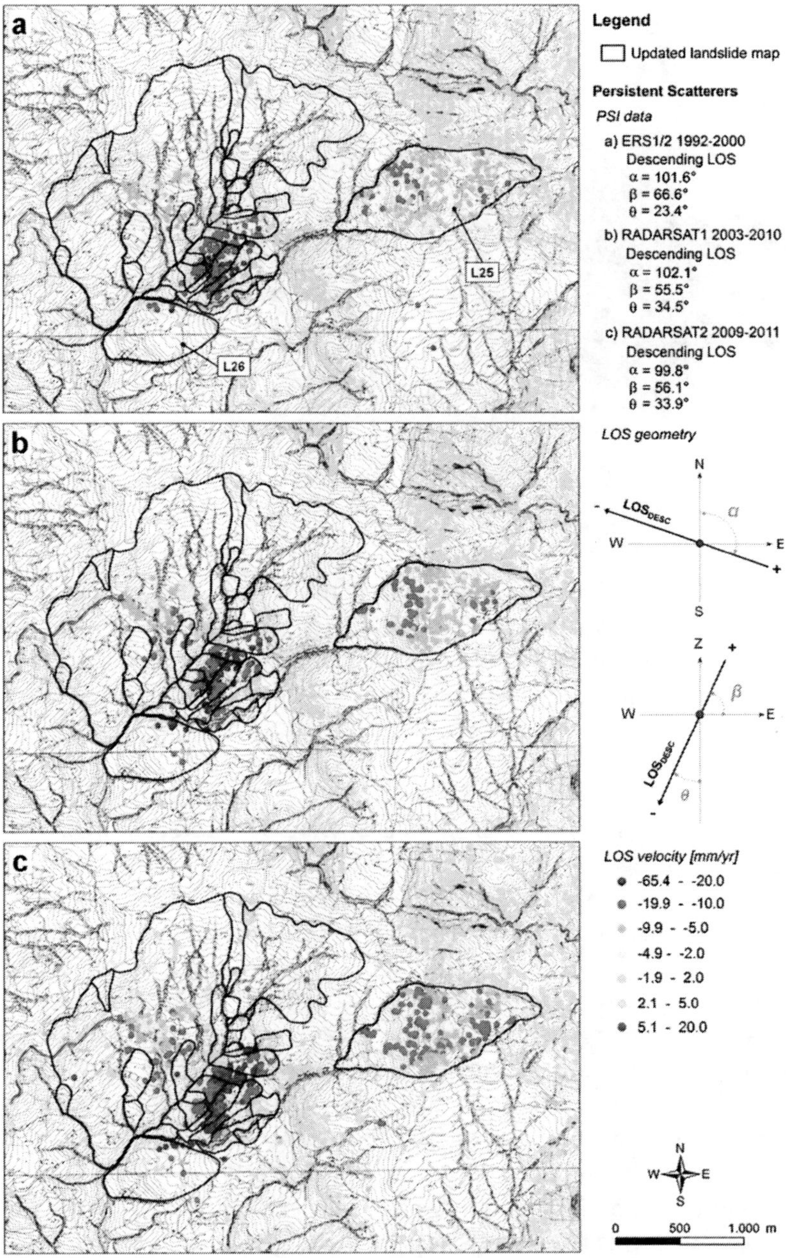

Figure 7. Updated landslide inventory and PSI-derived deformation in 1992–2001 (**a**): ERS1/2 data, descending), 2003–2010 (**b**): RADARSAT1 data, Standard Beam mode S3, descending), and 2009–2011 (**c**): RADARSAT2 data, Standard Beam mode S3, descending) in the area of Verbicaro, overlapped on a 1:10,000 topographic map

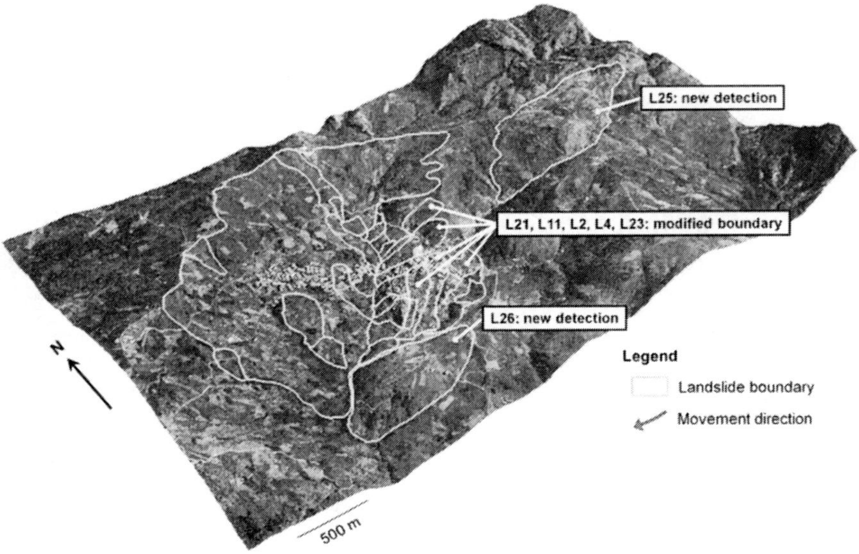

Figure 8. 3D view of the updated landslide mapping in the Verbicaro village, captured from SW at ~45 ° angle. Heights of the DEM are amplified by a factor of 1.3. Direction of landslide movements and main modifications to the preexisting inventory are also shown

Two additional phenomena, L25 and L26, were also detected by exploiting the information gathered from optical and radar data; both can likely be attributed to the typology of complex movements. L25 affects the eastern sector of the test area and involves a sparsely vegetated SW-facing slope less than 1 km far from the village of Verbicaro and characterized by weathered bare rocks made of well-stratified limestones, calcareous breccias, and conglomerates, partially covered by alluvional deposits. Several blocks of sizes variable between few cubic centimeters and decimeters can be observed in this area (Figure 10). Their movement is attributable to the typologies of falls, topples or slides, depending on local factors such as the geometrical proportions of each block (height to width ratio) and the slope inclination. A significant amount of PS in all the three available stacks was retrieved within the boundary of L25: 200 ERS1/2, 420 Radarsat1, and 540 Radarsat2 targets. We observe similar spatial distributions of LOS velocities within the landslide, with higher values concentrated in the foot (up to −21.3 mm/year in 2009–2011) where the slope is steeper (up to 40 °) and thus closer to the LOS geometry; lower rates (−5 to −10 mm/year in all the time intervals) are observed in the main body where the morphology is gentler. L26 involves a more densely vegetated WSW-facing slope in the southwestern portion of the suburban area of Verbicaro, which lies on metamorphic rocks and, in its northern portion, limestones. Its morphology is gentler than that of L25, with slopes averagely 13 °, but never exceeding 25 °. The presence of vegetation here hampers the identification of PS and induces the three data stacks to be populated by less than 80 PS in total. Velocities are higher in the northern portion of the landslide body where the orientation of the slope is closer to the descending LOS geometry and reach no more than 25–30 mm/year in the direction away from the sensors in all the available stacks.

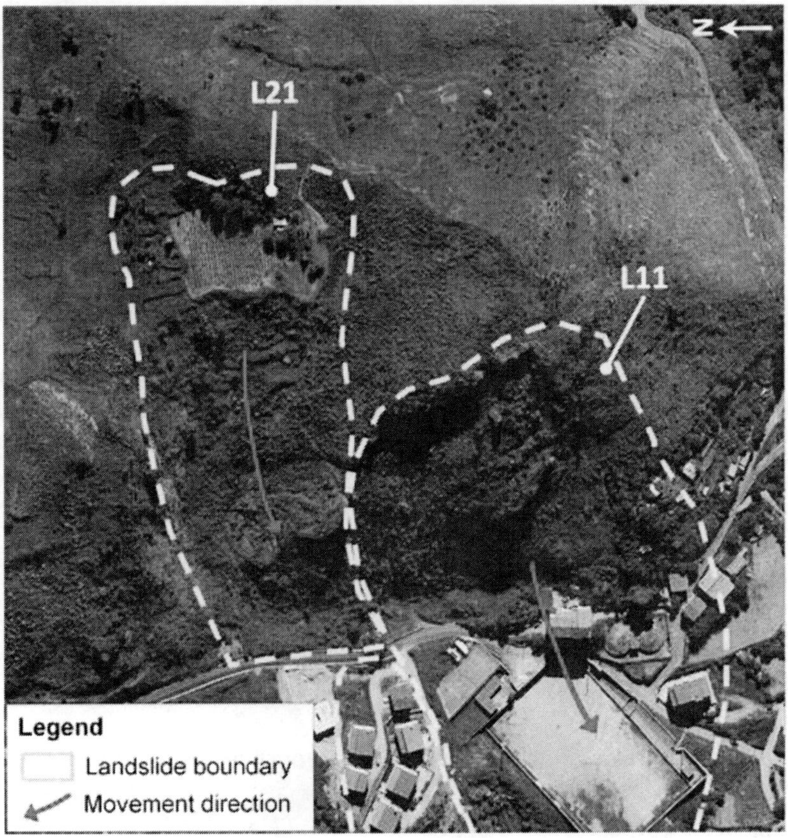

Figure 9. Aerial view of landslides L21 and L11after the updating of their boundaries through photo- and radar interpretation (modified from Bing Maps© Microsoft)

Figure 10. 3D view of landslide L25 (**a**) and two Street Views (© Google) captured in 2011 within its boundaries, (**b**) and (**c**)

In some other sectors of the test area, moving PS were not considered as part of nonmapped landslides because geomorphologic and geological evidences did not show actual presence of landslide processes. Moreover, some single unstable targets of the ERS1/2 and Radarsat1 stacks were attributable to settlements of single structures and not indicative of landslide occurrences.

5.3 Implementation

The representative velocities of the 26 landslides (L1 to L26) in the historical ($V_{H,LOS}$) and present ($V_{P,LOS}$) intervals were evaluated as the average yearly velocity of the available ERS1/2 and Radarsat1/2 PS data included within each landslide area. Given the availability of two different stacks of recent PSI data (Radarsat1 and Radarsat2) and considering their slightly different acquisition geometries, we preliminarily assigned two different representative velocities to each landslide, $V_{RSAT1,LOS}$ [2003–2010] and $V_{RSAT2,LOS}$ [2009–2011]. For both historical and recent PSI data, the calculation of the representative velocities was performed only for those unstable areas where a sufficient number of PS data was retrieved (at least 4 PS within the landslide boundary).

Using a simplified geomorphologic scheme, we assumed the motions affecting the area to be purely translational mechanisms; thus, we converted LOS deformation data in the descending mode (V_{LOS}) to the most probable direction of displacement, i.e., the direction of the maximum slope (steepest slope direction). The conversion to slope velocity (V_{SLOPE}) was performed—for each landslide—using the equations:

$$V_{SLOPE} = \rho_{L2S} \cdot V_{LOS} \tag{1}$$

$$\rho_{L2S} = (E_{LOS} \cdot E_{SLOPE} + N_{LOS} \cdot N_{SLOPE} + Z_{LOS} \cdot Z_{SLOPE})^{-1} \tag{2}$$

where the value ρ_{L2S} represents the conversion factor of LOS to SLOPE values, and*E*, *N*, and *Z* are the directional cosines of the LOS and the SLOPE vectors in the east, north, and zenith directions respectively.

The azimuth (α) and slope (β) angles of the steepest slope direction (SLOPE) were evaluated, respectively, as the average of the aspects and slopes of the DTM cells included within the updated landslide boundaries (Table 1) and then used to estimate the directional cosines of the SLOPE, employing the formulas:

$$E_i = sin\alpha_i \cdot cos\beta_i \tag{3}$$

$$N_i = cos\alpha_i \cdot cos\beta_i \tag{4}$$

$$Z_i = sin\beta_i \tag{5}$$

Considering the cell resolution of both the ERS1/2 and Radarsat1/2 imagery (i.e., 20–30 m on the ground), the spatial resolution of the employed DTM (20 m) was considered appropriate for this calculation.

For the implementation of the matrix approach, we used the ERS1/2-based velocity estimate as the historical representative velocity ($V_{H,SLOPE}$), while as the present velocity ($V_{P,SLOPE}$), the highest between $v_{RSAT1,SLOPE}$ and $v_{RSAT2,}$

$_{SLOPE}$ and the respective reference year [D_{P2}] (i.e., 2010 or 2011) was prudentially selected. Given the standard deviations of the three sets of PSI data available for Verbicaro (mentioned above), the LOS value of ±2 mm/year can be generally considered as an appropriate threshold discriminating moving from nonmoving targets (green PS in Figure 6). The activity threshold V_{ACT} is set by taking into account the average reprojection factor (ρ_{LS}) of all the landslides mapped in Verbicaro with respect to the three descending LOS of ERS1/2, Radarsat1, and Radarsat2. This average factor has the value of 2.5 for Verbicaro; it means that, for this area, the LOS directions are able to detect approximately 40 % of slope-oriented displacements; hence, an increment of 150 % is needed to convert LOS to SLOPE estimates. The V_{ACT} to be applied to the SLOPE-projected velocities was thereby set as ±5 mm/year.

V_{INT} of ±13 mm/year was selected to discriminate extremely slow from very slow phenomena; the official threshold of 16 mm/year (Cruden and Varnes 1996) was precautionarily reduced by ~20 % to account not only for the difference between peak values and temporally averaged velocities (PSI-derived estimates are averages over long observation periods) but also for the spatial averaging that was performed to assess the representative velocity of each analyzed phenomenon. The application of identical thresholds for both ERS1/2 and Radarsat1/2 data do not represent major source of errors in the present application, since LOS velocities of all the PSI stacks were preliminarily reprojected along the slope directions and so cannot be considered influenced by the slightly different orientations of the employed LOS.

6. RESULTS

Active complex landslides, L1 and L4, and rototranslational slides, L10, L11, L13, L14, and L21, showed sufficient number of both historical and present PS data within their boundaries (Table 1, Figure 11). According to the activity matrix shown in Figure 2, specifically devoted to premapped phenomena, only the present representative velocities were considered for the application of the PSI-based matrix approach. All these landslides revealed $V_{P,SLOPE}$ exceeding the activity threshold of ±5 mm/year and were thereby reattributed to the state of active (*A*), with new reference year of 2011 (L1 to L4, L21) or 2010 (L11, L13, L14)—depending on the Radarsat stack showing the highest rates. Similar to these phenomena, L22 to L24, previously classified as dormant complex landslides, were updated using the matrix of Figure 2; in particular, while for L22 and L24, the $V_{P,SLOPE}$ exceeded the activity threshold and induced them to be classified as reactivated (*R*), for L23 we observed a representative velocity lower than the threshold suggesting the stabilization of this phenomenon. However, following the conservative principle of not lowering the activity before confirming the actual lowering through on-site validation, we attributed to L23 the ambiguous state of dormant/stabilized (*D/S*). As for the assessment of the intensity of these premapped phenomena, all were classified as very slow (VS), since their $V_{P,SLOPE}$ were all higher than the intensity threshold of ±13 mm/year (Figure 4); an exception was the L23 which was attributed to the class of negligible (*N*), since its velocity $V_{P,SLOPE}$ was of only a few millimeters per year (Table 1).

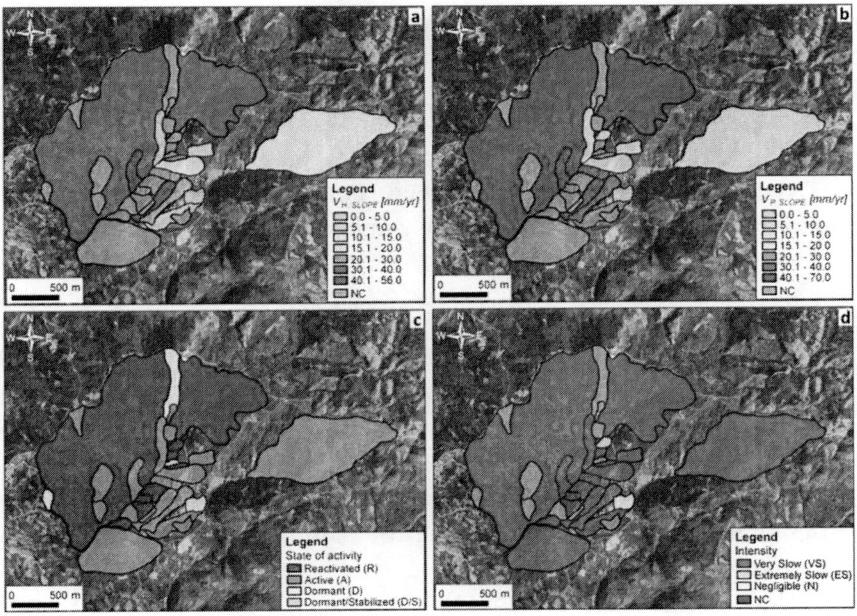

Figure 11. Historical (**a**) and present (**b**) representative velocities, and states of activity (**c**) and intensity (**d**) of landslides mapped in Verbicaro, overlaid on a color orthophoto. Summary and statistics of PSI-derived data and main modifications to the preexisting landslide inventory are shown in Table 1

Complex landslide, L2, active in 2007, and rototranslational slides, L18 and L20, dormant in 2007, showed typical situations occurring when one or more stacks of PSI data are not available or insufficiently distributed within the landslide boundaries. For L2 and L20, PSI-derived estimates of ground deformation were only available (or reliable) for the present scenario (i.e., Radarsat1/2 stacks; Figure 11b). In this particular case, given that these phenomena belong to the premapped landslides and that the use of historical PSI data is unnecessary (Figure 2), no difference in the application of the methodology was applied with respect to the previously described cases: for L2, the active (*A*) state was confirmed and the very slow intensity (VS) was attributed, given that its $V_{P,SLOPE}$ exceeded both the V_{ACT} and the V_{INT}. On the other hand, for L20, the state of activity was updated to reactivated (*R*) and the intensity to extremely slow (ES) in 2011, as its $V_{P,SLOPE}$ was between V_{ACT} and V_{INT} (Figs. 4 and 11). Phenomenon L18 showed a particular scenario where only one of the present stacks of PSI-derived estimates is available, in particular, the Radarsat1 one; Radarsat2 PS were less than four within the landslide boundary, hence considered not reliable for the methodology. In this case, the implementation of the matrix approach was done using the Radarsat stack of estimates in the period 2003–2010, and this conditioned the reference year of the updating which was set to 2010 (Table 1). For all the other premapped landslides (i.e., L3, L5 to L9, L12, L15 to L17, and L19), partial lack or total absence of PSI data was observed within the respective boundaries and thereby the employment of the matrix of Figure 2 and the scale of Figure 4 caused these phenomena neither to be updated in terms of state of activity (their reference year was left as 2007) nor to be attributed an intensity (NC phenomena; Figure 11c and d).

Summing up, for 13 of the 24 premapped landslides, the activity was updated to 2011 or 2010 and the intensity was also assessed with same reference year; on the other hand, for the remaining 11 phenomena, the activity recorded in the IFFI inventory was left unchanged and no evaluation of the intensity was performed due to the absence or insufficiency of PS data within the boundaries of these landslides (NC). A total of four reactivated (*R*), 16 active (*A*), three dormant (*D*), and one dormant/stabilized (*D/S*) phenomena were finally identified. Considering the intensity, the 13 updated landslides were distributed as follows: 11 very slow (VS), one extremely slow (ES), and one negligible (*N*).

The two newly identified phenomena, L25 and L26, were covered by PSI data in all the available stacks, i.e., ERS1/2, Radarsat1, and Radarsat2 (Table 1). Both these unstable areas showed historical and present velocities exceeding the activity threshold of ±5 mm/year and the intensity threshold of ±13 mm/year. L25 average motions in the slope direction were

12.8 and 16.0 mm/year in 1992–2000 and 2009–2011 respectively, while L26 showed slightly higher rates, i.e., 20.8 and 24.5 mm/year in 1992–2000 and 2003–2010. These phenomena were thereby attributed the state of active (*A*) and the intensity of very slow (VS); L25 had a new reference year of 2011 and L26, with year 2010 (Figure 11).

Given the areal extension of L25 and the variable morphology of this unstable area (Figure 10), the use of a unique representative velocity for the whole phenomenon can be considered inappropriate. Further studies could be focused on a segmentation of the unstable area into subsectors of homogeneous morphology for which sector-specific velocities might be calculated. Alternatively, a cell-by-cell projection of PS LOS velocities along the slope direction of each DTM cell might be performed to account for the local differences in the morphology of this unstable area. Considering the scope of this work and the scale of application of the PSI-based matrix approach (generally, regional), these further steps are not tackled in this paper but postponed to future experimentations.

7. DISCUSSION

Major influencing factors of the PSI-based matrix approach, its current limitations, and their respective solutions emerged through both the application to the test area of Verbicaro and the previously cited PSI-based landslide investigations, and can be summarized as follows.

7.1 Partial or total lack of PS within the landslide boundary

Total lack or insufficiency of targets within the landslide boundaries induce uncertainties in the activity and intensity evaluations, or even prevent the analyzed phenomena to be classified and/or updated (e.g., landslides L5 to L9 in Verbicaro). The different factors that influence the identification of radar targets during a PSI analysis are as follows.

7.1.1 Geometrical visibility

Combination of local topography and LOS orientation may cause significant underestimation of land motions, or even hamper the identification of PS within the observed scene, due to foreshortening, layover, and shadowing effects (Colesanti and Wasowski 2006). The geometrical visibility of a slope depends on its aspect and inclination (with respect to the sensor acquisition geometry, i.e., the LOS orientation) and can also vary within different portions of the same landslide (see, Cascini et al.2010). Use of appropriate geometry for the investigated area is, therefore, fundamental for each PSI-based analysis. Generally, ascending acquisition modes are suitable for E-facing slopes where the landslide movements are likely to be oriented in the E direction, while descending modes are used for W-facing slopes. Availability of both ascending and descending data obviously increases the number of the areas investigable through PSI data. Different look angles of available SAR sensors and the flexibility of some satellites to vary their acquisition geometry (e.g., Radarsat and TerraSAR-X) need also to be taken into account while planning a PSI investigation.

A feasibility analysis of the geometrical visibility and target density that can be achieved through a PSI analysis for a given area of interest can be performed by exploiting one of the different criteria developed in the last years. Colombo et al. (2007) forecast the PS density of an area by cross-combining the effects of radar geometry (i.e., layover and shadowing), local topography (i.e., slope and aspect) and land use; Notti et al. (2010) summarize the previous visibility factors introducing the RC-index; Cascini et al. (2010) discuss and implement the approach of a priori DInSAR landslide visibility mapping.

7.1.2 Vegetation

Complementarily to the exploitation of optical photointerpretation, landslides observable with radar interpretation are concentrated primarily in urbanized/built-up areas. Vegetation cover causes temporal decorrelation and makes difficult the identification of stable radar targets. Deeper penetration of vegetation may be achieved using higher radar wavelengths, such as those of the L band (e.g., 23 cm for ALOS PALSAR data).

Use of corner reflectors may improve the density of monitored targets in semivegetated, agricultural, and also urban areas by installing on the ground artificial benchmarks which act as PS (e.g., planar, dihedral, and trihedral reflectors; Savio et al. 2005); however, this solution cannot overcome the lack of targets in archived SAR scenes clearly because the identification of corner reflectors is achieved only after their installation on the ground.

7.1.3 Rapid movements

Intensity classes higher than 'very slow' cannot be considered yet in the PSI-based matrix approach. This is due to the acquisition parameters and characteristics of the currently available SAR sensors that do not enable the estimation of deformations faster than few tens of centimeters per year, which would compromise the results of the PSI processing. Aliasing effects, due to the ambiguous nature of the 2π wrapping of interferometric phases, limit to a quarter of the wavelength the maximum displacement between two successive acquisitions and two close PS of the same dataset (e.g., Hanssen 2005; Crosetto et al. 2010). Velocities compromising the PSI processing depend on the employed SAR wavelengths and satellite revisiting times; they are about 15 cm/year for ERS/Envisat data (C band), 20 cm/year for Radarsat (C band), 25 cm/year for TerraSAR-X (X band), and 45 cm/year for ALOS/JERS (L band). Consequently, higher motion rates cannot be tracked and/or interpreted correctly. Future improvements of the PSI-based approach could be achieved by exploiting future SAR missions employing higher temporal frequencies, or even conventional InSAR (single interferograms) or Ground-Based InSAR (GBInSAR; e.g., Antonello et al. 2004) techniques, which can extend the velocity range of applicability of the matrix approach to include faster phenomena.

7.2 Temporal coverage of PSI data

7.2.1 Modification of temporal scale for activity assessment

Depending on the temporal coverage of PSI-derived deformation data, the matrix approach may modify the temporal scale for the evaluation of the states of activity, changing the time interval for discriminating different states from a single cycle of seasons (i.e., 12 months; WP/WLI 1993) to two or more years (e.g., 2 years, using 2009–2011 Radarsat2 data for the analysis of recent deformation velocities in Verbicaro). The lengthening of the reference time interval makes the evaluation of the activity more precautionary for the subsequent evaluation of hazard, since the behavior of each landslide is analyzed within a wider time frame with respect to the 12-month-based observation range proposed by the WP/WLI in 1993. However, changes of the deformation behavior recorded at the end of a time period and indicative of movement reactivation may be blurred by the long history of motion recorded in the PS time series and may not be clearly recognized by simply using PS yearly velocities. Future development of advanced tools to better analyze PS deformation series will most likely contribute to better characterization and early detection of landslide accelerations or reactivations recorded within the monitored intervals.

7.2.2 Overlapping of PSI data with referenced landslide information

When the interval spanned by the available PSI-derived data temporally overlap with the date of completion of the preexisting inventory (e.g., 2003–2010 PSI data for the updating of a 2007 inventory, as for landslide L18 in Verbicaro), the PSI velocities do not represent just the deformational scenario subsequent to the time reference of the inventory itself (i.e. 2008–2010, for the above-mentioned example); they rather include some contributions from the scenario antecedent to this reference (i.e., 2003–2007). Previous PSI-based landslide studies and the application on Verbicaro used simple interpretation of full PS time series of ground deformation, regardless of this overlapping. Alternatively, advanced postprocessing PSI products obtained through the temporal subsampling of PSI deformation histories—as those recently exploited by Cigna et al. (2011)—may be considered to solve this temporal inconsistency. In particular, subsampled yearly deformation velocities might be derived by analyzing long-term PS time series and recalculating average rates in the time period subsequent to the reference year of the available preexisting inventory, i.e., separating the deformation velocities pre- and postinventory (for the above-mentioned example, 2003–2007 and 2008–2010 data, respectively). In some cases, however, this approach is strongly influenced by some parameters employed during the PSI processing, such as the model chosen to unwrap phase variations through time (generally, linear); thus, the whole PS time series (both pre- and postinventory) may be already intrinsically influenced by the entire deformation history of the analyzed time span.

7.2.3 Choice of activity and intensity thresholds

The choice of activity and intensity thresholds is a key point for the PSI-based matrix analysis and needs to be performed taking into account the following facets:

- Typology of observed deformation processes (e.g., geometry and expected velocity);
- Technical characteristics of available PSI data (e.g., LOS orientation);
- Postprocessing steps carried out during the evaluation of landslide representative velocities (e.g., reprojection or not of LOS estimates to most probable directions of movement; V_{ACT} and V_{INT} will be lower for LOS values than for reprojected data, as shown for the Verbicaro case study).

Usually, V_{INT} is precautionarily assumed lower than the official velocity threshold discriminating extremely from very slow landslides (i.e., 16 mm/year; Cruden and Varnes 1996). This choice is mainly driven by the frequent assumption of average velocities as the representative rates of movement for the analyzed phenomena; maximum deformation rates (peak velocities; Cruden and Varnes 1996) may significantly exceed the averaged patterns of the whole landslide, since displacement rates usually differ (spatially) within the same landslide area. Moreover, the temporal variations of ground displacements should also be considered; PSI-derived data are evaluated as averages of observations over long time periods (several years) and they are, therefore, expected to be lower than peak velocities (as also recently observed by Cascini et al. 2010).

If neither a reprojection of LOS velocities along the slope direction nor the combination of ascending and descending estimates are applied, the use of lowered thresholds might affect the results of the implementation for those phenomena moving approximately close to the direction of the employed LOS (Bianchini et al. 2012). For these landslides, the exploitation of low thresholds results in an overestimation of the landslide velocities—and consequently, the respective intensity—, since these phenomena are classified as faster than they really are.

With regard to V_{ACT}, the choice of the minimum value exceeding the precision of the PSI technique, i.e., a few millimeter per year, is usually performed (e.g., Cascini et al.2010; Meisina et al. 2006, 2008). The implementation of the methodology for Verbicaro showed also how this value may be converted to account for the projection of LOS velocity along the slope directions. Differently from other applications, this conversion entails the use of higher thresholds for discriminating states of activity and intensities in order to account for the lower underestimation of real velocities occurring with the use of SLOPE-projected values with respect to that attributed to LOS velocities.

Another approach to calibrate the activity threshold might use training data—for which ground truth or in situ data are available—and then apply the determined threshold to the entire zone of interest. To this aim, historical PSI data (e.g., ERS 1992–2001 data) and past field-based activity determinations (past landslide inventory; e.g., map referenced 2001) may also be used, respectively, as training inputs and outputs for the determination of the activity using the PSI data. Operatively, for some training areas, the updating of the landslide activity can be simulated, starting from the determination of landslide representative velocities by using PSI data and then testing different thresholds to determine the landslide states of activity. The appropriate activity threshold for the area can then be chosen as the value producing the highest number of correspondences between simulated activity determinations (based on PSI data) and field-based evaluations.

As pointed out by Bianchini et al. (2012), the use of significantly low activity thresholds (e.g., a few millimeter per year) reduces the probability of discarding movements potentially critical, guaranteeing that even the slower motions are detected and classified. However, it causes some gravitational movements, such erosion, seasonal, or continuous downhill creep, to be included in the updated inventory maps.

7.2.4 Need of field validation

The PSI-based matrix approach, as any landslide study supported by remote sensing data, should always be validated through field checks and surveys in order to confirm, or even improve, all the information and evaluations performed through satellite radar interpretation to verify damage on buildings and infrastructure and other geomorphologic indicators and features induced by ground instability (e.g., Meisina et al. 2008; Righini et al. 2011). The need of field checks is also induced by those ambiguous classifications of the states of activity included in the activity matrices (e.g., what happened for landslide L23 in Verbicaro), for which on-site inspections are required to confirm an actual lowering of the landslide activity. Anyhow, it is clear that PSI-based landslide studies facilitate the scanning of very large areas to identify hotspot zones on which on-site checks and surveys have to be prioritarily carried out, focusing field investigations on the areas of (potentially) higher risk and thus significantly reducing efforts, times, and costs of on site activities (Bianchini et al.2012).

8. CONCLUSIONS

We presented the definition, procedures, discussion, and one example of application of the 'PSI-based matrix approach', a methodology employed for the assessment or updating of the state of activity and intensity of extremely to very slow landslide processes. For the case study of Verbicaro, we exploited 1992–2011 PSI data, and the methodology allowed the assessment of the state of activity and the intensity for 13 of the 24 premapped phenomena and also for the two newly identified phenomena, L25 and L26, with the year 2011 or 2010 as temporal reference of these estimations. For the remaining 11 landslides, the activity recorded in 2007 in the IFFI inventory was confirmed, and no evaluation of the intensity was performed due to the absence or insufficiency of PS data within the boundaries of these phenomena. Considering both premapped and newly identified phenomena, four reactivated (*R*), 18 active (*A*), three dormant (*D*), and one dormant/stabilized (*D/S*) landslides were finally identified; as for intensity, the distribution showed 13 very slow (VS), one extremely slow (ES), one negligible (*N*), and 11 not classified (NC).

Both the application for the test area of Verbicaro and past landslide investigations brought to light the major influencing factors of the PSI-based matrix approach, such as the lack of PS data within the landslide boundaries, the temporal coverage of the available data, and the need of field checks and validation as well as the operative procedures to choose the activity and intensity thresholds. Further improvements of the methodology will surely follow in the upcoming years as a consequence of the use of high spatial and temporal resolution imagery acquired by new satellites as well as the development of advanced approaches for the postprocessing analysis of PS data and in-depth exploitation of the information stored within the PS time series.

ACKNOWLEDGMENTS

This work was carried out within the SAFER (Services and Applications For Emergency Response) project, funded by the European Commission within the 7th Framework Programme under the Global Monitoring for Environment and Security (EC GMES FP7) initiative, with grant agreement no. 218802. Radarsat1/2 data were provided by the ESA-managed GSC-DA, funded by the FP7/2007–2013 under EC-ESA Grant Agreement no. 223001. The authors would like to thank Gaia Righini for her support during the selection and acquisition of the radar imagery, and e-GEOS for processing Radarsat1/2 data with the PSP-DIFSAR technique in the framework of the SAFER project. ERS PSI products were made available through the PST-A (Piano Straordinario di Telerilevamento Ambientale) project and the WMS service of the Italian Ministry for the Environment, Land and Sea. Access to IFFI (Inventario dei Fenomeni Franosi in Italia) was performed through the WMS service of the Italian Institute for Environmental Protection and Research (www.isprambiente.gov.it), while PAI (Piano Stralcio di Bacino per l'Assetto Idrogeologico) was made available by the Italian Civil Protection Department.

REFERENCES

1. Amodio Morelli L, Bonardi G, Colonna V, Dietrich D, Giunta G, Ippolito F, Liguori V, Lorenzoni S, Paglionico A, Perrone V, Piccarretta G, Russo M, Scandone P, Zanettin Lorenzoni E, Zuppetta A (1976) L'arco Calabro-Peloritano nell'orogene Appenninico-Maghrebide. Mem Soc Geol Ital 17:1–60

2. Antonello G, Casagli N, Farina P, Leva D, Nico G, Sieber AJ, Tarchi D (2004) Ground-based SAR interferometry for monitoring mass movements. Landslides 1(1):21–28

3. Bianchini S, Cigna F, Righini G, Proietti C, Casagli N (2012) Landslide hotspot mapping by means of persistent scatterer interferometry. Environ Earth Sci.

4. Bovenga F, Nutricato R, Refice A, Wasowski J (2006) Application of multi-temporal differential interferometry to slope instability detection in urban/peri-urban areas. Eng Geol 88:219–240

5. Canuti P, Focardi P (1986) Slope stability and landslides investigation in Tuscany. Mem Soc Geol Ital 31:307–315

6. Carrara A, Merenda L (1976) Landslide inventory in northern Calabria, southern Italy. Geol Soc Am Bull 87:1153–1162

7. Cascini L, Fornaro G, Peduto D (2010) Advanced low- and full-resolution DInSAR map generation for slow-moving landslide analysis at different scales. Eng Geol 112:29–42

8. Cigna F, Del Ventisette C, Liguori V, Casagli N (2011) Advanced radar-interpretation of InSAR time series for mapping and characterization of geological processes. Nat Hazards Earth Syst Sci 11(3):865–881

9. Colesanti C, Wasowski J (2006) Investigating landslides with space-borne synthetic aperture radar (SAR) interferometry. Eng Geol 88:173–199

10. Colombo A, Mallen L, Pispico R, Giannico C, Bianchi M, Savio G (2007) Mappatura regionale delle aree monitorabili mediante l'uso della tecnica PS. In: Proceedings of 10° National Conference ASITA, Bolzano, Italy, 14–17 November 2006

11. Costantini M, Iodice A, Magnapane L, Pietranera L (2000) Monitoring terrain movements by means of sparse SAR differential interferometric measurements. In: Proceedings of IGARSS 2000, 20th IEEE International Geoscience and Remote Sensing Symposium, Honolulu, Hawaii, USA, 24–28 July 2000, pp 3225–3227

12. Crosetto M, Monserrat O, Iglesias R, Crippa B (2010) Persistent scatterer interferometry: potential, limits and initial C- and X-band comparison. Photogramm Eng Remote Sens 76(9):1061–1069

13. Cruden DM, Varnes DJ (1996) Landslide types and processes. In: Turner AK, Schuster RL (eds) Landslides: investigation and mitigation, Sp. Rep. 247, Transportation Research Board, National Research Council. National Academy, Washington DC, pp 36–75

14. Cucci L, Cinti FR (1998) Regional uplift and local tectonic deformation recorded by the Quaternary marine terraces on the Ionian coast of the northern Calabria (southern Italy). Tectonophys 292:67–83

15. Farina P, Colombo D, Fumagalli A, Marks F, Moretti S (2006) Permanent scatterers for landslide investigations: outcomes from the ESA-SLAM project. Eng Geol 88:200–217

16. Fell R, Corominas J, Bonnard C, Cascini L, Leroi E, Savage WZ (2008) Guidelines for landslide susceptibility, hazard and risk zoning for land use planning. Eng Geol 102:85–98

17. Ferranti L, Santoro E, Mazzella ME, Monaco C, Morelli D (2009) Active transpression in the northern Calabria Apennines, southern Italy. Tectonophys 476:226–251

18. Ferretti A, Prati C, Rocca F (2001) Permanent scatterers in SAR interferometry. IEEE Trans Geosci Remote Sens 39(1):8–20

19. Gonzalez-Diez A, Remondo J, Diaz de Teràn R, Cendrero A (1999) A methodological approach for the analysis of the temporal occurrence and triggering factors of landslides. Geomorphol 30:95–113

20. Greif V, Vlcko J (2011) Monitoring of post-failure landslide deformation by the PS-InSAR technique at Lubietova in Central Slovakia. Environ Earth Sci.

21. Guzzetti F, Carrara A, Cardinali M, Reichenbach P (1999) Landslide hazard evaluation: a review of current techniques and their application in a multi-scale study, Central Italy. Geomorphol 31:181–216

22. Hanssen RF (2005) Satellite radar interferometry for deformation monitoring: a priori assessment of feasibility and accuracy. Int J Appl Earth Obs 6:253–260

23. Herrera G, Notti D, García-Davalillo JC, Mora O, Cooksley G, Sánchez M, Arnaud A, Crosetto M (2010) Landslides analysis with C- and X-band satellite SAR data: the Portalet landslide area. Landslides 7:1–12

24. Hilley GE, Bürgmann R, Ferretti A, Novali F, Rocca F (2004) Dynamics of slow-moving landslides from permanent scatterer analysis. Science 304(5679):1952–1955

25. Hungr O (1997) Some methods of landslide intensity mapping. In: Cruden D, Fell R (eds) Landslide risk assessment, Proceedings of the International Workshop on Landslide Risk Assessment, Honolulu, USA, February 19–21, 1997. Balkema, Rotterdam, pp 215–226

26. Iannace A, Bonardi G, D'Errico M, Mazzoli S, Perrone V, Vitale S (2005) Structural setting and tectonic evolution of the Apennine Units of northern Calabria. C R Geoscience 337:1541–1550

27. Lateltin O, Haemmig C, Raetzo H, Bonnard C (2005) Landslide risk management in Switzerland. Landslides 2(4):313–320

28. Meisina C, Zucca F, Fossati D, Ceriani M, Allievi J (2006) Ground deformation monitoring by using the permanent scatterers technique: the example of the Oltrepo Pavese (Lombardia, Italy). Eng Geol 88:240–259

29. Meisina C, Zucca F, Notti D, Colombo A, Cucchi A, Savio G, Giannico C, Bianchi M (2008) Geological interpretation of PSInSAR data at regional scale. Sensors 8(11):7469–7492

30. Notti D, Davalillo JC, Herrera G, Mora O (2010) Assessment of the performance of X-band satellite radar data for landslide mapping and monitoring: Upper Tena Valley case study. Nat Hazards Earth Syst Sci 10:1865–1875

31. Parise M, Wasowski J (1999) Landslide activity maps for landslide hazard evaluation: three case studies from Southern Italy. Nat Hazards 20(2):159–183

32. Righini G, Pancioli V, Casagli N (2011) Updating landslide inventory maps using persistent scatterer interferometry (PSI). Int J Remote Sens.

33. Savio G, Ferretti A, Novali F, Musazzi S, Prati C, Rocca F (2005) PSInSAR™ validation by means of a blind experiment using dihedral reflectors. In: Proceedings of FRINGE 2005, Frascati, Italy, 28 November–2 December 2005, pp 1–6

34. Uzielli M, Nadim F, Lacasse S, Kaynia AM (2008) A conceptual framework for quantitative estimation of physical vulnerability to landslides. Eng Geol 102:251–256

35. Van Dijk JP, Bello M, Brancaleoni GP, Cantarella G, Costa V, Frixa A, Golfetto F, Merlini S, Riva M, Torricelli S, Toscano C, Zerilli A (2000) A regional structural model for the northern sector of the Calabrian Arc (Southern Italy). Tectonophys 324:267–320

36. Werner C, Wegmuller U, Strozzi T, Wiesmann A (2003) Interferometric point target analysis for deformation mapping. In: Proceedings of IGARSS 2003, 23rd IEEE International Geoscience and Remote Sensing Symposium, Toulouse, France, 21–25 July 2003, pp 4362–4364

37. Wieczorek GF (1984) Preparing a detailed landslide-inventory map for hazard evaluation and reduction. IAEG Bull 21(3):337–342

38. WP/WLI—Working Party on World Landslide Inventory (1993) Multilingual glossary for landslides. The Canadian Geotechnical Society, BiTech, Richmond BC

Chapter 9

Web3DGIS-Based System for Reservoir Landslide Monitoring and Early Warning

Huang Huang [1,2], Jianhua Ni [1,2], Yu Zhang [3], Tianlu Qian [1,2], Dingtao Shen [1,2,3] and Jiechen Wang [1,2,4,*]

[1]Jiangsu Provincial Key Laboratory of Geographic Information Science and Technology, Nanjing 210023, China

[2]Department of Geographic Information Science, Nanjing University, Nanjing 210023, China

[3]Changjiang River Scientific Research Institute, Changjiang Water Resources Commission, Wuhan 430010, China

[4]Jiangsu Center for Collaborative Innovation in Geographical Information Resource Development and Application, Nanjing 210023, China

ABSTRACT

Landslides are the most frequent type of natural disaster, and they bring about large-scale damage and are a threat to human lives and infrastructure; therefore, the ability to conduct real-time monitoring and early warning is important. In this study, a Web3DGIS (Web3D geographic information systems) system for monitoring and forecasting landslides was developed using the Danjiangkou Reservoir area as a case study. The development of this technique involved system construction, functional design, organizing and managing multi-source spatial data, and implementing a forecasting plan and landslide-forecasting model. By integrating sensor technologies, spatial information technologies, 3D visualization technologies, and a landslide-forecasting model, the results of this study provide a tool for real-time monitoring at potential landslide sites. When relevant data from these sites reach threshold values, the model automatically initiates forecasting procedures, and sends information to disaster prevention sectors for emergency management.

KEYWORDS

landslide monitoring; landslide early-warning; natural disasters; Web3DGIS

1. INTRODUCTION

In recent years, landslides, debris flows, and other natural disasters have been attributed to changes in the global climate, heavy localized rainfall, strong tropical windstorms, and other extreme weather situations. In China, official statistics from the Geographical Research Bureau show that in 2014 there were 10,907 natural disasters, causing 349 deaths and 5.41 billion yuan of damage in direct economic loss. Among these events, landslides were the most frequent, comprising ~74.5% of the total number of natural disasters, and the most severe in terms of damage. Given the threat to human life and infrastructure, minimizing the impact of landslides remains critical. Solutions that have been proposed include transforming the shapes of landslide-risk slopes, structural supports for soil and rock features in landslide areas, and reducing the effects of groundwater and surface water on slope stability. However, while these approaches could decrease the occurrence of landslides, large-scale control remains difficult owing to limitations from construction, available technologies, and cost. Therefore, the ability to conduct real-time monitoring and disaster early warning is critical.

Landslide stability is affected by both internal and external factors, whose mechanisms vary concurrently; therefore, the main approach to real-time monitoring involves monitoring the state of these inducing factors. Monitoring is focused on either slope surface or subsurface conditions (e.g., groundwater levels and physical composition). Global Positioning System (GPS) techniques were first applied to monitor landslide deformation in the 1990s [1,2,3,4], and Zumberge in Jet Propulsion Laboratory put forward the precise point positioning (PPP) technique, anoptimized GPS technique. The PPP technique has taken a dominating role in landslide monitoring [5,6,7]. In the mid-1990s, Dowding, Kane, and others, introduced Time Domain Reflectometry (TDR) techniques into obtaining landslide deformation data by burying coaxial cables in landslide areas and measuring their changing state [8,9]. In the late 1990s and early 2000s, Interferometric synthetic aperture radar, close range photogrammetry measurements, and other techniques were also applied to surface deformation monitoring [10,11,12,13]. Accompanied with the development of monitoring techniques, the monitoring sensors experienced rapid development, from simple and relatively independent monitoring instruments (e.g., rainfall gauge, wire extensometers) to professional landslide sensors, which can transfer monitoring data to scene sites set around the landslides (e.g., osmometer, inclinometer), and then to the multi-sensor networks with the ability of real-time remote data transmission [14,15]. The efficiency and accuracy of monitoring sensors have been further improved to provide a sound statistical basis for the development of landslide early warning systems. In addition, modern mathematical methods, including grey system and fuzzy math, have also been gradually applied to landslide early warning endeavors [16,17,18,19].

The development of geographical information systems (GIS), and network and sensor techniques has provided strong support for the integration, management, and analysis of spatial statistical data regarding landslides. By integrating the above techniques with landslide early warning models, real-time early warning of landslides becomes possible [20,21,22,23,24,25], as does the design and development of practical monitoring systems [26,27,28] and disaster handling decision-making and support systems [29,30,31,32]. Some progress has been made on the development of an early warning system; however, there are issues that remain to be resolved: (1) expression and analysis of three-dimensional spatial information; (2) real-time collection, transfers, storage, analysis and integration of multi-spot and multi-type sensor statistics; and (3) connecting the landslide early warning modal with a decision-making model and with information publishing functions, so as to effectively achieve the functional integration of the system.

In this study, the Danjiangkou Reservoir area was used as a case study to develop a system monitoring and modeling landslides based on Web3DGIS

(Web3D Geographic Information Systems), which features three-dimensional spatial expression and function analysis of the ground surface environmental and monitoring spot data. The developed method, which is based on a real-time data transport network among sensors, on-spot monitoring sites, and monitoring early warning centers, is able to collect and store real-time monitoring data, and to integrate monitoring, early warning, and information publishing functions.

2. TARGET AREA

Danjiangkou Reservoir, China (Figure 1), is the largest artificial freshwater lake in Asia, and forms a water source for the South-to-North water diversion project in China. The project's central line was officially launched in late 2003, introducing water from Danjiangkou Reservoir to the Huanghuaihai Plain, which hosts over 20 cities ofvarying size. During construction of the water diversion project, the height of the Danjiangkou dam was extended, and its regular water-storage level was raised from 157 to 170 m, storage capacity increased from 17.45 to 29.05 billion m^3, and the water area expanded from 700 to 1022.75 km^2. When the reservoir water level rose, original drying landslide fronts were immersed, and stability of rock and soil mass were reduced. Furthermore, hydrodynamic pressure caused by the rapid water level change induced deformation. According to the geological survey results, before the water level rose, there were about 23 landslides with ascale of over 100,000 m^3; five years after the water level rose, the number of landslides with ascale of over 100,000 m^3 was about 30. To protect human life and infrastructure, the regional management and disaster prevention agencies require real-time monitoring of potential landslide sites, and an early warning system for landslide events in the Danjiangkou Reservoir area.

3. SYSTEM STRUCTURE AND DESIGN

The software structure of the system developed in this study was based on the monitoring database, which is supported by a Brower/Server model with WebGIS and Web3D technologies (Figure 2). Using SQL (Structure Query Language) server, Visual Studio and skyline platform (a 3D GIS application development platform), a multi-structure for application-information and data was constructed, and the service was oriented to different applications, offering convenience for reuse and the expansion of system functions.

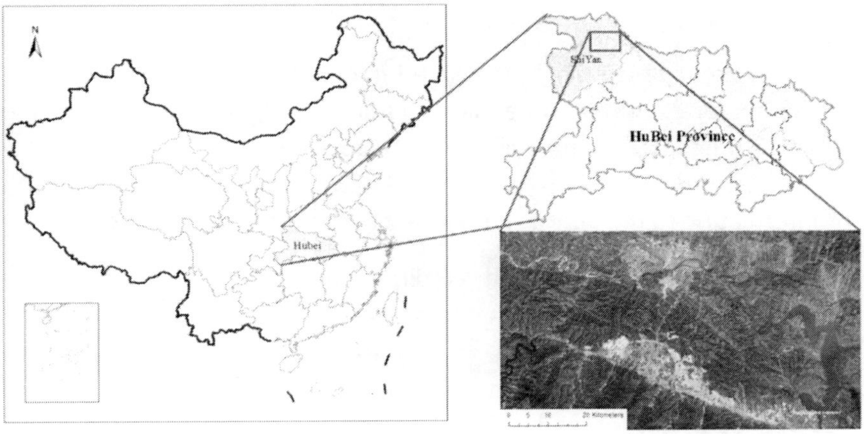

Figure 1. Location of the study area within the northwestern region of Hubei Province, China.

Data storage and management services were integrated into the system, including those for real-time monitoring data, historical monitoring data, rock and soil mass characteristics data, emergency information, and basic geographical data. The assembly level of the system offers components relevant for the operational sector, including the analysis of landslide monitoring data and an early warning module. After the assembly level integration, the service sector offers simple but practical operational services, primarily including three-dimensional spatial analysis, real-time monitoring services, landslide data consulting and management, early warning analysis, a disaster publishing service, and a system management service. The performance sector is responsible for human–computer interaction services and a thematic map-making output service.

According to the operational procedures and application requirements for a landslide early warning service, the system has five main functions:

1 Collection and management of multiple datasets pertaining to landslides, including fundamental geographical data, real-time and

historical monitoring data, acquirement, transfer, preprocessing and management of emergency service data.

2 Query research and spatial statistical analysis, including comprehensive querying of each monitoring site (e.g., feature, section, and sphere) and statistical analysis of feature data.

3 Landslide early warning and related information publishing. Based on the features of the potential landslide location and its real-time monitoring data, a model for landslide forecasting and emergency warning that is able to publish potential disaster evaluation information to the public and relevant departments is selected.

4 Expression and analysis of three-dimensional information, including a three-dimensional depiction of the target area and key monitoring locations. Focused on data relating to flow fields and groundwater level, a combination of functions and needs was developed to address landslide prediction and the evaluation of relevant three-dimensional spatial analysis.

5 Support functions, including user management, operational right management, data management and maintenance, and other regular functions.

4. KEY TECHNOLOGIES

4.1. Multi-Data Organization and Management

Landslide monitoring and early warning systems involves a large volume of data from different sources and of various types, and with complex relationships between them. They include both static and dynamic real-time monitoring data, lattice and arrow data, and non-spatial data on charts, texts, photos, and videos. In order to guarantee system functionality, a reasonable data model was needed, in which data were effectively organized and managed. This required a number of key technologies to be applied to the construction of the landslide special data pool, and to the organization, transport, and expression of three-dimensional scene data.

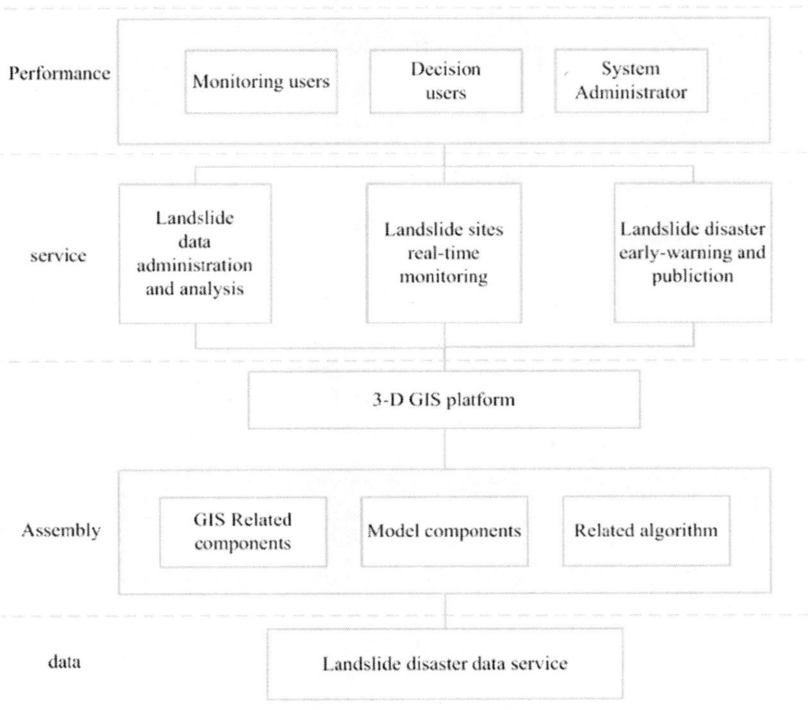

Figure 2. Software structure of the landslide monitoring and early warning system.

4.1.1. Special Data Pool for Landslide Monitoring

With regard to data content, the special topics data of landslide monitoring were placed into one of the following five categories (Table 1), while the technological procedures were integrated into the database construction (Figure 3).

With the spatial database and spatial data engine, the spatial data form an object oriented database management system, which has integrated management over spatial data and attribute data. For the design of spatial databases, previous studies have investigated landslide features in-depth using an object-oriented information model and a design method for a structured UML static view. Using this approach, they obtained the relationships between the patterns of landslide information management, countermeasures, and limitations. The spatial data of potential landslides include dotted and polygon maps, unified serial numbers (id) for monitoring sites, influential areas, and boundaries.

Table 1. Landslide Monitoring Data Categories.

Categories		Data Contents
Static data	Basic geographical data	Basic arrow data on geological information, including big scale administration division, lake and water systems, residential area, and lattice spatial data for the digital elevation model
	Remote sensing image data	Aerial image and satellite-borne remote sensing image data
	Survey data about potential landslide sites	Plane graph, profile, location map of landslides, rock and soil mass characteristics data
	Documentary information and other data	Relevant statistical materials recorded about historical disasters, photos, and technical reports.
Dynamic data	Sensor monitoring data from potential landslide sites	Ground surface displacement, depth displacement, groundwater level, inclination and rainfall

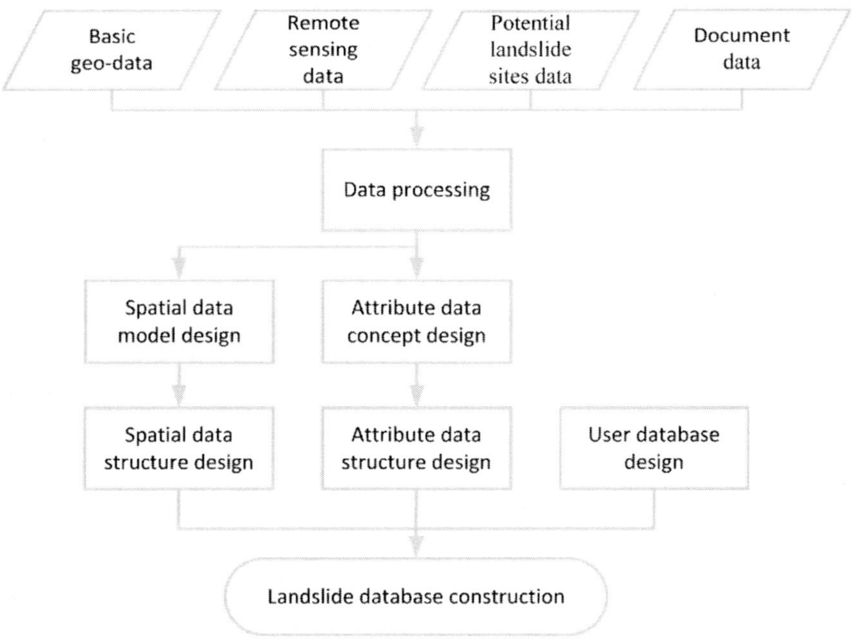

Figure 3. Flow-chart of database construction.

The attribute data of potential landslides involve large volumes and contain complex relationships, primarily for those relating to real-time monitoring, forecasting and early warning, historical information, and monitoring management over relevant staff. The real-time monitoring data include landslide monitoring information, real-time monitoring sites, and sensors, among which various relationships exist (Figure 4).

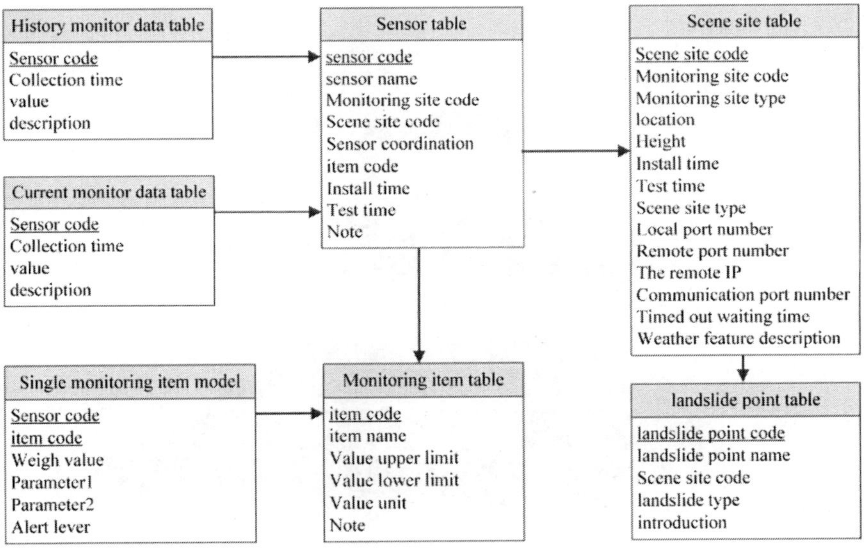

Figure 4. Entity Relationship Diagram of real-time monitoring data model.

4.1.2. Three-Dimensional Scene Data Organization

In order to establish an accurate image of the target area, researchers need to integrate remote image data with the digital elevation model, overlapped with multiple geological factors, so as to allow functions including browse, cruise, consultancy, measurement, and analysis in a three-dimensional space. The main data for three-dimensional scene construction are from highly accurate digital evaluation models and high-resolution digital orthophoto maps, whose data are normally restored in the form of a lattice. As a direct three-dimensional display can severely affect system effectiveness and due to the difficulties of achieving real time-mapping of the three-dimensional scene, this study combined the tile pyramid model method with a linear quad-tree index in order to preprocess and reorganize the data needed for the scene [33].

The pyramid model contains multiple resolution levels (Figure 5). When constructing the pyramid, the original lattice data (e.g., the digital evaluation model or remote image) is first set to the bottom (No. 0 level) of the pyramid tile matrix. Based on the No. 0 level, every 2 × 2 lattice unit combines together into one unit to form the 1 level lattice matrix. This process is then repeated to form the whole tile pyramid [34].

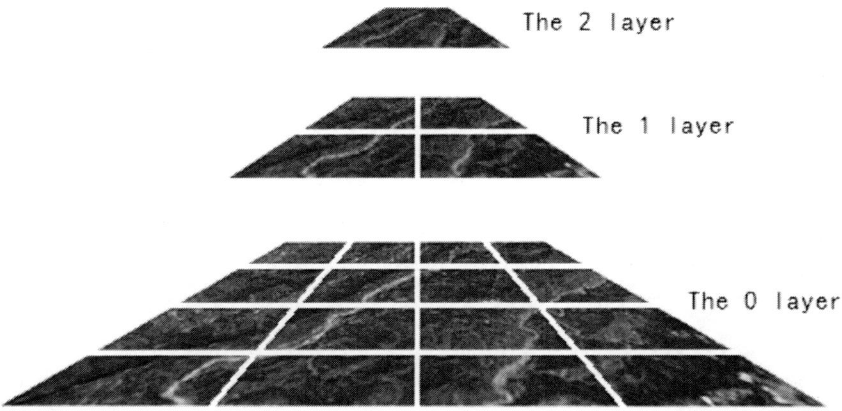

Figure 5. Image tile pyramid model.

A quad-tree is a tree structure with up to 4 branches in each non-leaf node [33]. It is also a hierarchical data structure, featuring spatial interconnection rescission. The quad-tree structure offers convenience for the index and management of tile data. Based on tile pyramid, the construction of quad-tree index can be divided into three steps: logical partitioning, node coding, and physical partitioning. After partitioning, a topological relationship between tiles is built, which includes two aspects: a neighboring link of the same level and higher, and lower relationships (e.g., that of parents and their children).

The above method can, to a great extent, satisfy the need to visualize a three-dimensional scene in a large area; therefore, it can greatly improve the display effectiveness of the system. This is primarily because cutting out topographical data from outside of the view and using a fast tile search method based on resolution rate greatly reduces the amount of data described in a scene. While mapping the scene, for the specific display proportion, data from the digital elevation model and the image texture of a specific pyramid level (the closest to the display resolution rate) are directly used in the rapid representation of the three-dimensional surface.

After considering the Browser/Server structure of the system, users can set a client-oriented browser and a quick three-dimensional display of related data. To ensure a fluent display of the three-dimensional scene, this study adopted several measures, including a seamless network and the effective transmission of three-dimensional data. Once the initial image is accepted by the user in a low-resolution state, the client displays the three-dimensional data, without waiting for the full transmission of data. Using a distribution system and load balance technology, different servers in different areas can share the burden of three-dimensional data collection; thus, automatically distributing load among all applicable servers.

4.2. Landslide Forecasting Model

A landslide-forecasting model must determine the time that a landslide reaches its accelerated deformation-instability phase. The stability of a landslide is affected by internal factors (e.g., landslide shape, soil/rock mechanical parameters) and external factors (e.g., groundwater level, rainfall, and human construction work), which not only have different effects on the landslide, but are also constantly changing.

At present, models for forecasting the time of landslide instability include the Saito Model, creep-spline Combined Model, Verhulst Model, the grey Model (1, 1) and so on [35,36,37,38]. Each model describes the relationship between instability time and deformation values and velocity, as estimated by the surface conditions, and is able to predict when a landslide will enter the accelerating deformation-instability phase. However, for large areas, we have to set many landslides monitoring sites independently, so using a single forecasting model is not sufficient. In this study, three models were selected: a landslide deformation forecasting, non-linear dynamics model; a grey forecasting model that varies with different periods; and a multi-factor grey forecasting model. These models are respectively responsible for short-term forecasting of relatively unstable landslides, long-middle term forecasting of relatively stable landslides, and forecasting landslides easily affected by external factors. In detail:

1 Non-linear dynamics model. By nature, landslides are a non-linear dynamic systems controlled by soil and rock conditions (e.g., topography, groundwater, rainfalls, and human engineering). The non-linear dynamics model has long been used [39,40], and has been developed to reach a high level of accuracy in the short-term prediction of landslides. This study applied the non-linear dynamics model proposed by Qin *et al.*[19], which makes predictions based on three major factors: stress, landslide displacement, and groundwater.

2 Grey forecasting model. This model, proposed by Deng, is a method for forecasting systems with uncertain factors [38]. By collecting slices of incomplete information, this model was built to consider a long-term general description of a system's development. Luan *et al.* [41] used this method to monitor and forecast ground surface deformation, and since then it has been commonly used in landslide forecasting [42,43,44]. In this study, the grey model was combined with a time sequence analysis model, such that after amendment of the residual model in time sequence, it was able to forecast trends of uniform motion-accelerating motion, accelerating motion-uniform motion, slow change-uniform motion, and slow change-accelerating motion. Through extraction of the items, which shock periodically, the model can greatly offset influences from external factors.

3 Multi-factor grey forecasting model. This model, from the perspective of both internal and external factors that can affect landslide deformation, regards the whole landslide as the production of multiple affecting factors. The description of internal affecting factors are modeled after a tendency grey model that varies with different periods, while external affecting factors can be divided into several independent but related linear factors. By combining the model with internal and external factors, a multiple and combined model is constructed and can be used for grey forecasting [45]. The multi grey model considers both external and internal factors relating to deformation, and when these change, the deformation value can be estimated as a fixed quantity. The model is suitable for forecasting landslide deformation affected by external factors (e.g., rainfall and groundwater level).

5. SYSTEM COMPLETION

5.1. Sensors Deployment

The landslide monitoring and early warning system developed in this study was deployed at the Danjiangkou Reservoir. Based on past disaster information, 31 potential landslide sites were selected and the responsible department constructed monitoring sites, which were fitted with a rainfall gauge, inclinometer, water level gauge, osmometer and displacement gauge whose real-time information was transmitted to the monitoring and early warning center using GPRS wireless signal transmitter. The deployment of multi-sensors is shown in Figure 6.

(a) (b)

Figure 6. (**a**) The sensors network in one of the landslide; and (**b**) the deployment of sensors.

5.2. Software System Completion

The integrated structure of the early warning software system includes a Brower/Server structure, subdivided into the server end and customer end, designed and developed in a .NET structure (a Microsoft XML web services development platform). The server end was completed in the Java programming language and its database is managed by SQL Serve. The server end was mainly developed through JavaScript and FLEX on the skyline three-dimensional digital earth platform, which is able to rapidly construct a simulated three-dimensional scene using large volumes of remote image data, digital elevation data, and two- and three-dimensional data. The system, topography of the integrated target area, and the image data with multiple attributes are used to construct the three-dimensional model, which then allows for various functions of contour line analysis, slope analysis, ground surface runoff analysis, section analysis, and other three-dimensional space analysis and expression procedures. The system interface is shown in Figure 7.

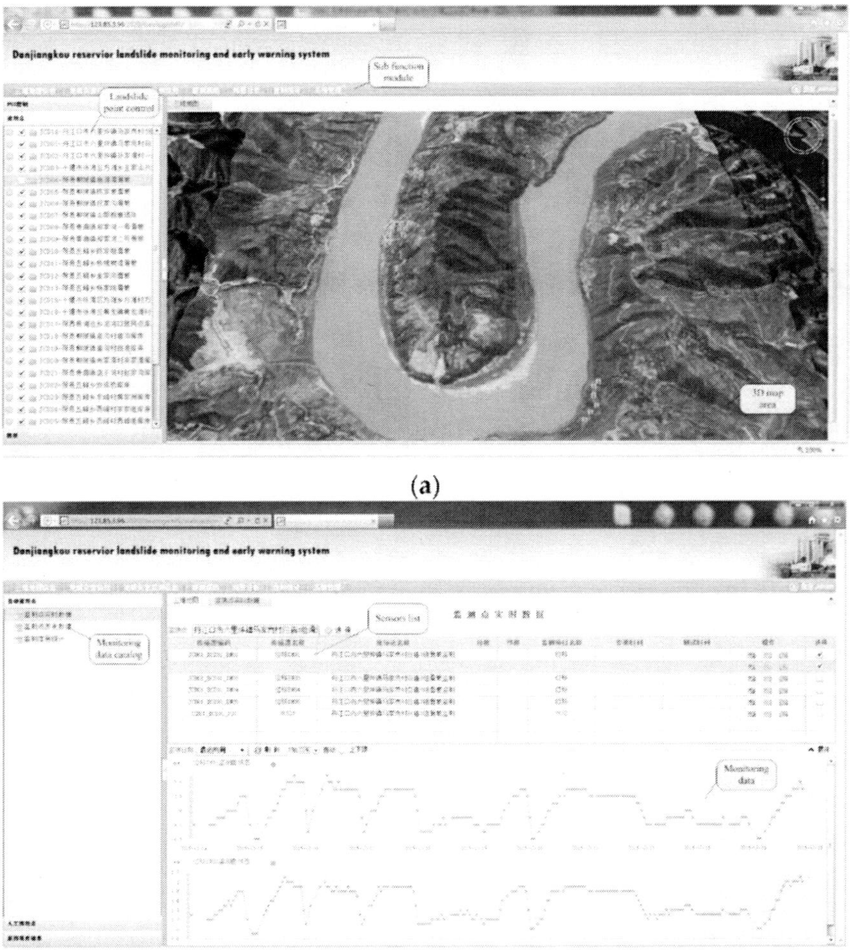

(a)

(b)

Figure 7. System interface: (**a**) the 3D map page; (**b**) the monitoring data page.

5.3. Early Warning Completion

For every single landslide monitoring sites, we analyzed the relationship between rainfall, deformation value/velocity and landslide, then set the threshold for starting landslide forecasting at this site according to the analysis result. When certain data streams change and meet the threshold value, the system automatically initiates its landslide early warning procedure.

By combining past and new real-time monitoring data with deformation early warning conditions (e.g., accumulated accelerated speed, rainfall, and groundwater level), the system can automatically select an appropriate early warning model and calculate the landslide transformation trend and instability time. After observing the calculation results, the system can automatically transmit relevant forecasting information to target users (e.g., disaster control experts and monitoring supervisors). After checking the monitoring data with experts, the system finally determines whether or not to publish the warning information, at which point it will transmit the relevant emergency plan and action suggestions to target people at disaster sites. In summary, this model integrates both the early warning results and expert experience, so as to ensure that the published information is as accurate as possible in order to avoid negative social influences from model flaws (Figure 8).

6. CONCLUSIONS

This study focused on the development of a landslide monitoring and early warning system in the Danjiangkou Reservoir area, China. The proposed real-time disaster forecasting system is based onWeb3DGIS and uses analysis results from landslide monitoring data and linked forecasting models. After reviewing local conditions and historical landslide data, we set up 31 landslide monitoring sites around the Danjiangkou Reservoir in order to collect data on surface displacement, rainfall, and groundwater. After receiving the data, the system consults the early warning model and information from experts before deciding whether to issue timely disaster information to decision-makers and emergency personnel. The system incorporates many technologies, including those of the sensors, WebGIS, three-dimensional visualization, and landslide forecasting. The final systems offers: (1) a comprehensive combination of multi data and landslide forecasting models that support forecasting accuracy; (2) real-time data collection by different sensors and timely wireless data transmission; and (3) Web3DGIS technology for three-dimensional analysis and expression of spatial information, allowing it to accurately identify the environmental conditions at the monitoring site, and to support the analysis of relevant three-dimensional space.

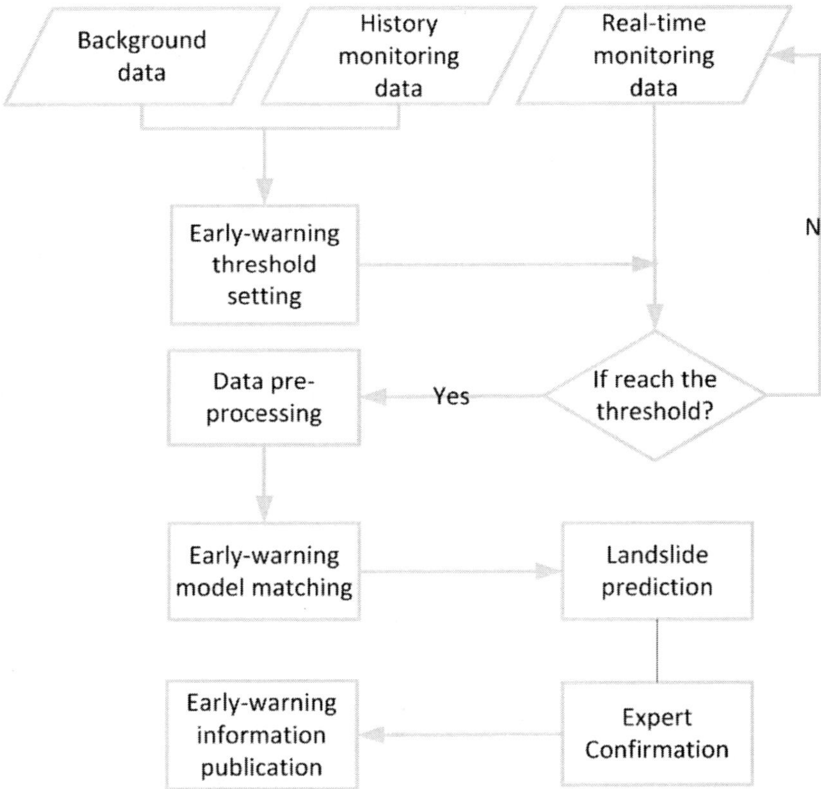

Figure 8. Flow-chart of landslide early warning system.

ACKNOWLEDGMENTS

This work was supported by the National Natural Science Foundation of China (Grant Nos. 41501558 and 41571377).

AUTHOR CONTRIBUTIONS

Huang Huang, Jianhua Ni and Yu Zhang conducted the primary experiments, cartography and analyzed the results. Yu Zhang and Tianlu Qian offered data support for this work. Jiechen Wang and Dingtao Shen provided the original idea for this paper.

REFERENCES

1. Mantovani, F.; Soeters, R.; van Westen, C.J. Remote sensing techniques for landslide studies and hazard zonation in Europe. *Geomorphology* **1996**, *15*, 213–225.

2. Gili, J.A.; Corominas, J.; Rius, J. Using Global Positioning System techniques in landslide monitoring. *Eng. Geol.* **2000**, *55*, 167–192

3. Squarzoni, C.; Delacourt, C.; Allemand, P. Differential single-frequency GPS monitoring of the La Valette landslide (French Alps). *Eng. Geol.* **2005**, *79*, 215–229.

4. Xiao, R.; He, X. Real-time landslide monitoring of Pubugou hydropower resettlement zone using continuous GPS. *Nat. Hazards* **2013**, *69*, 1647–1660.

5. Palmerini, G. Capabilities of the GNSS precise point positioning technique for landslide monitoring. *Disaster Adv.* **2012**, *5*, 509–513.

6. Wang, G.Q. Millimeter-accuracy GPS landslide monitoring using Precise Point Positioning with Single Receiver Phase Ambiguity (PPP-SRPA) resolution: A case study in Puerto Rico. *J. Geod. Sci.* **2013**, *3*, 22–31.

7. Yigit, C. Experimental assessment of post-processed kinematic Precise Point Positioning method for structural health monitoring. *Geomat. Nat. Hazards Risk* **2014**, *7*.

8. Dowding, C.H.; Huang, F.C. Early detection of rock movement with time domain reflectometry. *J. Geotech. Eng.* **1994**, *120*, 1413–1427.

9. Kane, W.F.; Perez, H.; Anderson, N.O. *Development of Time Domain Reflectometry System to Monitor Landslide Activity*; Final Report,

FHWA/CA/TL-96/09; Federal Highway Administration: Washington, DC, USA, 1996.

10. Fruneau, B.; Achache, J.; Delacourt, C. Observation and modelling of the Saint-Etienne-de-Tinée landslide using SAR interferometry. *Tectonophysics* **1996**, *265*, 181–190.

11. Carnec, C.; Delacourt, C. Three years of mining subsidence monitored by SAR interferometry, near Gardanne, France. *J. Appl. Geophys.* **2000**, *43*, 43–54.

12. Ohnishi, Y.; Nishiyama, S.; Yano, T.; Matsuyama, H.; Amano, K. A study of the application of digital photogrammetry to slope monitoring systems. *Int. J. Rock Mech. Min. Sci.* **2006**, *43*, 756–766.

13. Atzeni, C.; Barla, M.; Pieraccini, M.; Antolini, F. Early Warning Monitoring of natural and engineering slopes with Ground-based Synthetic Aperture Radar. *Rock Mech. Rock Eng.* **2015**,*48*, 235–236.

14. Angeli, M.G.; Pasuto, A.; Silvano, S. A critical review of landslide monitoring experiences.*Eng. Geol.* **2000**, *55*, 133–147.

15. Ramesh, M.V. Real-time wireless sensor network for landslide detection. In Proceedings of the Third International Conference on Sensor Technologies and Applications, SENSORCOMM'09, Athens, Greece, 18–23 June 2009; pp. 405–409.

16. Xia, Y. Grey System Prediction Model of Landslide and Its Application. *Chin. Sci. Abstr. Ser. B* **1995**, *14*, 74.

17. Chen, H.W.; Chang, N.B. Prediction analysis of solid waste generation based on grey fuzzy dynamic modeling. *Resour. Conserv. Recycl.* **2000**, *29*, 1–18.

18. Chung, C.J.F.; Fabbri, A.G. Probabilistic prediction models for landslide hazard mapping.*Photogramm. Eng. Remote Sens.* **1999**, *65*, 1389–1399.

19. Qin, S.; Jiao, J.J.; Wang, S. A nonlinear dynamical model of landslide evolution.*Geomorphology* **2002**, *43*, 77–85.

20. Carrara, A.; Guzzetti, F.; Cardinali, M.; Reichenbach, P. Use of GIS technology in the prediction and monitoring of landslide hazard. *Nat. Hazards* **1999**, *20*, 117–135.

21. Zan, L.; Latini, G.; Piscina, E.; Polloni, G.; Baldelli, P. Landslides early warning monitoring system. In Proceedings of the 2002 IEEE International Geoscience and Remote Sensing Symposium, IGARSS'02, Honolulu, Hawaii, USA, 25–30 July 2002; Volume 1, pp. 188–190.

22. Yin, Y.; Wang, H.; Gao, Y.; Li, X. Real-time monitoring and early warning of landslides at relocated Wushan Town, the Three Gorges Reservoir, China. *Landslides* **2010**, *7*, 339–349.

23. Gao, H.X.; Yin, K.L. Study on spatial prediction and time forecast of landslide. *Nat. Hazards***2014**, *70*, 1735–1748.

24. Barla, M.; Antolini, F. An integrated methodology for landslides early warning systems.*Landslides* **2015**, 1–14.

25. Costanzo, S.; Di Massa, G.; Costanzo, A.; Morrone, L.; Raffo, A.; Spadafora, F.; Borgia, A.; Formetta, G.; Capparelli, G.; Versace, P. Low-cost radars integrated into a landslide early warning systems. *Adv. Intell. Syst. Comput.* **2015**, *354*, 11–19.

26. Huang, R.; Huang, J.; Ju, N.; He, C.; Li, W. WebGIS-based information management system for landslides triggered by Wenchuan earthquake. *Nat. Hazards* **2013**, *65*, 1507–1517.

27. Mao, S.J.; Liu, Q.X.; Li, M. Design and development of safety production management information system based on a digital coalmine. *Procedia Earth Planet. Sci.* **2009**, *1*, 1121–1127.

28. Intrieri, E.; Gigli, G.; Mugnai, F.; Fanti, R.; Casagli, N. Design and implementation of a landslide early warning system. *Eng. Geol.* **2012**, *147*, 124–136.

29. Zerger, A.; Smith, D.I. Impediments to using GIS for real-time disaster decision support.*Comput. Environ. Urban Syst.* **2003**, *27*, 123–141.

30. Yu, F.C.; Chen, C.Y.; Lin, S.C.; Lin, Y.C.; Wu, S.Y.; Cheung, K.W. A web-based decision support system for slopeland hazard warning. *Environ. Monit. Assess.* **2007**, *127*, 419–428.

31. Chang, K.T.; Wan, S.; Lei, T.C. Development of a spatial decision support system for monitoring earthquake-induced landslides based on aerial photographs and the finite element method. *Int. J. Appl. Earth Obs. Geoinform.* **2010**, *12*, 448–456.

32. Giardino, M.; Perotti, L.; Lanfranco, M.; Perrone, G. GIS and geomatics for disaster management and emergency relief: A proactive response to natural hazards. *Appl. Geomat.* **2012**, *4*, 33–46.

33. Finkel, R.A.; Bentley, J.L. Quad trees a data structure for retrieval on composite keys. *Acta Inform.* **1974**, *4*, 1–9.

34. Heeger, D.J.; Bergen, J.R. Pyramid-based texture analysis/synthesis. In Proceedings of the 22nd Annual Conference on Computer Graphics and Interactive Techniques, Los Angeles, CA, USA, 9–13 August 1995; pp. 229–238.

35. Saito, M. Forecasting time of slope failure by tertiary creep. In Proceedings of the 7th International Conference of Soil Mechanics and Foundation Engineering, Mexico City, Mexico, 5–7 May 1969; pp. 677–683.

36. Zhou, C.B.; Zhang, H. A creep-spline combined model and its application to the time prediction of landslides. *J. Nat. Disasters* **1996**, *5*, 60–67.

37. Li, T.B.; Chen, M.D. Time prediction of landslide using verhulst inverse-function model. *J. Geol. Hazards Environ. Preserv.* **1996**, *3*, 13–17.

38. Deng, J.L. Control problems of grey systems. *Syst. Control Lett.* **1982**, *1*, 288–294.

39. Haigh, M.J. *Dynamic Systems Approaches in Landslide Hazard Research*; Zeitschrift für Geomorphologie; Supplementbände; Schweizerbart science publishers: Stuttgart, Germany, 1988; Volume 67, pp. 79–81.

40. Phillips, J.D. Instability and chaos in hillslope evolution. *Am. J. Sci.* **1993**, *293*, 25.

41. Luan, Y.Z.; Jiang, Y.; Jia, M.H. Forecasting the ground subsidence in mine area by grey system. *Site Investig. Sci. Technol.* **1990**, *5*, 45–48.

42. Li, X.H.; Jin, X.G.; Kang, H.M.; Lu, Y.; Yang, X.H. Application of GM (1,1) Majorized Model to Simulation-forecast of Landslide. *J. Mt. Sci.* **2001**, *19*, 265–269.

43. Lu, P.; Rosenbaum, M.S. Artificial neural networks and grey systems for the prediction of slope stability. *Nat. Hazards* **2003**, *30*, 383–398.

44. Li, X.; Kong, J.; Wang, Z. Landslide displacement prediction based on combining method with optimal weight. *Nat. Hazards* **2012**, *61*, 635–646.

45. Wang, X.; Niu, R. Landslide intelligent prediction using object-oriented method. *Soil Dyn. Earthq. Eng.* **2010**, *30*, 1478–1486.

Chapter 10

Integrating Expert Knowledge with Statistical Analysis for Landslide Susceptibility Assessment at Regional Scale

Christos Chalkias [1,*], Christos Polykretis [1], Maria Ferentinou [2†] and Efthimios Karymbalis [1]

[1]Department of Geography, Harokopio University, El. Venizelou 70, Athens 17671, Greece

[2]Department of Geological Sciences, University of KwaZulu-Natal, Westville, Private Bag X54001, South Africa

ABSTRACT

In this paper, an integration landslide susceptibility model by combining expert-based and bivariate statistical analysis (Landslide Susceptibility Index—LSI) approaches is presented. Factors related with the occurrence of landslides—such as elevation, slope angle, slope aspect, lithology, land cover, Mean Annual Precipitation (MAP) and Peak Ground Acceleration (PGA)—were analyzed within a GIS environment. This integrated model produced a landslide susceptibility map which categorized the study area according to the probability level of landslide occurrence. The accuracy of the final map was evaluated by Receiver Operating Characteristics (ROC) analysis depending on an independent (validation) dataset of landslide events. The prediction ability was found to be 76% revealing that the integration of statistical analysis with human expertise can provide an acceptable landslide susceptibility assessment at regional scale.

KEYWORDS

integration modeling; landslide susceptibility; expert-based fuzzy weighting; bivariate statistics; GIS; Greece

1. INTRODUCTION

Natural hazards have been studied continuously, and systematically, and during the last three decades, their frequency seems to be on the rise [1]. The need for reliable hazard maps is uncontested, in order to mitigate their effects in both natural and manmade environments. From 1900 to 2014, nearly 6% of global natural hazards were associated with landslides [2]. Mountainous or rural environments are subject to landslides as a major natural hazard which causes substantial economic, human, and environmental losses throughout the world each year [3].

Landslide Susceptibility (LS) is defined as the likelihood of a landslide occurring in an area on the basis of local terrain conditions [4]. LS zonation refers to the division of the land in homogeneous areas or domains according to the degree of actual or potential hazard [5]. In the last years, LS mapping has become a valuable tool for assessing disaster phenomena. The cartographic products of LS modeling can be used for developing mitigation plans and selecting the more suitable locations for construction.

Given the powerful abilities of Geographic Information Systems (GIS) technology to deal with a large amount of spatial data, LS is analyzed through effective spatial analysis tools. Many GIS-based approaches have been developed for LS assessment. In general, they are divided into two groups: qualitative and quantitative methods. The most important difference between these two types of methods is their degree of objectivity. The qualitative methods, used extensively during the 1970s and 1980s, depend on the knowledge and previous experience of the experts, resulting in a high degree of subjectivity. They include the geomorphological analysis [6,7] and the use of index or parameter maps.

The use of index or parameter maps is subdivided into two methods: the combination or overlay of index maps [8,9] and the logical analytical models [10,11]. Some qualitative approaches, however, incorporate the idea of ranking and weighting the parameters involved, and may be semi-quantitative in nature [12]. Examples of this kind of approach are the Analytical Hierarchy Process (AHP) [13,14], the Weighted Linear Combination (WLC) [15], and their combination [16].

The quantitative methods are based on numerical expressions of the relationship between controlling factors and landslide occurrence. They include statistical analysis, geotechnical engineering approaches [17,18], and soft computing techniques such as Artificial Neural Network (ANN) [19,20,21,22], neuro-fuzzy [23,24], support vector machine [16,23] and decision tree [23,25] methods. In statistical analysis methods, the combination of parameters that have led to landslides in the past are determined statistically, and quantitative predictions are made for areas not affected by landslides but where similar background conditions exist [26]. Several researchers have applied these methods, which involve either bivariate [27,28,29,30] or multivariate analysis [16,31,32,33]. In bivariate analysis, each individual landslide-influencing factor is combined with a landslide inventory map, and weight values based on landslide densities are calculated for its corresponding categories. A well-known and widely used bivariate analysis method for LS zonation is the Landslide Susceptibility Index (LSI) [14,22,34,35]. It is increasingly recognized that in both qualitative and quantitative models, the results are prone to the inherent uncertainties related to various analysis parameters such as the errors and variability in model choice, weighting factors, data used, and human judgment [15].

In recent years, another soft computing technique, known as fuzzy logic, has been applied to map LS. Considering the determination of weightings (fuzzy membership values), this technique is either expert-based (semi-quantitative) approach or data-driven (quantitative) approach or a combination of both [36]. In expert-based approaches [37,38], the weightings are assigned based on the field knowledge of the experts, whereas in data-driven approaches [39,40] the weightings are determined according to correlations between landslide locations and landslide-conditioning factors. Such an expert-based approach, referred as Trapezoidal Fuzzy Number Weighting (TFNW) [12,41,42], was used in this study.

In Greece, previous studies concerning LS assessment have been implemented on moderate scale (1:25,000–1:500,000) using semi-quantitative methods such as AHP, WLC and TFNW [12,43,44,45], statistical analysis [22,35,46,47], ANN [22,48] and integration methods [49], and on national scale (<1:1,000,000) using statistical analysis [50,51], ANNs [50] and fuzzy logic [52].

The main aim of this paper was to improve our previous research works [12,35], integrating two different and already tested LS assessment methods: LSI and TFNW. Thus, a GIS-based integration method was developed and presented at regional scale (1:500,000) for Peloponnese peninsula, Greece. Furthermore, a validation analysis was implemented to estimate the prediction ability of the proposed method.

2. STUDY AREA

Peloponnese was selected as the study area for the proposed method. It is one of the nine geographical departments of Greece and is located in its southern part (Figure 1). Peloponnese constitutes the largest peninsula of the country with an extent of 21,439 km^2. It is divided into seven administrative prefectures: Achaia, Argolida, Arkadia, Ilia, Korinthia, Lakonia and Messinia (Figure 1), with a population of 1,086,935 inhabitants [53]. Peloponnese has a typical Mediterranean climate with a hot and relatively dry summer, and a wet season during autumn, winter and spring [54]. The climate as well as the geotectonic status, the lithology and the seismicity have mainly formed the geomorphology of the study area. In general, its geomorphology can be characterized as complex, with a mountainous interior, many coastal cliffs in the south, and basins, coastal beaches, lakes and inland basins on the west and southeast coasts.

Peloponnese is a region highly vulnerable to the occurrence of severe natural disasters such as earthquakes, floods and forest fires. However, due to its complex geomorphology, all causal and triggering factors of landslide occurrence present a high spatial variability. Considering the morphological and geological conditions of Peloponnese, it is evident that slope stability is one of the most severe and frequent hazards for public safety as well as land-use planning and management of the area.

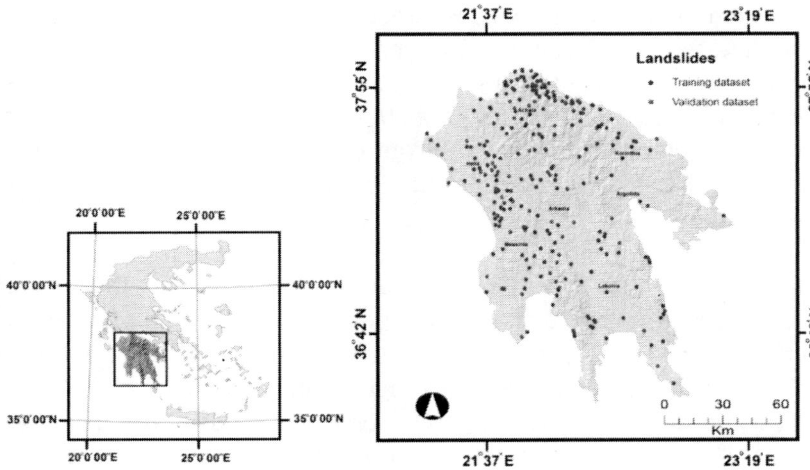

Figure 1. The location of the study area (Peloponnese peninsula) and the landslide inventory.

3. MATERIALS AND METHODS

In order to accomplish the LS analyses in the study area, a spatial database was designed and developed, and spatial analysis tools were implemented within GIS environment with the use of an ArcGIS (version 9.3) software package. This database comprises two main parts: (a) the landslide inventory dataset and (b) the datasets for the landslide-influencing factors.

3.1. Landslide Inventory

In the present study, two main landslide inventories were used: (a) an inventory maintained by Institute of Geology and Mineral Exploration (IGME) covering a time period from 1910 to 1995 [55] and (b) an inventory, developed on the basis of fieldwork and aerial photography [56] interpretation, which contains landslide events occurring from 1995 to 2003. The final landslide dataset consists of 282 landslides (217 and 65 landslides from the first and second inventory, respectively) presented as point features in the centroid of each pixel (Figure 1). This dataset was randomly split into two separate groups: a training dataset and a validation dataset. The training dataset (75% of the landslide inventory) was used for the implementation of the model, whereas the validation dataset (25% of the landslide inventory) was used during the verification of the results produced from the model (see among others [57,58,59]). Our intention was to keep the same sampling strategy (*i.e.*, the same percentages for training and validation datasets) for both the integration model and the single models [12,35] combined for its creation in order to increase their comparability.

The landslides under investigation have more or less the same characteristics: lateral based and downslope movement of soils or rocks. In line with [60], in this study, the term landslide is used for translational and rotational earth slides, which were recorded in landslide inventories. These events vary consistently in volume, from some thousands of m^3 to several million m^3 [51], and depicting small to extremely large magnitude according to [61] classification. Subaqueous and liquefaction events are not examined.

3.2. Landslide-Influencing Factors

Although there are no standard guidelines for selecting the landslide-influencing factors, the nature of the study area, the scale of the analysis, the data availability and the general literature guidelines were taken into account [26,62,63,64,65]. In this study, land cover, lithology, elevation, slope angle, slope aspect, MAP and PGA have been selected as the landslide-influencing factors.

Slope stability is highly influenced by land cover. The land cover layer was classified by following the level-1 classification scheme of the CORINE Coordinate of Information on the Environment) program data [66]. Agricultural, forest and semi-natural areas cover the main part of Peloponnese, whereas urban is the dominant land cover in the coastal zone of the peninsula.

Lithology plays a key role in landslide activity since different lithologic units have different slope stability performances [67]. The lithological layer of the study area was created from the Geological Map of Greece [68]. The main lithological formations are carbonate rocks (44%): limestones, dolomites and marbles, and Neogene sediments (22%): usually marls, sandstones and mudstones. The central and western part of Peloponese belongs to the External Hellenides, which is characterized by Mesozoic–early Cenozoic sedimentary units (mainly carbonate rocks, schists-cherts and flysch) [69]. The most critical landslide-prone formations are flysch and neogene sediments, while schists and cherts significantly contribute in landslide phenomena [51].

Among all parameters for LS zonation, elevation, slope angle and slope aspect have been recognized as the most important [70,71,72]. The elevation dataset is useful to classify the local relief and locate points of maximum and minimum heights within terrains. Furthermore, it is well supported in the literature [43,46] that slope gradients have a large impact on landsliding in Peloponnese. In the study area, the slopes vary from gentle to very steep. Most of the landslides occur on gentle natural slopes where the translational type predominates. The aspect parameter is also related to differential weathering, exposure to sunlight and drying winds and soil moisture [62]. The Digital Elevation Model (DEM) was the key to generating the topographic factors related to the landslide activity of the area under investigation. We used the DEM from SRTM (Shuttle Radar Topography Mission) database. Elevation (0–2367 m), slope angle (0°–54°) and slope aspect were extracted from this DEM.

Previous studies emphasized the need of incorporating “dynamic” factors (MAP, PGA) with other “static” factors for LS zonation mapping in areas, whereas these factors are playing an important role not only in the reactivation of old landslides but in the development of new ones [73,74]. It is, therefore, imperative to incorporate these parameters, while carrying out LS analysis in seismically active and frequently rainfall-influenced regions. The precipitation data (339–1655 mm) used in this study refers to the mean annual precipitation during the period from 1950 to 1974 (source: Public Power Company, Athens, Greece). MAP is the average of the available long-term records [75]. The study area belongs to a seismically active zone with tectonism expressed through faults, thrust zones and folds. The seismic factor (0.05–3.95 m/s^2) was produced from the map of expected peak ground acceleration with a 475-year return period (10% probability of exceedance in 50 years). This map, from which the corresponding layer was derived, was launched in 1992 from the Technical Chamber of Greece. PGA is the absolute maximum amplitude of recorded ground acceleration [76]. Heavy rainfall and earthquakes have triggered several landslide events, mainly in northern and western areas of the peninsula [43,77,78]. Rainfall triggered landslides usually are rapid short moving events, while slow short-moving type also occur including extensive instability zones [79]. On the other hand, seismic triggered landslides occur in the vicinity of active faults, and usually related to other secondary seismic events, like soil liquefaction, subsidence of the coastal strip, and rock falls [80,81].

Most of the final layers were in raster format (grid), while others were converted from vector (point, line or polygon) to raster format (with cell size 250 m × 250 m for land cover, lithology, MAP and PGA factors, and 90 m × 90 m for elevation, slope angle and slope aspect factors).

3.3. Methodology

A GIS-based integration approach was developed to create a LS map using two different methods: LSI and TFNW. Such integration can adjust the parameters included in this study by means of the precise LS assessment and effectively reduce the subjectivity and uncertainty resulting from the use of single method. The LSI method was proposed by [82] and is based on the relationship between the spatial distribution of landslides and each conditioning factor [62]. Here, in order to quantify the impact of various categories for the standardization of each factor, we used the TFNW. The details of this approach have been described by [41]. The main steps of this methodology are presented below:

a. Categorization of all landslide-influencing factors. In this step, the "Natural Breaks (Jenks)" categorization (five categories) was implemented for the "dynamic" factors (MAP and PGA), whereas for the other two factors with continuous values (elevation and slope angle) the categorization was executed in a manual way based on their presented values. "Natural Breaks (Jenks)" method is a data classification method that minimizes variance within groups of data and maximizes variance between groups of data [83]. For land cover, lithology, and aspect, all the categories of the nominal scale were preserved.

b. Calculation of LSI for all factor categories (Table 1) based on Equation (1) [26]:

c. The maximum value, the range and the standard deviation of LSI values for all factors were calculated in order to interpret the importance of each factor. The range is the difference between the maximum and minimum LSI values. The standard deviation measures the spread of the LSI values about the mean value.

d. Definition of linguistic variables and fuzzy numbers for LS categories in order to incorporate uncertainty in the analysis. All fuzzy numbers were expressed as (a_k, b_k, c_k, d_k). The definition of these fuzzy numbers is presented in [12].

e. Three independent experts, with scientific background in the fields of geosciences and engineering geology, and with experience and scientific knowledge of the study area, were invited to assign a linguistic importance weighting for every category of each factor. From these linguistic judgments we obtained the corresponding fuzzy numbers. Such judgments are inevitably subjective, but, by proposing several possible scenarios, followed by the systematic testing and elimination of options, as a result of additional investigation and discussion, it is possible to develop reliable estimates. Experimental evidence suggests that group judgments appear to be more accurate than judgments of a typical group member [84]. In this case, we considered a homogeneous group of experts with equal degree of importance for each one. Accordingly, the overall expert-based judgment is the mean value of the three judgment values. The sum of these numbers is still a fuzzy number. Thus, we proceeded to the computation of the aggregated fuzzy weights of individual categories (Table 1).

f. A fourth "statistical expert" judgment was added in the above list on the basis of the statistical analysis (Table 1). This statistical interpretation was based on the LSI values for each category according to the following rules:

- In a category with LSI value between –0.15 and 0.15, a "Moderate" susceptibility was assigned.
- Categories with negative (<–0.15) LSI values were defined as zones of "Very Low" or "Low" susceptibility according to their ascending order.
- Categories with positive (>0.15) LSI values were defined as zones of "Very High" or "High" susceptibility according to their descending order.
- Accordingly, the range of LSI values was interpreted for each factor according to the following rules:

- In a factor with range between 1.00 and 1.50, a "Moderate" susceptibility was assigned.

- Factors with range <1 were characterized with "Very Low" or "Low" susceptibility according to their ascending order.

- Factors with range >1.50 were characterized with "Very High" or "High" susceptibility according to their descending order.

It is worth mentioning that in our case, we assigned an equal degree of importance between the "statistical expert" and the group of experts.

- After the defuzzification of the fuzzy weights of individual LS categories, we proceed to the computation of the normalized weights and the construction of the weight vector (Table 1). More details of this step could be found in [12].

By implementing the same procedure described previously, we used expert-based and statistical-based linguistic variables to calculate the weight of importance for each conditioning factor (Table 2).

- The last step is the aggregation of relative values, and the generation of the final LS map (Figure 2). This step was implemented by using the WLC method [85]. We classified the final LS map into five discrete categories: "Very Low", "Low", "Moderate", "High" and "Very High" landslide susceptibility according to the standard deviation classification method. This method uses the mean value to generate class breaks by adding or subtracting one standard deviation at a time [70]. Moreover, in order to maintain five categories, we embedded low and high outliers into "Very Low" and "Very High" susceptibility categories, respectively.

$$LSI_{i,j} = ln\left(\frac{N_{i,j}}{A_{i,j}} / \frac{N_T}{A_T}\right) \qquad (1)$$

with $N_{i,j}$ the number of landslides in category j of the factor i, $A_{i,j}$ the area of this category, N_T the total number of landslides and A_T the total area under investigation. Thus, the calculation of the landslide density in each category and of the area for each category was implemented using GIS-based overlay functions.

Table 1. Categories, landslide densities, susceptibilities (LSI), fuzzy values and weights for landslide related factors.

Layers (Factors)	Categories (Classes)	Landslide Density	LSI	(Experts), STAT Fuzzy Value	LSI/TFNW Weight
Land cover	Artificial surfaces	0.09	2.04	(H, M, M), VH	0.45
	Agricultural areas	0.66	0.42	(M, H, H), H	0.42
	Forest and semi-natural land	0.25	−0.79	(L, L, M), VL	0.13
Lithology	Phyllites/Gneiss (metamorphic)	0.07	0.16	(L, M, L), H	0.14
	Limestones—Marbles	0.20	−0.77	(L, L, L), VL	0.05
	Schists (metamorphic)	0.01	−0.51	(M, M, M), VL	0.08
	Neogene	0.44	0.69	(H, VH, M), VH	0.22
	Quaternary	0.12	−0.06	(H, VH, M), M	0.16
	Flysch	0.09	0.06	(VH, VH, VH), M	0.19
	Cherts—Schists	0.06	0.36	(M, L, M), H	0.15
MAP	<663 mm	0.03	−1.69	(L, VL, M), VL	0.07
	663–884 mm	0.30	0.14	(M, L, H), M	0.19
	885–1079 mm	0.43	0.29	(M, M, H), H	0.25
	1080–1295 mm	0.19	−0.03	(H, H, VH), M	0.25
	>1295 mm	0.05	−0.22	(VH, VH, VH), L	0.23
PGA	<1.16 m/s^2	0.02	−0.14	(L, VL, L), M	0.16
	1.16–1.88 m/s^2	0.01	−0.98	(L, L, L), VL	0.08
	1.89–2.53 m/s^2	0.17	−0.81	(M, M, M), VL	0.13
	2.54–3.11 m/s^2	0.67	0.53	(H, H, H), VH	0.37
	>3.11 m/s^2	0.13	−0.28	(VH, VH, VH), L	0.27
Elevation	<234 m	0.36	0.17	(L, VL, L), H	0.25
	234–524 m	0.27	0.11	(L, L, M), M	0.22
	525–851 m	0.22	−0.03	(H, M, M), M	0.28
	>851 m	0.15	−0.43	(M, H, H), L	0.24
Slope angle	<5°	0.31	−0.16	(VL, VL, L), L	0.09
	5°–10°	0.30	0.19	(L, L, L), H	0.21
	11°–15°	0.22	0.23	(M, M, M), H	0.25
	16°–20°	0.12	0.08	(H, H, H), M	0.25
	>20°	0.05	−0.77	(VH, VH, VH), VL	0.21
Slope aspect	North	0.31	0.31	(H, H, M), H	0.31
	East	0.21	−0.15	(M, M, M), M	0.23
	South	0.16	−0.47	(M, M, M), L	0.18
	West	0.32	0.16	(H, H, H), M	0.28

LSI: Landslide Susceptibility Index, LSI/TFNW: The integration method, VL: Very Low, L: Low, M: Moderate, H: High, VH: Very High.

4. RESULTS—VALIDATION

As mentioned previously, we developed an integration LS model according to the LSI/TFNW approach. This approach allowed for the assessment of landslide susceptibility by integrating expert-based and statistical modelling methods.

The results from the LSI/TFNW model are presented in Table 1 and Table 2. The most important causal factors were found to be MAP, slope angle and lithology with weight values 0.20, 0.18 and 0.17, respectively. Elevation was the weakest factor in the scale of importance (weight value = 0.08).

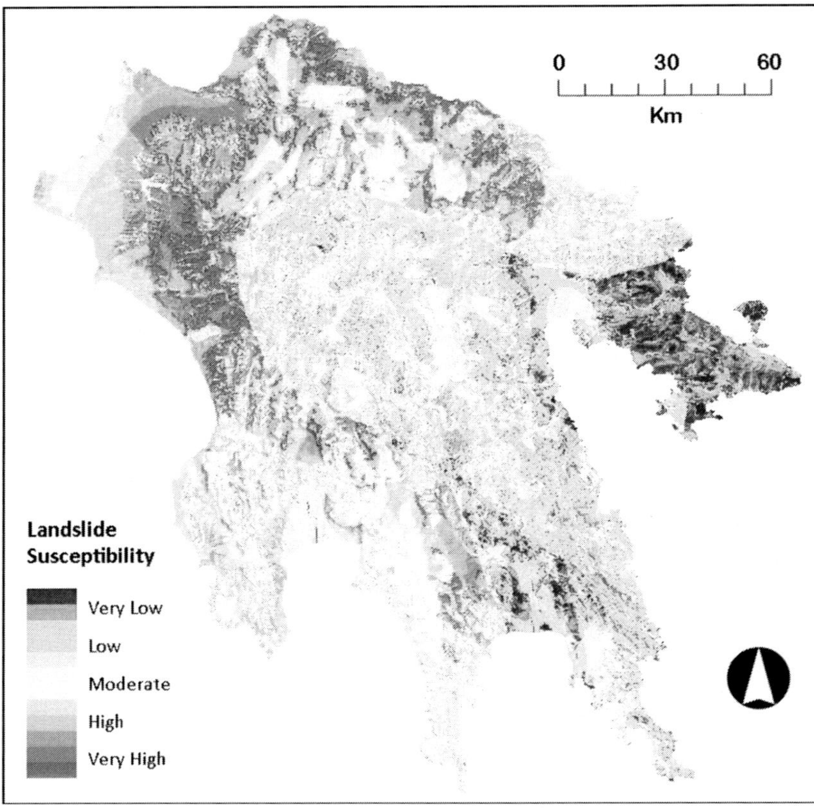

Figure 2. The landslide susceptibility map produced from the integration LSI/TFNW model.

Table 2. Weight values for each factor according to the integration LSI/TFNW method.

Parameter	Parameter Importance (Expert Judgment), STAT	LSI/TFNW Weight
Land cover	(M, M, H), M	0.12
Lithology	(VH, VH, H), H	0.17
MAP	(VH, VH, VH), VH	0.20
PGA	(H, H, H), H	0.16
Elevation	(M, L, H), L	0.08
Slope angle	(VH, VH, VH), H	0.18
Slope aspect	(H, M, M), L	0.09

The output LS map (Figure 2) from the integration model shows that 26% and 7% of the study area were classified as "High" and "Very High" susceptibility zones, respectively. The same map also shows that the zones of "High" and "Very High" susceptibility are mainly located around the northern (coastal) and western part of Peloponnese, with pockets of "High" susceptibility in the rest part of the peninsula.

The overlay of the final LS map with the landslide training dataset indicated that 28%, 42% and 24% (total: 94%) of the landslide events fall within "Very High", "High" and "Moderate" LS zones (in total 66% of the study area), respectively. It is notable that according to the used model only 6% of the dataset falls in "Low" susceptibility zone, with no landslides into the "Very Low" susceptibly zone.

We have also proceeded with a standard validation analysis using data from landslide events, not included in the initial spatial database (validation dataset), in order to estimate the overall performance of the model in the study area. Validation is essential to know the predictive value of the model [86]. For the validation of the output, the Receiver Operating Characteristics (ROC) curve was drawn, and the Area Under Curve (AUC) value was calculated for the proposed model. ROC analysis is considered as a powerful method for the validation of LS models [87,88]. In practice, the AUC performs very well and is often used when a general measure of predictiveness is desired [89]. The AUC values range from 0.5 to 1.0. The ideal model yields a value close to 1.0 (perfect fit), whereas a value close to 0.5 indicates an inaccurate model (random fit).

Figure 3 shows the ROC curve of LSI/TFNW model for the validation dataset. The AUC value of 0.76 indicates a reasonable prediction ability of the model.

Previous studies [12,35] in the same area with the use of the single methods combined in this integration model (TFNW and LSI, respectively), produced slightly worse results (Figure 3). The AUC values for these models were 0.70 and 0.74, respectively, based on the standard deviation classification method for their final LS maps.

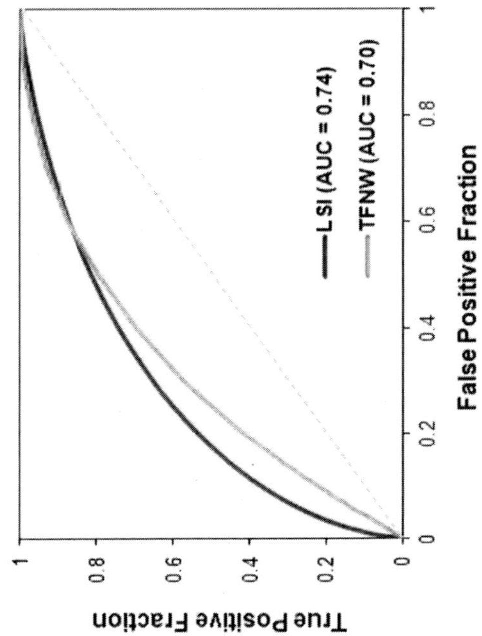

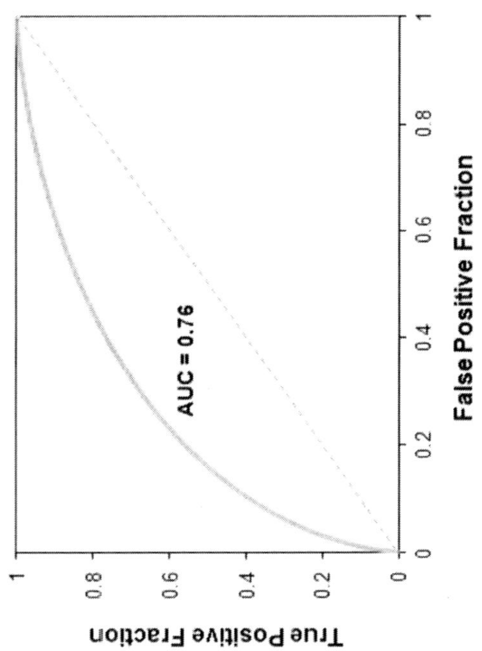

Figure 3. ROC curves for the integration LSI/TFNW model and the combined single models, LSI and TFNW.

Specifically, comparing the final map of the proposed integrated method with the landslide susceptibility map produced from TFNW model [12], it is shown that in both maps 14% (3039 km^2), 9% (2001 km^2) and 1% (301 km^2) of the study area are defined as zones of "Moderate", "High" and "Very High" susceptibility, respectively. On the other hand, the differences between the LS maps for these three susceptibility zones are presented to cover 32% (6773 km^2) of the study area. The differences for the remaining susceptibility zones ("Very Low" and "Low") cover a restricted part of the study area (7% or 1377 km^2).

In addition, comparing the final LS map of this study with this one created from LSI model [35], it is derived that in both maps 21% (4427 km^2), 18% (3829 km^2) and 4% (881 km^2) of the study area belong to "Moderate", "High" and "Very High", respectively, susceptibility zones. On the contrary, the coverage of differences between the final maps for these three susceptibility zones is found to be 16% (3291 km^2) of the study area. Finally, the coverage of differences for the remaining susceptibility zones is more or less similar with the relative coverage of the aforementioned comparison between integration and TFNW models.

5. DISCUSSION

The idea of implementing an integration method for LS zonation is not new. Recent examples are the efforts of [90,91,92]. In the first study, AHP and fuzzy sets were combined to determine LS levels for areas of the Metro Vancouver region, British Columbia, Canada. The parameters were standardized to a common measurement scale using fuzzy set membership functions and the weight value for each parameter was calculated using AHP. In the second study, the authors produced a LS zonation map in the Chehelchay Basin (northeastern Iran). They applied Geographically-Weighted Principal Component Analysis (GWPCA) to calculate LS using a fuzzy gamma operator. Finally, the third study focused on the methodology of the urgent LS assessment right after the 2013 Lushan earthquake (southwest of Sichuan Province, China). An assessment approach was developed integrating bivariate statistical analysis (LSI) and AHP methods for the assignment of factor class weights and factor weights, respectively.

This study combined an expert-based semi-quantitative approach (TFNW) with a data-driven statistical method (LSI) to prepare a LS map at a regional scale on the Peloponnese peninsula, Greece. The main motivation is to create an integration model that preserves and integrates the advantages of the combined models at the specific scale of analysis. To achieve this objective, seven conditioning factors (elevation, slope angle, slope aspect, lithology, land cover, MAP and PGA) were taken into consideration.

The LSI model is a data-driven bivariate statistical approach. The bivariate statistical technique is one of the most preferable models at this scale [93]. In this LS modeling, all the problems of landslide inventory maps will automatically be inherited to the final susceptibility maps [94]. In general, a reliable, accurate and complete landslide inventory map will provide high-quality statistical models [95]. In many countries—including Greece—this kind of extended landslide inventory is not available.

On the other hand, for the expert-based fuzzy weighting model, the landslide inventory map is not needed. This kind of analysis is purely subjective. To some extent, opinions may change for every individual expert and thus may be subjected to cognitive limitations with uncertainty and subjectivity. However, methods that depend on expert opinions are often useful for regional assessments [96]. A significant part of the semi-quantitative LS research follows a similar strategy as the proposed one by inviting a limited number of experts (see among others [42,97]). The main issue is not to invite as many scientists as possible but to invite experts with detailed knowledge of the landslide events in the area under investigation and to ensure an overall consensus of their evaluation about the importance of the factors involved.

Recently, ROC analysis has been widely used in the international literature to evaluate the prediction capability of the LS mapping methods. In this study, the empiric ROC area for the integration model was estimated to be 0.76 for the validation dataset. Thus, there is 76% agreement between prepared LS map and landslide locations of the validation dataset, which is a reasonable result, taking into consideration the scale of analysis. Moreover, this model proved to be slightly more effective for the overall LS assessment of the study area as it recorded the highest percentage in our validation test in comparison with LSI and TFNW methods (74% and 70%, respectively). Therefore, the reliability of the proposed integration model is directly related to the reliability of both data-driven and expert-based methods combined. In other words, it seems that a combination of two quite satisfactory methods according to the proposed integration model generates even better results. However, this preliminary finding needs more research to be confirmed through the implementation of the proposed model in more studies areas with various geographical and geotectonic settings.

The implementation of the integration model in the study area revealed that there are different zones within Peloponnese which seem to configure various landslide susceptibility clusters. The "High" susceptibility areas are concentrated in the western Peloponnese with few scattered pockets in the remaining parts of the peninsula. Additionally, a linear concentration of "High" susceptibility values is located in the northern coastal zone of the study area. According to the final LS map, the "Very High" susceptibility zone covers a significant part of the study area (7% of the total area).

The comparison of the proposed method with the expert-based and the statistical methods showed that the integrated method illustrates more emphatically the zones of "High" or "Low" susceptibility in both of the individual methods. Accordingly, the main part the areas of "High" and "Very High" susceptibility in the hybrid map (classes "4" and "5"), is within the class "5" ("Very high" susceptibility) in the statistical and the expert-based LS maps (99% and 67%, respectively). Concerning the spatial distribution of the areas with high LS values in the integrated method, despite the preservation of the general spatial pattern (high LS values in the northern and the western part of Peloponnese) it has to be noticed that there are some pockets of high LS in areas with limited existence of previous landslides. This fact is due to the expert-based classification and it may be useful to identify high LS zones in areas with lack of landslide events.

The most important factors for the zonation of LS in the study area are MAP, slope angle and lithology. It seems that the incorporation of "dynamic" factors (MAP, PGA) in this regional analysis was more or less beneficial to the LS assessment. Among the categories of these two factors, the categories with high levels of annual precipitation (885–1079 mm) and seismic acceleration (2.54–3.11 m/s^2) have the highest LSI values combining the highest landslide density and coverage area. In contrast with aforementioned categories of MAP and PGA factors, the category with 11°–15° slope angle presents the highest LSI value combining a lower landslide density and coverage area than the categories with low levels of slope angle (<10°).

Some limitations and assumptions of the proposed method have to be pointed out. First of all, as the analysis is based on medium-scale datasets, the results are unsuitable for detailed site-oriented specific analysis. At large scales, more exhaustive datasets and detailed geotechnical information are required. A second limitation is related to the landslide inventory dataset. At regional scale, this dataset does not include the total amount of the landslide events within the area under investigation. In this paper, we considered equal importance between the statistical and expert-based models for the construction of the integration model. Although this is a reasonable starting point, alternative importance ratios should be further tested. Finally, another limitation of this study is that the proposed method does not take into account the mutual relationships between causal factors. Future work will focus on this issue by combining expert-based with multivariate statistical models. Furthermore, it should be mentioned that, according to our analysis, the output LS map presents only the predicted spatial distribution of landslides. It does not present the temporal probability of landsliding.

Therefore, the results from this paper should be used in the first stage of hazard mapping, which is crucial in an essential part of quantitative risk analysis [98,99]. Despite its limitations, the proposed method can produce reliable landslide susceptibility maps at regional scale. This is very useful information for local authorities and decision makers in order to choose suitable locations for future land-use planning and implementation of developments, and target their mitigation strategies.

6. CONCLUSIONS

Statistical analysis and expert-based methods are often used to evaluate LS zonation. Considering that these general approaches have their pros and cons when applied at regional scale, we investigate the performance of an integration model. The results obtained show that it is possible to produce satisfactory LS maps with the proposed hybrid model. It seems that the overall assessment provided by the experts efficiently incorporates the knowledge of the study area and of the factors involved in the LS zonation. Thus, to some extent, the expertise of specialists can contribute to the enhancement of the standard statistical method especially in moderate and small scale analyses in areas with incomplete landslide event archives.

Further research may include the validation of the proposed model in other areas characterized by different geomorphological characteristics, and its comparison with other advanced LS models (such as logistic regression and ANN models) at regional scale.

ACKNOWLEDGMENTS

Authors thank the three experts for their contribution in this paper. Also, authors are grateful to the anonymous reviewers for their useful comments and constructive manuscript reviewing.

AUTHOR CONTRIBUTIONS

Christos Chalkias conceived of the study, designed the analysis, and participated in its implementation. Christos Polykretis participated in the spatial database creation and the implementation of statistical analysis, and drafted the manuscript. Maria Ferentinou and Efthimios Karymbalis participated in the design and implementation of the study and in the preparation of the validation dataset. Christos Chalkias, Christos Polykretis, Maria Ferentinou and Efthimios Karymbalis participated in the finalization of the manuscript. All authors read and approved the final manuscript.

REFERENCES

1. Smith, K.; Petley, D.N. *Environmental Hazards: Assessing risk and reducing disaster*, 5th ed.; Routledge: Abingdon, UK, 2009.

2. Centre for Research on the Epidemiology of Disasters (CRED). *EM-DAT: International Disaster Database*; Université Catholique de Louvain: Brussels, Belgium, 2014.

3. Aleotti, P.; Chowdhury, R. Landslide hazard assessment: Summary review and new perspectives. *Bull. Eng. Geol. Environ.* **1999**, *58*, 21–44.

4. Brabb, E.E. Innovative approaches to landslide hazard mapping. In Proceedings of the 4th International Symposium on Landslides, Toronto, ON, Canada, 16–21 September 1984; Volume 1, pp. 307–324.

5. Varnes, D.J. *Landslide Hazard Zonation: A Review of Principles and Practice*; UNESCO: Paris, France, 1984; Volume 3.

6. Santangelo, M.; Cardinali, M.; Rossi, M.; Mondini, A.C.; Guzzetti, F. Remote landslide mapping using a laser rangefinder binocular and GPS. *Nat. Hazards Earth Syst. Sci.* **2010**, *10*, 2539–2546.

7. Listo, F.L.R.; Vieira, B.C. Mapping of risk and susceptibility of shallow-landslide in the city of Sao Paulo, Brazil. *Geomorphology* **2012**, *169–170*, 30–44.

8. Castellanos Abella, E.A.; van Westen, C.J. Qualitative landslide susceptibility assessment bymulticriteria analysis: A case study from San Antonio del Sur, Guantánamo, Cuba.*Geomorphology* **2008**, *94*, 453–466.

9. Avtar, R.; Singh, C.K.; Singh, G.; Verma, R.L.; Mukherjee, S.; Sawada, H. Landslide susceptibility zonation study using remote sensing and GIS technology in the Ken-Betwa River Link area, India. *Bull. Eng. Geol. Environ.* **2011**, *70*, 595–606.

10. Intarawichian, N; Dasananda, S. Analytical hierarchy process for landslide susceptibility mapping in lower Mae Chaem watershed, northern Thailand. *Suranaree J. Sci. Technol.* **2010**,*17*, 277–292.

11. Demir, G.; Aytekin, M.; Akgün, A.; Ikizler, S.B.; Tatar, O. A comparison of landslide susceptibility mapping of the eastern part of the North Anatolian Fault Zone (Turkey) by likelihood-frequency ratio and analytic hierarchy process methods. *Nat. Hazards* **2013**, *65*, 1481–1506.

12. Chalkias, C.; Ferentinou, M.; Polykretis, C. GIS supported landslide susceptibility modeling at regional scale: An expert-based fuzzy weighting method. *ISPRS Int. J. Geo-Inf.* **2014**, *3*, 523–539.

13. Kayastha, P.; Dhital, M.R.; De Smedt, F. Application of the Analytical Hierarchy Process (AHP) for landslide susceptibility mapping: A case study from the Tinau watershed, west Nepal. *Comput. Geosci.* **2013**, *52*, 398–408.

14. Pourghasemi, H.R.; Moradi, H.R.; Fatemi Aghda, S.M. Landslide susceptibility mapping by binary logistic regression, analytical hierarchy process, and statistical index models and assessment of their performances. *Nat. Hazards* **2013**, *69*, 749–779.

15. Feizizadeh, B.; Blaschke, T. An uncertainty and sensitivity analysis approach for GIS-based multicriteria landslide susceptibility mapping. *Int. J. Geogr. Inf. Sci.* **2014**, *28*, 610–638.

16. Kavzoglu, T.; Sahin, E.K.; Colkesen, I. Landslide susceptibility mapping using GIS-based multi-criteria decision analysis, support vector machines, and logistic regression. *Landslides***2014**, *11*, 425–439.

17. Michel, G.P.; Kobiyama, M.; Goerl, R.F. Comparative analysis of SHALSTAB and SINMAP for landslide susceptibility mapping in the Cunha River basin, southern Brazil. *J. Soils Sediments* **2014**, *14*, 1266–1277.

18. Dahal, R.K.; Bhandary, N.P.; Hasegawa, S.; Yatabe, R. Topo-stress based probabilistic model for shallow landslide susceptibility zonation in the Nepal Himalaya. *Environ. Earth Sci.* **2014**,*71*, 3879–3892.

19. Wu, X.; Niu, R.; Ren, F.; Peng, L. Landslide susceptibility mapping using rough sets and backpropagation neural networks in the Three Gorges, China. *Environ. Earth Sci.* **2013**, *70*, 1307–1318.

20. Bi, R.; Schleier, M.; Rohn, J.; Ehret, D.; Xiang, W. Landslide susceptibility analysis based on ArcGIS and Artificial Neural Network for a large catchment in Three Gorges region, China.*Environ. Earth Sci.* **2014**, *72*, 1925–1938.

21. Conforti, M.; Pascale, S.; Robustelli, G.; Sdao, F. Evaluation of prediction capability of the artificial neural networks for mapping landslide susceptibility in the Turbolo River catchment (northern Calabria, Italy). *Catena* **2014**, *113*, 236–250.

22. Polykretis, C.; Ferentinou, M.; Chalkias, C. A comparative study of landslide susceptibility mapping using landslide susceptibility index and artificial neural networks in the Krios River and Krathis River catchments (northern Peloponnesus, Greece). *Bull. Eng. Geol. Environ.* **2015**, *74*, 27–45.

23. Pradhan, B. A comparative study on the predictive ability of the decision tree, support vector machine and neuro-fuzzy models in landslide susceptibility mapping using GIS.*Comput. Geosci.* **2013**, *51*, 350–365.

24. Sdao, F.; Lioi, D.S.; Pascale, S.; Caniani, D.; Mancini, I.M. Landslide susceptibility assessment by using a neuro-fuzzy model: A case study in the Rupestrian heritage rich area of Matera.*Nat. Hazards Earth Syst. Sci.* **2013**, *13*, 395–407.

25. Park, I.; Lee, S. Spatial prediction of landslide susceptibility using a decision tree approach: A case study of the Pyeongchang area, Korea. *Int. J. Remote Sens.* **2014**, *35*, 6089–6112.

26. He, Y.; Beighley, R.E. GIS-based regional landslide susceptibility mapping: A case study in southern California. *Earth Surf. Process. Landf.* **2008**, *33*, 380–393.

27. Pourghasemi, H.R.; Pradhan, B.; Gokceoglu, C.; Mohammadi, M.; Moradi, H.R. Application of weights-of-evidence and certainty factor models and their comparison in landslide susceptibility mapping at Haraz watershed. *Iran. Arab. J. Geosci.* **2013**, *6*, 2351–2365.

28. Epifânio, B.; Zêzere, J.L.; Neves, M. Susceptibility assessment to different types of landslides in the coastal cliffs of Lourinhã (Central Portugal). *J. Sea Res.* **2014**, *93*, 150–159.

29. Sujatha, E.R.; Kumaravel, P.; Rajamanickam, G.V. Assessing landslide susceptibility using Bayesian probability-based weight of evidence model. *Bull. Eng. Geol. Environ.* **2014**, *73*, 147–161.

30. Youssef, A.M.; Al-Kathery, M.; Pradhan, B. Landslide susceptibility mapping at Al-Hasher Area, Jizan (Saudi Arabia) using GIS-based frequency ratio and index of entropy models.*Geosci. J.* **2015**, *19*, 113–134.

31. Bathrellos, G.D.; Kalivas, D.P.; Skilodimou, H.D. GIS-based landslide susceptibility mapping models applied to natural and urban planning in Trikala, Central Greece. *Estud. Geol.* **2009**,*65*, 49–65.

32. Bai, S.B.; Wang, J.; Thiebes, B.; Cheng, C.; Chang, Z.Y. Susceptibility assessments of the Wenchuan earthquake-triggered landslides in Longnan using logistic regression. *Environ. Earth Sci.* **2014**, *71*, 731–743.

33. Reichenbach, P.; Busca, C.; Mondini, A.C.; Rossi, M. The influence of land use change on landslide susceptibility zonation: The Briga catchment test site (Messina, Italy). *Environ. Manag.* **2014**, *54*, 1372–1384.

34. Bourenane, H.; Bouhadad, Y.; Guettouche, M.S.; Braham, M. GIS-based landslide susceptibility zonation using bivariate statistical and expert approaches in the city of Constantine (Northeast Algeria). *Bull. Eng. Geol. Environ.* **2015**, *74*, 337–355.

35. Chalkias, C.; Ferentinou, M.; Polykretis, C. GIS-based landslide susceptibility mapping on the Peloponnese Peninsula, Greece. *Geosciences* **2014**, *4*, 176–190.

36. Tien Bui, D.; Pradhan, B.; Lofman, O.; Revhaug, I.; Dick, O.B. Spatial prediction of landslide hazards in Hoa Binh province (Vietnam): A comparative assessment of the efficacy of evidential belief functions and fuzzy logic models. *Catena* **2012**, *96*, 28–40.

37. Sujatha, E.R.; Rajamanickam, V. Landslide susceptibility mapping of Tevankarai Ar sub-watershed, Kodaikkanal taluk, India, using weighted similar choice fuzzy model. *Nat. Hazards* **2011**, *59*, 401–425.

38. Zhu, A.X.; Wang, R.; Qiao, J.; Qin, C.Z.; Chen, Y.; Liu, J.; Du, F.; Lin, Y.; Zhu, T. An expert knowledge-based approach to landslide susceptibility mapping using GIS and fuzzy logic.*Geomorphology* **2014**, *214*, 128–138.

39. Pourghasemi, H.R.; Pradhan, B.; Gokceoglu, C. Application of fuzzy logic and Analytical Hierarchy Process (AHP) to landslide susceptibility mapping at Haraz watershed, Iran. *Nat. Hazards* **2012**, *63*, 965–996.

40. Kayastha, P.; Bijukchhen, S.M.; Dhital, M.R.; De Smedt, F. GIS based landslide susceptibility mapping using a fuzzy logic approach: A case study from Ghurmi-Dhad Khola area, eastern Nepal. *J. Geol. Soc. India* **2013**, *82*, 249–261.

41. Wang, W.D.; Xie, C.M.; Du, X.G. Landslides susceptibility mapping based on geographical information system, GuiZhou, south-west China. *Environ. Geol.* **2009**, *58*, 33–43.

42. Wang, W.D.; Guo, J.; Fang, L.G.; Chang, X.S. A subjective and objective integrated weighting method for landslides susceptibility mapping based on GIS. *Environ. Earth Sci.* **2012**, *65*, 1705–1714.

43. Rozos, D.; Bathrellos, G.D.; Skillodimou, H.D. Comparison of the implementation of rock engineering system and analytic hierarchy process methods, upon landslide susceptibility mapping, using GIS: A

case study from the Eastern Achaia County of Peloponnesus, Greece.*Environ. Earth Sci.* **2011**, *63*, 49–63.

44. Kritikos, T.; Davies, T.R.H. GIS-based multi-criteria decision analysis for landslide susceptibility mapping at northern Evia, Greece. *Z. Deutsch. Ges. Geowiss.* **2011**, *162*, 421–434.

45. Kouli, M.; Loupasakis, C.; Soupios, P.; Rozos, D.; Vallianatos, F. Landslide susceptibility mapping by comparing the WLC and WofE multi-criteria methods in the West Crete Island, Greece. *Environ. Earth Sci.* **2014**, *72*, 5197–5219.

46. Papathanassiou, G.; Valkaniotis, S.; Ganas, A.; Pavlides, S. GIS-based statistical analysis of the spatial distribution of earthquake-induced landslides in the island of Lefkada, Ionian Islands, Greece. *Landslides* **2013**, *10*, 771–783.

47. Chalkias, C.; Kalogirou, S.; Ferentinou, M. Landslide susceptibility, Peloponnese Peninsula in South Greece. *J. Maps* **2014**, *10*, 211–222.

48. Tsangaratos, P.; Benardos, A. Estimating landslide susceptibility through an artificial neural network classifier. *Nat. Hazards* **2014**, *74*, 1489–1516.

49. Gemitzi, A.; Falakakis, G.; Eskioglou, P.; Petalas, C. Evaluating landslide susceptibility using environmental factors, fuzzy membership functions and GIS. *Glob. NEST J.* **2011**, *13*, 28–40.

50. Ferentinou, M.; Chalkias, C. Mapping mass movement susceptibility across Greece with GIS, ANN and statistical methods. *Landslide Sci. Pract.* **2013**, 321–327.

51. Sabatakakis, N.; Koukis, G.; Vassiliades, E.; Lainas, S. Landslide susceptibility zonation in Greece. *Nat. Hazards* **2013**, *65*, 523–543.

52. Gournellos, T.; Chalkias, C.; Tsagas, D. Landslide susceptibility zonation of Greece using fuzzy logic and GIS. *Geographies* **2006**, *12*, 114–126.

53. Hellenic Statistical Authority (ELSTAT). *Population and Housing Census: Resident Population*; ELSTAT: Piraeus, Greece, 2011.

54. Pnevmatikos, J.D.; Katsoulis, B.D. The changing rainfall regime in Greece and its impact on climatological means. *Meteorol. Appl.* **2006**, *13*, 331–345.

55. Koukis, G.; Tsiambaos, G.; Sabatakakis, N. Landslides movements in Greece: Engineering geologicalcharacteristics and environmental consequences. In Proceedings of the 7th International Symposium of Engineering Geology and Environment, Athens, Greece, 23–27 June 1997; Volume 1, pp. 789–792.

56. Hellenic Military Geographical Service (HMGS). *Aerial Photography of Various Areas of Greece, Scale 1:40,000*; HMGS: Athens, Greece, 1969.

57. Che, V.B.; Kervyn, M.; Suh, C.E.; Fontijn, K.; Ernst, G.G.J.; del Marmol, M.A.; Trefois, P.; Jacobs, P. Landslide susceptibility assessment in Limbe (SW Cameroon): A field calibrated seed cell and information value method. *Catena* **2012**, *92*, 83–98.

58. De Guidi, G.; Scudero, S. Landslide susceptibility assessment in the Peloritani Mts. (Sicily, Italy) and clues for tectonic control of relief processes. *Nat. Hazards Earth Syst. Sci.* **2013**, *13*, 949–963.

59. Youssef, A.M.; Pradhan, B.; Pourghasemi, H.R.; Abdullahi, S. Landslide susceptibility assessment at Wadi Jawrah Basin, Jizan region, Saudi Arabia using two bivariate models in GIS. *Geosci. J.* **2015**, *19*, 449–469.

60. Cruden, D.M.; Varnes, D.J. Landslide types and processes. In *Landslides: Investigation and Mitigation, Special Report*; Turner, A.K., Schuster, R.L., Eds.; National Academy Press: Washington, DC, USA, 1996.

61. Fell, R. Landslide risk assessment and acceptable risk. *Can. Geotech. J.* **1994**, *31*, 261–272.

62. Magliulo, P.; Lisio, A.; Russo, F.; Zelano, A. Geomorphology and landslide susceptibility assessment using GIS and bivariate statistics: A case study in southern Italy. *Nat. Hazards***2008**, *47*, 411–435.

63. Das, O.; Sonmez, H.; Gokceoglu, C.; Nefeslioglu, H. A study on the effect of the ground acceleration on landslide susceptibility maps. *Geophys. Res. Abstr.* **2010**, *12*, EGU2010-5885-1.

64. Yalcin, A. GIS-based landslide susceptibility mapping using analytical process and bivariate statistics in Ardesen (Turkey): Comparisons of results and confirmations. *Catena* **2008**, *72*, 1–12.

65. Hasekioğullari, G.D.; Ercanoglu, M. A new approach to use AHP in landslide susceptibility mapping: A case study at Yenice (Karabuk, NW Turkey). *Nat. Hazards* **2012**, *63*, 1157–1179.

66. Heymann, Y.; Steenmans, C.; Croisille, G.; Bossard, M. *CORINE Land Cover: Technical Guide*; Office for Official Publications of the European Communities: Luxembourg City, Luxembourg, 1994.

67. Carrara, A.; Cardinali, M.; Detti, R.; Guzzetti, F.; Pasqui, V.; Reichenbach, P. GIS techniques and statistical models in evaluating landslide hazard. *Earth Surf. Process. Landf.* **1991**, *16*, 427–445.

68. Institute of Geology and Mineral Exploration (IGME). *Geological Map of Greece, Scale 1:500,000*, 2nd ed.; IGME: Athens, Greece, 1983.

69. Underhill, J.R. Late Cenozoic deformation of the Hellenide foreland, western Greece. *Geol. Soc. Am. Bull.* **1989**, *101*, 613–634.

70. Ayalew, L.; Yamagishi, H. The application of GIS-based logistic regression for landslide susceptibility mapping in the Kakuda-Yahiko Mountains Central Japan. *Geomorphology* **2005**,*65*, 15–31.

71. Rahmani, M.; Hosieni, S.A.; Molla Aghajanzadeh, A.; Mashayekhan, A.; Mahalleh, S.M.M. Landslide hazard zonation using the MCE method (Case study: Agh Mashhad, Mazandaran Province, Iran). *Int. J. AgriSci.* **2013**, *3*, 240–244.

72. Ahmed, M.F.; Rogers, J.D.; Ismail, E.H. A regional level preliminary landslide susceptibility study of the upper Indus river basin. *Eur. J. Remote Sens.* **2014**, *47*, 343–373.

73. Sato, H.P.; Hasegawa, H.; Fujiwara, S.; Tobita, M.; Koarai, M.; Une, H.; Iwahashi, J. Interpretation of landslide distribution triggered by the 2005 Northern Pakistan earthquake using SPOT 5 imagery. *Landslides* **2007**, *4*, 113–122.

74. Owen, L.A.; Kamp, U.; Khattak, G.A.; Harp, E.L.; Keefer, D.K.; Bauer, M.A. Landslides triggered by the 8 October 2005 Kashmir earthquake. *Geomorphology* **2008**, *94*, 1–9.

75. Huade, G.; Wilson, J.L.; Makhnin, O. Geostatistical mapping of mountain precipitation incorporating autosearched effects of terrain and climatic characteristics. *J. Hydrometeorol.***2005**, *6*, 1018–1031.

76. Arjun, C.R.; Kumar, A. Artificial neural network-based estimation of peak ground acceleration. *ISET J. Earthq. Technol.* **2009**, *46*, 19–28.

77. Stournaras, G.; Tsimpidis, G.; Tsoumanis, P.; Yannatos, G.; Guillande, R. Instability Phenomena in the Neogenne Deposits of Northern Peloponnese. *Bull. Eng. Geol. Environ.***1998**, *57*, 65–68.

78. Papathanassiou, G. Estimating slope failure potential in an earthquake prone area: A case study at Skolis Mountain, NW Peloponnesus, Greece. *Bull. Eng. Geol. Environ.* **2012**, *71*, 187–194.

79. Dounias, G.; Fikiris, I.F.; Kollios, A.; Kabbadas, M.; Marinos, P. The Nemea Landslide. Trigger Mechanism–Remedial Measures. In Proceedings of the 5th Panellenic Conference of Geotechnics and Geoenvironmental Engineering, Xanthi, Greece, 31 May–2 June 2006.

80. Papadopoulos, G.A.; Plessa, A. Magnitude-distance relations for earthquake-induced landslides in Greece. *Eng. Geol.* **2000**, *58*, 377–386.

81. Chousianitis, K.; Del Gaudio, V.; Sabatakakis, N.; Kavoura, K.; Drakatos, G.; Bathrellos, G.D.; Skilodimou, H.D. Assessment of earthquake-induced landslide hazard in Greece: From Arias Intensity to spatial distribution of slope resistance demand. *Bull. Seismol. Soc. Am.***2016**.

82. Van Westen, C.J. Statistical landslide hazard analysis. In *ILWIS 2.1 for Windows Application Guide*; ITC Publication: Utrecht, The Netherlands, 1997; pp. 73–84.

83. Jenks, G.F. The data model concept in statistical mapping. *Int. Yearb. Cartogr.* **1967**, *7*, 186–190.

84. Lee, E.M.; Jones, D.K.C. *Landslide Risk Assessment*; Thomas Telford: London, UK, 2004; p. 161.

85. Malczewski, J. *GIS and Multicriteria Decision Analysis*; John Wiley & Sons: New York, NY, USA, 1999.

86. Remondo, J.; González-Díez, A.; Díaz de Terán, J.R.; Cendrero, A. Landslide susceptibility models utilizing spatial data analysis techniques. A case study from the Lower Deba Valley, Guipúzcoa (Spain). *Nat. Hazards* **2003**, *30*, 267–279.

87. Nandi, A.; Shakoor, A. Preparation of a landslide susceptibility map of summit county, Ohio, USA, using numerical models. In Proceedings of the 10th IAEG Congress, Nottingham, UK, 6–10 September 2006.

88. Frattini, P.; Crosta, G.; Carrara, A. Techniques for evaluating the performance of landslide susceptibility models. *Eng. Geol.* **2010**, *111*, 62–72.

89. Fawcett, T. An introduction to ROC analysis. *Pattern Recognit. Lett.* **2006**, *27*, 861–874.

90. Dragićević, S.; Lai, T.; Balram, S. GIS-based multicriteria evaluation with multiscale analysis to characterize urban landslide susceptibility in data-scarce environments. *Habitat Int.* **2015**,*45*, 114–125.

91. Faraji Sabokbar, H.; Shadman Roodposhti, M.; Tazik, E. Landslide susceptibility mapping using geographically-weighted principal component analysis. *Geomorphology* **2014**, *226*, 15–24.

92. Yang, Z.H.; Lan, H.X.; Gao, X.; Li, L.P.; Meng, Y.S.; Wu, Y.M. Urgent landslide susceptibility assessment in the 2013 Lushan earthquake-impacted area, Sichuan Province, China. *Nat. Hazards* **2015**, *75*, 2467–2487.

93. Van Westen, C.J.; Rengers, N.; Terlien, M.T.J.; Soeters, R. Prediction of the occurrence of slope instability phenomena through GIS-based hazard zonation. *Geol. Rundsch.* **1997**, *86*, 404–414.

94. Ayalew, L.; Yamagishi, H.; Maruib, H.; Takami, K. Landslides in Sado Island of Japan: Part II. GIS-based susceptibility mapping with comparisons of results from two methods and verifications. *Eng. Geol.* **2005**, *81*, 432–445.

95. Tien Bui, D.; Lofman, O.; Revhaug, I.; Dick, O. Landslide susceptibility analysis in the Hoa Binh province of Vietnam using statistical index and logistic regression. *Nat. Hazards* **2011**,*59*, 1413–1444.

96. Van Westen, C.J.; Rengers, N.; Soeters, R. Use of geomorphological information in indirect landslide susceptibility assessment. *Nat. Hazards* **2003**, *30*, 399–419.

97. Saboya, F., Jr.; da Glória Alves, M.; Dias Pinto, W. Assessment of failure susceptibility of soil slopes using fuzzy logic. *Eng. Geol.* **2006**, *86*, 211–224.

98. Bathrellos, G.D.; Gaki-Papanastassiou, K.; Skilodimou, H.D.; Papanastassiou, D.; Chousianitis, K.G. Potential suitability for urban planning and industry development by using natural hazard maps and geological—Geomorphological parameters. *Environ. Earth Sci.* **2012**, *66*, 537–548.

99. Rozos, D.; Skilodimou, H.D.; Loupasakis, C.; Bathrellos, G.D. Application of the revised universal soil loss equation model on landslide prevention. An example from N. Euboea (Evia) Island, Greece. *Environ. Earth Sci.* **2013**, *70*, 3255–3266.

Index